GENERIC TECHNIQUES IN SYSTEMS RELIABILITY ASSESSMENT

NATO ADVANCED STUDY INSTITUTES SERIES

Proceedings of the Advanced Study Institute Programme, which aims at the dissemination of advanced knowledge and the formation of contacts among scientists from different countries.

The series is published by an international board of publishers in conjunction with NATO Scientific Affairs Division

A	Life Sciences	Plenum Publishing Corporation
B	Physics	London and New York
C	Mathematical and Physical Sciences	D. Reidel Publishing Company Dordrecht and Boston
D	Behavioural and Social Sciences	Sijthoff International Publishing Company Leyden, The Neth. and Reading, Mass., USA
E	Applied Sciences	Noordhoff International Publishing Leyden, The Neth. and Reading, Mass., USA

Series E: Applied Science - No. 5

GENERIC TECHNIQUES IN SYSTEMS RELIABILITY ASSESSMENT

edited by

E.J. HENLEY
Cullen College of Engineering
University of Houston, USA

and

J.W. LYNN
Dept. of Electrical Engineering and Electronics
The University of Liverpool, UK

NOORDHOFF - LEYDEN - 1976

Proceedings of the NATO Advanced Study Institute on Generic Techniques in Systems Reliability Assessment, The University of Liverpool, UK, July 17-28, 1973.
The Institute was co-directed by Prof. E. J. Henley and Dr. J. W. Lynn.
As Scientific Directors acted:
Mr. Eric Green, UK Atomic Energy Authority
Prof. T. Regulinski, Air Force Institute, Wright Patterson Air Force Base, USA
Prof. M. L. Shooman, Polytechnic Institute of Brooklyn, USA
Mr. W. Vinck, European Economic Communities, Belgium.

ISBN-13: 978-94-010-1555-4 e-ISBN-13: 978-94-010-1553-0
DOI: 10.1007/978-94-010-1553-0

Copyright © 1976 by Noordhoff International Publishing, a division of A. W. Sijthoff International Publishing Company bv, Leyden, The Netherlands.
Softcover reprint of the hardcover 1st edition 1976

All rights reserved. No part of this publication may be reproduced, stored in a retrieval system, or transmitted, in any form or by any means, electronic, mechanical, photocopying, recording, or otherwise, without the prior permission of the copyright owner.

PREFACE

The NATO Advanced Study Institute is, to quote the Notes for Applicants "primarily a high level teaching activity at which a carefully defined subject is presented in a systematic and coherently structured programme. The subject is treated in considerable depth by eminent lecturers........."

The NATO ASI on Generic Techniques in Systems Reliability Assessment was held at the University of Liverpool and the Proceedings are presented in the present volume. Regrettably many of the papers are in shortened version. This was an interdisciplinary assembly designed to focus on the synthesis of generic reliability concepts and technology and to discuss relevant teaching and research in universities and colleges. One important objective was, of course, to give opportunity for interchange of information on advanced techniques in reliability in various fields.

The Institute was held in Dale Hall, one of the halls of residence of the University of Liverpool, England, from 17th to 28th July, 1973. Sixty-four engineers from twelve countries attended, namely, 27 from U.K., 14 from U.S.A. 8 from Italy, 4 from West Germany, 2 from France, 2 from the Netherlands, 2 from Sweden and one each from Belgium, Canada, Denmark, India and Norway. Seven of these had their wives and some brought their children also. The technical affiliations which were represented were 23 universities, 22 national laboratories, 11 industry, 5 military, 3 consultants.

On the first day the Institute was privileged to have present Mr. M. di Lullo of NATO, who addressed the meeting and discussed the work of the Scientific Affairs Division. During nine working days while visits were arranged for the families, more than forty lectures were given, covering overviews, mathematics, techniques and reliability specifications, data banks, software and computing, optimisation, nuclear and chemical plant, electronics, power systems, human factors, mechanical structures and design. Each morning and afternoon session began with an invited lecture, followed by presentation of shorter papers. The Institute finished with discussions by the members in eight groups, each of which considered the material presented in relation to special interests. It is difficult to convey in the printed page the atmosphere of group discussion. The topics covered a wide range but every group was concerned with several specific questions, for example, reliability in university courses, common language and precise definitions, the use of reliability specifications in industry, aspects of research, mathematics, data processing, and so on. The conclusions which were reached by the groups are included here, along with the papers presented in the formal sessions.

The amount of material presented for publication far exceeded the space available. Because of the nature of the meeting, however, the editors felt that every member who had contributed a paper should be given the opportunity of publishing a shortened version, each paper thus helping to give some integration over the whole field. The editors would like at this point to express their appreciation of the co-operation they received from authors in this respect and of the advice and assistance of the publishers.

The work of organising the ASI was undertaken by a Scientific Committee, whose members wish to express their thanks to all who so ably contributed by presenting papers or by taking part in the discussions. Thanks are also due to the Scientific Affairs Division of NATO for the award of the Institute funds, to the Universities of Liverpool, England, and Houston, Texas, U.S.A. for facilities provided, to the Safety and Reliability Directorate of the U.K.A.E.A. Risley, for technical advice and to Noordhoff Publishing, Netherlands, for undertaking the task of publishing the proceedings.

E.J.H.
J.W.L.

TABLE OF CONTENTS

Preface	V
D.H. Allen and G.D.M. Pearson	
A Preliminary Model for Achieving Optimum Reliability Investment in Chemical Plants	1
H. Ascher	
Development of Systems Reliability Models from Basic Physical/Mathematical Concepts	11
S.A. Austin	
A Technique for Selectively Reducing Small Sample Error in Maintainability Demonstrations when Apportionment is Based upon Predicted Reliability	29
I. Bazovsky, Jr.	
On Definitions of Software Reliability and Computing Reliability	37
I. Bazovsky, Jr.	
On Some Objective Functions for Reliability Optimization	45
R.G. Bennetts	
A Comment on Reliability Evaluation of Software	55
R. Billinton	
Power System Reliability Indices	61
A. Carnino	
Reliability Techniques in Assessment of Nuclear Reactor Safety	83
G.J. Dasani, J.M. Kontoleon and J.W. Lynn	
Digital Analysis of an Electronic Amplifier for Reliability Studies	93
A.R. Eames, C.D.H. Fothergill, R. Roughley and E.R. Woodcock	
Operation of a Reliability Data Bank	99
R.A. Evans	
Reliability Optimization	117
D.N. Farnan	
A Test Methodology Allowing Trade-Off of Reliability/Maintainability Requirements	125
J.B. Fussell	
Fault Tree Analysis: Concepts and Techniques	133

S.L. Gandhi and E.J. Henley
    Optimal Availability of a Complex System   163
A.E. Green
    A Review of System Reliability Assessment   183
H.H. Happ
    Security Assessment of Power Systems   207
J.A. Hess
    A Reliability/Maintainability Data Feedback System   223
F.P. Lees
    Design for Man-Machine System Reliability
    in Process Control   233
G. Manzoni, P.L. Noferi, L. Paris and M. Valtorta
    Risk Indices for Evaluating the Reliability
    of an Electrical System   255
P. Martin
    Reliability in Mechanical Design and Production   267
J. Munro
    The Analysis and Synthesis of Safe and
    Serviceable Structures   273
J.D. Murchland
    Fundamental Probability Relations
    for Repairable Items   293
S. Panichelli, L. Salvaderi and S. Scalcino
    Unavailability of Statistical Models for
    Power System Reliability Studies   295
G.J. Powers and F.C. Tompkins, Jr.
    Computer-Aided Synthesis of Fault Trees
    for Complex Processing Systems   307
J. Rasmussen
    The Role of the Man-Machine Interface
    in Systems Reliability   315
T.L. Regulinski
    Human Performance Reliability Modeling
    in Time Continuous Domain   325
H.V.K. Shetty and D.P. Sen Gupta
    Optimization of Redundant Systems:
    a Non-linear Programming Approach   337
M.L. Shooman
    Software Reliability: Analysis and Prediction   343
S.L. Surana and A. Brameller
    Application of Reliability Theory to Computation
    of System Operation Security in a Power Industry   375
A.D. Swain
    Shortcuts in Human Reliability Analysis   393
D.R. Towill
    Recent Developments in the Prediction of
    Human Operator Performance   411
C.R. Wakeman and M.A. Laughton
    Some Design Aspects of Electrical Power
    Distribution System Reliability   417

R.E.C. Weaver
 Reliability, Redundancy and Cascaded
 Priorities in Psysiological Systems 455
S.A. Wilson
 Distribution of Time-to-Damage Method
 in Reliability of Pipes 463
Group Reports 475
Attendance 491

A PRELIMINARY MODEL FOR ACHIEVING OPTIMUM RELIABILITY INVESTMENT
IN CHEMICAL PLANTS

D.H. Allen and G.D.M. Pearson

Chemical Engineering Department
University of Nottingham, U.K.

THE ECONOMIC PAYOFF FROM IMPROVING PLANT RELIABILITY

In the chemical industry an externally specified target reliability may often be required from a system for reasons of safety or environmental control, but another common situation is where the actual target level of system reliability is, or should be, decided mainly by its economic implications. Expenditure on reliability improvement is here subject to the law of diminishing returns. Further expenditure is only justified up to the point where the marginal payoff is still just attractive, and this is the point of optimum reliability investment. Thus it is necessary to identify the payoff from any proposal for improving reliability and to compare this with the minimum cost of achieving it. The payoff is the improvement in subsequent long-term cash flow during the plant's operation, i.e. the net addition to income or reduction in operating costs expected as a result of the improvement in plant reliability.

If a loss of product sales and sales income results from a failure, this lost income is an opportunity cost which could be avoided by improving reliability and preventing the failure or its adverse effects. The payoff from avoiding lost production can be very high for large chemical plants (see for example Finneran et alia (1)) and is frequently the over-riding incentive for improving reliability. Improving reliability also of course reduces failure correction costs. Besides the actual costs of maintenance, repair or replacement, the costs of shutting down plant and restarting it can be considerable (see for example Lenz (2)). These can be lengthy procedures during which chemical feedstocks and utilities are consumed but no saleable product is made. Improving reliability can therefore pay off by reducing the frequency of shutdowns and by reducing repair and replacement work needed. Many chemical plants operate continuously for long periods and some failures develop gradually and steadily increase running costs until they are put right. This latter situation and some other economic aspects of chemical plant reliability have been discussed in general terms by Allen (3) but are not considered further here.

The payoff from a specific proposal to improve reliability is obtained by establishing the anticipated economic performance of the plant both with and without the reliability improvement. The difference in performance (i.e. in subsequent operating cash flows) is the payoff attributable to the reliability improvement. It is therefore necessary to be able to evaluate plant operating cash flows for different component reliabilities, configurations and modes of operation before any optimization can be attempted.

EVALUATION OF THE PLANT OPERATING CASH FLOW RATE

The smoothed operating cash flow rate for a plant is the long-term, steady-state, average net rate of cash flow resulting from plant operation, in which the effects of failures are smoothed out. Thus component and plant reliabilities need to be expressed as constant long-term availabilities. A plant is considered to be composed of stages so chosen that while the reliability of stages can be found independently, complete failure of a stage interrupts production.

The smoothed operating cash flowrate is a function of running, standby and failure correction costs (excluding depreciation), the value added to the product, the plant availability and the full design production rate or capacity, assuming no sales limitation. Plant availability depends on the availabilities and configuration of component equipment and on the mode of plant operation.

At any instant, component equipment is assumed to be either available to operate normally or in a completely failed state, i.e. unavailable. An available component may be either onstream or offstream. When onstream it is assumed to incur a constant fixed cost and a variable cost which is linearly proportional to throughpout. When offstream, a component incurs a fixed cost to maintain it in a standby state. A failed component incurs a fixed shutdown, startup and repair or replacement cost.

A plant is assumed to be made up of a number of stages in series with each stage consisting of a number of identical components in parallel, each with the same steady-state availability.

Using the nomenclature defined at the end of this paper, the general equation for the smoothed operating cash flow rate for a plant is:

$$C_p = M_p(W-V_p) - (F_p + G_p + H_p) \quad (1)$$

Depending on the plant configuration and mode of operation, the main items in equation (1) can be expressed in terms of stage costs and availabilities. The derivation of these were included in the full paper presented at the Institute but only the resulting expressions are now reported here.

A SINGLE-STAGE, SINGLE-COMPONENT PLANT

$$C_p = [m(W-V_1)-F_1] A_o A_1 - G_1(1-A_o)A_1 - H_1(1-A_1) \quad (2)$$

If there are D separate feeds and the availability of feed d is A_d, then:

$$A_o = \prod_{d=1}^{D} A_d \quad (3)$$

A MULTISTAGE PLANT WITH ONE COMPONENT PER STAGE

$$C_p = [m(W - \sum_{i=1}^{I} V_i) - (\sum_{i=1}^{I} F_i - \sum_{i=1}^{I} G_i)]A_o \prod_{i=1}^{I} A_i - \sum_{i=1}^{I} G_i A_i - \sum_{i=1}^{I} H_i(1-A_i) \quad (4)$$

If all availabilities are unity, this of course simplifies to the basic equation for onstream cash flow:

$$C_p = m(W-V_p) - F_p \quad (5)$$

If any availability is zero the plant is shut down and the appropriate offstream costs are incurred. Equation (4) can be used to explore the sensitivity of the operating economics of a single-stream plant to the availabilities of its individual stages.

INTRODUCING STAGE REDUNDANCIES IN A PLANT - MODES OF OPERATION

An alternative to improving the reliability of single-component stages is of course to introduce stage redundancies by having multi-component stages. Three alternative ways of operating with multi-component stages are considered. In mode A operation, all available parallel components in a stage are onstream together except when another stage fails completely or when feed is unavailable. At any time the plant throughput is limited by the 'bottleneck' stage with the lowest capacity due to component failures, and all available components in each remaining stage are operated at reduced throughput, which is shared between them.

Mode B operation is similar to mode A except that when the plant is running below full throughput due to component failures, components in non-bottleneck stages are shut down wherever possible to keep those remaining at or near their full throughputs. Mode A operation may be adopted when component shutdown and startup procedures are time-consuming or expensive and they are sufficiently flexible in operation to run satisfactorily over a range of throughputs. Conversely, mode B may be adopted when component shutdown and startup procedures are quick and inexpensive or when their throughput flexibility is small.

The third way of operating is with standby redundancy. Only one component per stage is onstream at any time and takes the full stage throughput. Other available components in the stage are on standby, ready to be brought onstream when required. Thus the plant either operates at full throughput, or is completely shutdown when all components of a stage are in a failed state together, or when feed is unavailable. The expressions for the smoothed operating cash flow rate equation are now given for these three types of operation.

A SINGLE-STAGE, MULTI-COMPONENT PLANT, MODE A AND MODE B OPERATION

$$C_p = [m(W-V_1) - n_1(F_1-G_1)]A_o A_1 - n_1[G_1 A_1 + H_1(1-A_1)] \quad (6)$$

Full throughput is only achieved when feed is available and all components are also available and onstream. Standby costs only arise when a component is available but feed is unavailable. Putting $n_1 = 1$ in equation (6), equation (2) is obtained.

A MULTI-STAGE, MULTI-COMPONENT PLANT, MODE A OPERATION

For any of the availability states possible the corresponding available plant capacity is determined by the lowest available capacity of an individual stage, i.e. the 'bottleneck' stage. Thus the plant capacity for a specified plant availability state is :

$$M_T = M . \min \left(\frac{n_k - f_k}{n_k} \right)_{k=1}^{I} \quad (7)$$

It follows that :

$$M_p = MA_o \sum_{f_1=0}^{n_1-1} \cdots \sum_{f_I=0}^{n_I-1} \left[\min\left(\frac{n_k-f_k}{n_k}\right)_{k=1}^{I} \prod_{j=1}^{I} {}^{n_j}C_{f_j} \cdot (A_j)^{n_j-f_j} (1-A_j)^{f_j} \right] \quad (8)$$

$$V_p = \sum_{i=1}^{I} V_i \quad (9)$$

$$F_p = A_o \sum_{i=1}^{I} n_i F_i A_i \cdot \frac{\prod_{j=1}^{I}[1-(1-A_j)^{n_j}]}{[1-(1-A_i)^{n_i}]} \quad (10)$$

$$G_p = \sum_{i=1}^{I} n_i G_i A_i \left[1 - \frac{A_o \prod_{j=1}^{I}[1-(1-A_j)^{n_j}]}{[1-(1-A_i)^{n_i}]} \right] \quad (11)$$

$$H_p = \sum_{i=1}^{I} n_i H_i (1-A_i) \quad (12)$$

Equations (8) - (12) are substituted in equation (1) to obtain the complete expression for C_p.

A MULTI-STAGE, MULTI-COMPONENT PLANT, MODE B OPERATION

Plant throughput is not affected by mode A or mode B operation. Hence equations (8) and (9) apply for mode B. Similarly equation (12) also applies. F_p, however, depends on the minimum number of components in each stage needed to maintain the available throughput. Available plant capacity is given by equation (7). If λ_i components of stage i are needed to maintain the available plant capacity,

$$\lambda_i = n_i \cdot \text{gint} \left[\min\left(\frac{n_k - f_k}{n_k}\right)_{k=1}^{I} \right] \quad (13)$$

where gint(x) is the nearest integer $\geq x$.
It follows that :

$$F_p = A_o \sum_{f_1=0}^{n_1-1} \cdots \sum_{f_I=0}^{n_I-1} \left[\sum_{i=1}^{I} F_i \lambda_i \prod_{j=1}^{I} {}^{n_j} C_{f_j} (A_j)^{n_j-f_j} (1-A_j)^{f_j} \right] \quad (14)$$

$$G_p = \sum_{i=1}^{I} n_i G_i A_i - A_o \sum_{f_1=0}^{n_1-1} \cdots \sum_{f_I=0}^{n_I-1} \left[\sum_{i=1}^{I} G_i \lambda_i \prod_{j=1}^{I} {}^{n_j} C_{f_j} \cdot (A_j)^{n_j-f_j} (1-A_j)^{f_j} \right] \quad (15)$$

Equations (8), (9), (12), (14) and (15) are substituted in equation (1) to obtain the complete expression for C_p.

A SINGLE-STAGE, MULTI-COMPONENT PLANT, STANDBY OPERATION

As long as the plant is running, one available component is onstream and the remaining components are offstream. If no feed is available, all available components are on standby. Operating throughput is achieved when at least one component and the feed are available at the same time.

$$C_p = [M(W-V_1)-(F_1-G_1)]A_o[1-(1-A_1)^{n_1}] - n_1 G_1 A_1 - n_1 H_1 (1-A_1) \quad (16)$$

A MULTI-STAGE, MULTI-COMPONENT PLANT, STANDBY OPERATION

The equations for V_p and H_p are the same as those for mode A operation, namely (9) and (12).

$$M_p = MA_o \prod_{j=1}^{I} [1-(1-A_j)^{n_j}] \quad (17)$$

$$F_p = A_o \sum_{i=1}^{I} F_i \prod_{j=1}^{I} [1-(1-A_j)^{n_j}] \quad (18)$$

```
┌─────────────────────────────────────────────────────────┐
│ Establish initial base case plant configuration         │
└─────────────────────────────────────────────────────────┘
                          ▼
┌─────────────────────────────────────────────────────────┐
│ Establish initial base case reliability and economic data│
└─────────────────────────────────────────────────────────┘
                          ▼
┌─────────────────────────────────────────────────────────┐
│ Calculate initial base case $C_p$                       │
└─────────────────────────────────────────────────────────┘
                          ▼
   ┌──►┌──────────────────────────────────────────────────┐
   │   │ Identify alternative new proposals for plant     │
   │   │ reliability improvement                          │
   │   └──────────────────────────────────────────────────┘
   │                      ▼
   │   ┌──────────────────────────────────────────────────┐
   │   │ Establish reliability data and initial cost for  │
   │   │ each independent reliability improvement proposal│
   │   └──────────────────────────────────────────────────┘
   │                      ▼
   │   ┌──────────────────────────────────────────────────┐
   │   │ Incorporate each proposal in turn into the base  │
   │   │ case plant and calculate the corresponding new $C_p$│
   │   └──────────────────────────────────────────────────┘
   │                      ▼
   │   ┌──────────────────────────────────────────────────┐
   │   │ Calculate the payoff for each proposal as the    │
   │   │ difference between $C_p$'s, new case minus base case│
   │   └──────────────────────────────────────────────────┘
   │                      ▼
   │   ┌──────────────────────────────────────────────────┐
   │   │ From the initial cost and payoff of each proposal│
   │   │ calculate its NPV (or DCFR)                      │
   │   └──────────────────────────────────────────────────┘
   │                      ▼
   │   ┌──────────────────────────────────────────────────┐
   │   │ Select the proposal with the largest NPV (or DCFR)│
   │   │ Is this acceptable?                              │
   │   └──────────────────────────────────────────────────┘
   │                      ▼
   │                    ◇ ──── No ────┐
   │                  Yes             │
   │                    ▼             │
   │   ┌──────────────────────────────────────────────────┐
   │   │ Incorporate this proposal into the plant to form │
   │   │ a new base case for the next iteration           │
   │   └──────────────────────────────────────────────────┘
   │                      ▼
   │   ┌──────────────────────────────────────────────────┐
   │   │ Re-evaluate as necessary the feasibility of, and │
   │   │ the data on, existing proposals in relation to   │
   │   │ this new base case                               │
   │   └──────────────────────────────────────────────────┘
   └──────────────────────┘
                          ▼
┌─────────────────────────────────────────────────────────┐
│ Optimum reliability improvement level has already       │
│ been reached                                            │
└─────────────────────────────────────────────────────────┘
```

FIGURE 1. Iterative procedure for optimization of plant reliability improvement.

$$G_p = \sum_{i=1}^{I} G_i [n_i A_i - A_o \prod_{j=1}^{I} [1-(1-A_j)^{n_j}]] \qquad (19)$$

Equations (9),(12),(18)-(20) are substituted in equation (1) to obtain the complete expression for C_p.

A GENERAL ITERATIVE PROCEDURE FOR THE OPTIMIZATION OF RELIABILITY INVESTMENT

Because of the long-term nature of the effects of reliability improvement, discounting methods of evaluating the economic attractiveness of a proposal are appropriate, the most common ones being net present value (NPV) and discounted cash flow return (DCFR) over the anticipated life of the proposal. Assuming all the necessary reliability and economic data are available, the procedure in Figure 1 can be used to arrive at the policy for investment in reliability improvement which establishes the economic optimum reliability and also indicates the order of preference for successive reliability improvement proposals. Thus if for any reason, such as shortage of funds, the optimum level of reliability cannot be reached, this procedure ensures that the maximum benefit is obtained from whatever level of reliability investment is possible. The procedure involves putting forward independent alternative proposals for reliability improvement and care is needed in defining these so that real optimization and not sub-optimization is achieved. For repeat iterations, proposals are reappraised to take account of the change in base case. A modified procedure can be used where the target level of plant reliability is fixed, to ensure that it is achieved at minimum cost by selecting the appropriate reliability improvement proposals.

The biggest practical problem in applying this type of approach to reliability investment decisions is obtaining the data. However, it is only when the specific data needed is thus identified and the use to which it can be put is made apparent that the incentive to obtain it can be seen.

NOMENCLATURE

d	:	plant feed number
i,j,k	:	plant stage number
D	:	total number of plant feeds
I	:	total number of plant stages
M	:	full onstream plant capacity
W	:	value added to product
f_i	:	number of component failures in stage i

n_i : number of components in stage i
A_o : total availability of feed
A_d : availability of feed d
A_i : steady-state availability of each component in stage i
C_p : smoothed plant operating cash flow rate
F_i^p : onstream fixed cost rate of each component in stage i
F_p : total smoothed plant onstream fixed cost rate
G_i^p : offstream standby cost rate of each component in stage i
G_p : total smoothed plant offstream standby cost rate
H_i^p : offstream failure correction cost rate of each component in stage i
H_p : total smoothed plant offstream failure correction cost rate
M_T : smoothed available plant capacity
M_p : smoothed available production rate
V_i^p : onstream variable cost of each component in stage i
V_p : total plant onstream variable cost
$^{n_j}C_{f_j}$: $\dfrac{n_j!}{f_j!(n_j-f_j)!}$

λ_i : number of components of stage i needed to maintain available plant capacity.

REFERENCES

1. Finneran, J.A., Sweeney, N.J. and Huchinson, T.G. "Startup performance of large ammonia plants" Chemical Engineering Progress, 64 (8), 72 (1968)

2. Lenz, R.E., "Reliability design in process plants", Chemical Engineering Progress, 66 (12), 42 (1970)

3. Allen, D.H., "Economic aspects of plant reliability", Institution of Chemical Engineers Symposium on Design for Reliability, Manchester (April 1973), to be published in The Chemical Engineer (October 1973)

(Copies of the full paper, of which this is a shortened version, may be obtained from the authors)

DEVELOPMENT OF SYSTEMS RELIABILITY MODELS FROM BASIC PHYSICAL/
MATHEMATICAL CONCEPTS

Harold Ascher

Naval Research Laboratory
Washington, D. C. 20375

1. INTRODUCTION

Most papers on the subject of systems reliability models have
been written from an abstract, probabilistic point of view.
This paper attempts to show that that approach, even when
mathematically correct, has often been based on questionable or
invalid assumptions which lead to equally questionable or invalid
conclusions. Though it does utilize applicable theory, this
paper emphasizes the physical or real world factors which should
serve as guidelines for the development of realistic systems
reliability models. Section 2 presents definitions and basic
models for parts and sockets and some systems reliability models
are briefly described in section 3. Following this background
material, the next section discusses some recent graphical results
which indicate the slowness with which systems approach exponentiality, the model often assumed from system age zero. By considering realistic values for the mean-time-to-failure (MTTF) of
the parts comprising electronic and mechanical systems, it is
shown that the assumption of exponentiality is highly questionable,
even for many systems which have been operated for their entire
useful lives. Section 5 shows that the statistical approach has
great potential for improving our understanding of how real
systems behave. It is also stressed that, of the few statistical
treatments available in the literature, a large percentage are
invalid because trend tests were not performed properly. Hence,
the final section concludes that better application of existing
probabilistic models and statistical techniques is required
for meaningful assessment of systems reliability.

2. BASIC CONCEPTS

A. Definitions

The following definitions apply for the purposes of this paper.

(1) <u>Part</u>. An item which is not subject to disassembly and hence, is discarded the first time it fails.

(2) <u>Socket</u>. A circuit or equipment position which, at any given time, holds a part of a given type; as parts fail, they are replaced by new or good-as-new parts from the same population as the original part.

<u>Comment</u>: The term "good-as-new" should be interpreted in the reliability sense. That is, not only should the performance of the good-as-new part meet new part specifications, in addition, its distribution of time to failure must be identical to the new part's failure distribution.

(3) <u>System</u>. A collection of two or more sockets and their associated parts, interconnected to perform one or more functions.

(4) <u>Non-repairable system</u>. A system which is discarded the first time that it ceases to perform satisfactorily.

<u>Comment</u>: An example of a non-repairable system is an unmanned satellite which has no provision for remote switching of redundant circuits. It is noted that though a non-repairable system contains at least two parts, it is indistinguishable from a part for some purposes. For example, if an unmanned satellite fails and the function it performs is to be continued, another satellite must be orbited. If the replacement satellite is from the same population as the failed one then, in a sense, it is the replacement part in the "socket in the sky".

(5) <u>Repairable system</u>. A system which, after failure to perform at least one of its functions, can be restored to performing all of its required functions by the replacement of, at most, some of its constituent parts.

<u>Comment 1</u>: The above definition is worded to include the possibility that no parts are replaced. For example, the system might be repaired by an adjustment or by a well directed kick.

<u>Comment 2</u>: A system which has redundant paths which are repaired, but which is discarded as soon as it fails to perform at least one of its required functions satisfactorily, is <u>not</u> considered a repairable system.

B. Basic mathematical models for parts and sockets. Since these models are discussed in Ascher (1973) only the most basic notions concerning hazard functions and renewal rates will be covered here.

(1) <u>Hazard function of a part</u>. If $T \equiv$ random variable, time to failure, then

$$F_T(t) \equiv \Pr\{T \leq t\} \equiv \text{distribution function of } T$$

$$h_T(t) \equiv \frac{F_T'(t)}{1-F_T(t)} \equiv \frac{f_T(t)}{R_T(t)} \equiv \text{ hazard function of T}$$

$$h_T(t)\,dt = \Pr\{t < T \le t+dt \mid T > t\} \qquad (1)$$

Eq. (1) is interpreted as follows: $h_T(t)\,dt$ is the conditional probability that a part, from the population with distribution of time to failure $F_T(t)$, put into service at $t = 0$ and known to have operated until t, fails in $(t, t + dt)$.

(2) <u>Renewal rate of a socket.</u> A renewal process is defined as a sequence of independent, non-negative identically distributed random variables, not all 0 with probability 1. We will identify this sequence of random variables as the successive times-between-failures (interarrival times) in a socket[1]. Then, if we let $M(t)$ be the expected number of failures in the socket[2] in $(0,t)$, the renewal rate, $r(t)$, is defined as $M'(t)$. In this context, $r(t)\,dt$ can be interpreted as the (unconditional) probability that the part in the socket at time t, fails in the interval $(t, t+dt)$. Since the part in the socket at t may be the first, second, third, etc., part installed in the socket from the time the first one was installed at $t=0$, the failure in $(t, t+dt)$ will correspondingly be the first, second, third, etc., failure in the socket.

In the special case where the interarrival times are exponentially distributed, it can be shown that the renewal rate becomes a constant, λ, which is numerically equal to the hazard function of each part installed in the socket. Even in this special case, however, there is <u>no</u> equivalence between renewal rate and hazard function since the condition which results in numerical equality does not alter the fundamental differences in the way these two terms are defined. That is, the hazard function is still the ratio of the pdf to reliability function of a distribution whereas the renewal rate is still the time derivative of an expected number of failures. Therefore, $h_T(t)$ is still the intensity with which <u>one</u> part is tending to fail and $r(t)$ remains the rate of occurrence of failures of the parts successively installed in the socket. The prime reason that these concepts have been confused so often is that <u>each</u> has often been called

---

[1] It should be noted, however, that a renewal process is <u>not</u> automatically the correct model for a socket. For example, the stress on the part in the socket may change over the course of time, hence changing the distribution of time to failure even for nominally identical parts.

[2] It is assumed here and throughout the paper (except for a short discussion of system availability) that repair times are either instantaneous or measured on a different time scale.

failure rate. It is probably inevitable that such confusion occurs when distinct concepts, which become numerically equal in the special case usually assumed, are given the same name.

3. SYSTEMS RELIABILITY MODELS

A very large number of stochastic processes have been proposed to model systems reliability. Many of these models have resulted from consideration of complex internal structures. In this paper, only the following "external" models for repairable systems will be considered:
- A. Homogeneous Poisson Process (HPP)
- B. Renewal Process
- C. Non-homogeneous Poisson Process (NHPP)
- D. Branching (or Clustered) Poisson Process
- E. Non-homogeneous Branching Poisson Process
- F. The "Catastrophic" Model

A. Homogeneous Poisson Process (HPP)

In order to define a HPP we must first introduce the notion of a counting process. A stochastic process $\{N(t), t \geq 0\}$ is said to be a counting process if $N(t)$ represents the total number of events which have occurred in the interval $(0,t)$. The counting process $\{N(t), t \geq 0\}$ is said to be a HPP if
  (1) $N(0) = 0$
  (2) $\{N(t), t \geq 0\}$ has independent increments
  (3) The number of events (in our context, failures) in any interval of length $t_2 - t_1$ has a Poisson distribution with mean $\rho(t_2 - t_1)$. That is, for all $t_2 > t_1 \geq 0$,

$$\Pr\{N(t_2) - N(t_1) = m\} = \frac{e^{-\rho(t_2-t_1)} \{\rho(t_2-t_1)\}^m}{m!}$$

for $m \geq 0$.
From condition (3) it follows that

$$E\{N(t_2 - t_1)\} = \rho(t_2 - t_1)$$

where the constant, ρ, is the rate of occurrence of failures. The present author has called the Poisson process's rate of occurrence the "peril rate" in previous papers and this nomenclature will be retained here.

It can be shown that the successive times-between-failures of the HPP defined above are independent and identically distributed exponential random variables. Hence, the HPP is a special case of a renewal process and it is the appropriate model for a socket containing parts with exponentially distributed times-between-failures. This process also has been used widely as a model for the number of system failures over a given time interval. One reason for this choice is its mathematical tractability. In addition, if all the following assumptions are made, it is readily shown that the HPP is the correct model. The assumptions (henceforth, called Assumption Set A) which must be satisfied, at least to an adequate approximation, are that all parts must be in series and have constant hazard functions and that none of the following factors have significant effect:

(a) Secondary failures or damage to other parts caused by the original failure.

(b) Damage/failures caused by repair procedures or improper operation.

(c) Incomplete repairs (i.e., the real defect is not found until the ith repair attempt, $i > 1$).

(d) Many parts may be susceptible to the same stress experienced by the system, e.g., severe temperature cycling, power line fluctuations or on/off cycling. Moreover, factors like on/off cycling may be more important than total operating time.

(e) Drift failures and system instabilities.

(f) Reliability growth due to design improvements during the system life cycle.

(g) Overhauls-these are incompatible with the constant hazard assumption but may be performed anyhow.

(h) Unusual failure modes which may affect normally redundant channels.

(i) Deterioration in storage, shipping or installation.

(j) The system may have different modes of operation with different rates of occurrence of failures.

(k) The system may be down even if no parts are failed or conversely, one or more parts may fail without causing system failure.

It has been shown (Drenick (1960)) that the HPP is also the appropriate model under a slightly different set of assumptions. If we assume that the system has been operating for a very long period of time and is very complex then we can drop the requirement for each part to have a constant hazard function. All other assumptions of Set A still must hold in this set, henceforth called Set B. Unless indicated otherwise, Set B also applies to all other models discussed in this paper.

Many papers have presented much more complex models than the HPP. These models are based on complex redundancy and/or involved maintenance schemes and are often aptly described by the phrase (Evans (1971)), "solutions looking for problems". In spite of these papers, the HPP has been the "old standby" in the reliability field. While it certainly has a place as a first order model for many systems, it has been overemphasized, i.e., there is no valid reason to believe that Assumption Sets

A or B apply as often as the HPP is presented as a systems
reliability model. For example, it is empirically evident that
it does not model an automobile appropriately since, among other
reasons, a car does not have a constant rate of occurrence of
failures. The graphical studies of Section 4 will be used to
provide further evidence of the overemphasis on the HPP.

B. Renewal Process

This process was introduced earlier as the most plausible model
for a socket, though it was stressed that, even in that case, it
is not necessarily the appropriate model. Its plausibility as a
systems model is greatly reduced-particularly when its special
case, the HPP, is excluded from consideration. This is because
it is generally unrealistic to assume that a repair renews a system
in a reliability sense, even if performance is good-as-new, since
most of the parts comprising the system have the same age after
repair is completed as they did just before the repair commenced.
Hence, as shown in Ascher (1970), successive times-between-failures
in general will not be independent and identically distributed.
This argument, however, is inapplicable when Assumption Set A or
B applies but then the special case of a renewal process, the HPP,
is the appropriate model. The renewal process has been over-
utilized as a systems model because of:

(a) Confusion between good-as-new performance and good-as-new
reliability.

(b) The fact that the renewal process is one of the possible
generalizations of the HPP.

The overemphasis on the renewal process extends to the conventional wisdom concerning steady-state availability. The following formula for availability (A) is the one almost universally
used:

$$A = \frac{MUT}{MUT + MTTR} \quad (2)$$

where

MUT = Mean Up Time
MTTR = Mean Time to Repair.

While it is often recognized that Eq. (2) holds exactly only as
system operating time increases without bound from its initial
condition availability, it is seldom realized that it is also based
on the assumption that successive failures and repairs are from an
alternating renewal process. That is, this equation follows from
the assumption that up times are from a renewal process and down
times are from another renewal process, independent of the up time
process. While it is possible that other sufficient conditions
to assure the validity of Eq. (2) could be found, this equation
does not have the universal applicability so often ascribed to it.
For example, empirical considerations indicate that it does not
apply to an automobile which begins its life with low availability,
then improves, and subsequently degrades to poor availability

again. It is true that a "steady state" value is finally attained but this value is zero corresponding to the car's consignment to the scrap pile!

C. Non-homogeneous Poisson Process (NHPP)

The NHPP differs from the HPP only in that the peril rate varies with time rather than being a constant. That is, conditions (1) and (2) are retained and condition (3) is modified to be:

(3a) The number of failures in any interval (t_1, t_2) has a Poisson distribution with mean $\int_{t_1}^{t_2} \rho(t) dt$

That is, for all $t_2 > t_1 \geq 0$

$$\Pr\{N(t_2) - N(t_1) = m\} = \frac{e^{-\int_{t_1}^{t_2} \rho(t) dt} \left\{\int_{t_1}^{t_2} \rho(t) dt\right\}^m}{m!}$$

for $m \geq 0$.

From (3a) it follows that

$$E\{N(t_2) - N(t_1)\} = \int_{t_1}^{t_2} \rho(t) dt$$

The interpretation of $\rho(t)dt$ is that it is the probability that a system put into service at $t = 0$ and repaired in a "bad-as-old" sense (defined shortly) fails in the interval $(t, t+dt)$. It is hardly surprising that this is a similar interpretation to $r(t) dt$ since in the special case of the HPP $\rho(t) \equiv r(t) =$ a constant, usually denoted by λ. It is stressed that, just as for the renewal rate, the relationship between a constant peril rate and the hazard function of the corresponding exponential distribution of interarrival times is one of numerical equality rather than of equivalence.

The physical argument for the applicability of the NHPP stems from the fact that most systems are composed of so many parts that only a very small percentage are replaced in any one repair. Therefore, it is reasonable to assume that, even if a repair renews a system to its original performance specifications, from the reliability viewpoint it leaves the system with its full age. Since this is diametrically opposite to the good-as-new reliability imposed by the selection of the renewal process as a model, this assumption has been dubbed "bad-as-old" by the present author (Ascher, 1968).

In Ascher and Feingold (1969) it was shown that this assumption leads to the NHPP as the appropriate system reliability model. Starting from the approach of superimposing the renewal processes associated with the sockets comprising a system, Grigelionis (1964) and Blumenthal, Greenwood and Herbach (1973) (henceforth referenced as BGH (1973)) also showed that the NHPP was the appropriate model for the system's early-life. In section 4 we will return to a discussion of how long a system can be considered to be in this early-life state.

It is noted that, in spite of the phrase bad-as-old, the NHPP can be used to model repairable system burn-in or reliability growth. This is readily accomplished by using a peril rate which is a decreasing function of time. Lest the reader object to the phrase bad-as-old on these grounds, it is pointed out that a "good-as-new" system, i.e., one modeled by a renewal process, is more accurately termed "bad-as-new" if its distribution has a decreasing hazard function.

Downton (1969) made the "bad-as-old" assumption but he was concerned with the distributions of the successive times-between-failures. In particular, he showed that as the number of failures $\longrightarrow \infty$, the interarrival times approach exponentiality. This is a significant theoretical result but may have little practical impact. For instance, if the interarrival times tend to become too small as they approach exponentiality, the system will be discarded or overhauled before the limit is reached. If, on the other hand, reliability growth is occurring, knowledge of this fact may be more important than knowing the limiting distribution.

D. Branching (or Clustered) Poisson Process

This process is a generalization of the HPP where each failure modeled by the HPP generates a number of failures (possibly zero) modeled by a subsidiary process. Lewis (1964) showed that this process was an adequate representation of the failure patterns of three computers for which he had data. The physical basis for this model is the fact that failed parts are not always located and removed the first time they cause system failure. Hence, this model accounts for Assumption (c) of sets A and B and possibly for Assumptions (a) and (b) as well. Since Lewis (1964) gives a thorough probabilistic and statistical treatment of this model it will not receive further consideration here.

E. Non-homogeneous Branching Poisson Process

As implied by its name this process is similar to the previous one except that the main process is a NHPP. Lewis (1967) studied this model and as he pointed out, one application would be to a computer which is not always repaired at the first attempt and which, additionally, has a high initial rate of occurrence of failures. In this case, however, Lewis did not present data to demonstrate applicability of the model to real systems.

F. The "Catastrophic" Model

Blumenthal, Greenwood and Herbach (1970) considered a model where system failure could be caused either by the failure of a single part or by a catastrophe which caused all parts to fail simultaneously. In the latter case the system was assumed to be replaced with a new one and the failure process continued from that point. It was assumed that times between catastrophes were exponentially distributed. In spite of the name and description of this model it can be applied to any situation where a system is renewed in the reliability sense, e.g., by a thorough overhaul instead of a cataclysm. Of course, it is implausible to model overhauls with the HPP but generalizing the "catastrophic" model to include renewal processes would result in a highly useful model.

4. GRAPHICAL STUDIES

A. Introduction

In section 3.C it was indicated that Grigelionis (1964) and BGH (1973) have shown that the NHPP is the appropriate model for a system's early life. Since the HPP is the appropriate model for a fully aged system it is of interest to determine how the transition is made, what other model(s) may be needed in the transition region(s), and the amount of time that the system is appropriately modeled by each applicable stochastic process. BGH (1973) studied this problem and some graphical results they presented are particularly useful in assessing the impact of a system's age on its reliability. Three of the plots they presented, for gamma shape parameter $\alpha = 1/2, 3/2, 12$ respectively, are shown as Figures 1-3. The notation is defined as follows:

t_1 = system age
μ = mean time to failure (MTTF) of each part
n = number of parts (all in series) in the system
s is a dummy variable defined by

$$\lim_{\substack{t_1 \to \infty \\ n \to \infty}} R(\mu s/n; t_1, n) = \exp(-s) \qquad (3)$$

where $R(\mu s/n; t_1, n) =$
Pr $\{$ surviving interval $(t_1, t_1 + \mu s/n) |$ a failure at $t_1 \}$

Before these graphs are discussed the following points must be noted:

(a) the original graphs included curves for finite number of parts (n=64 and 256). Since these curves are, for practical purposes, the same as those for $n = \infty$ they are not reproduced here.

(b) notation has been changed, to be consistent with that used earlier in this paper.

The plots should be interpreted as follows: In order to compress the interval $(0, \infty)$ into a finite scale, the abscissa is plotted as $\exp(-t_1/\mu)$. Therefore, as t_1 varies from 0 to ∞ the abscissa goes from 1 to 0, i.e., the graphs must be read from

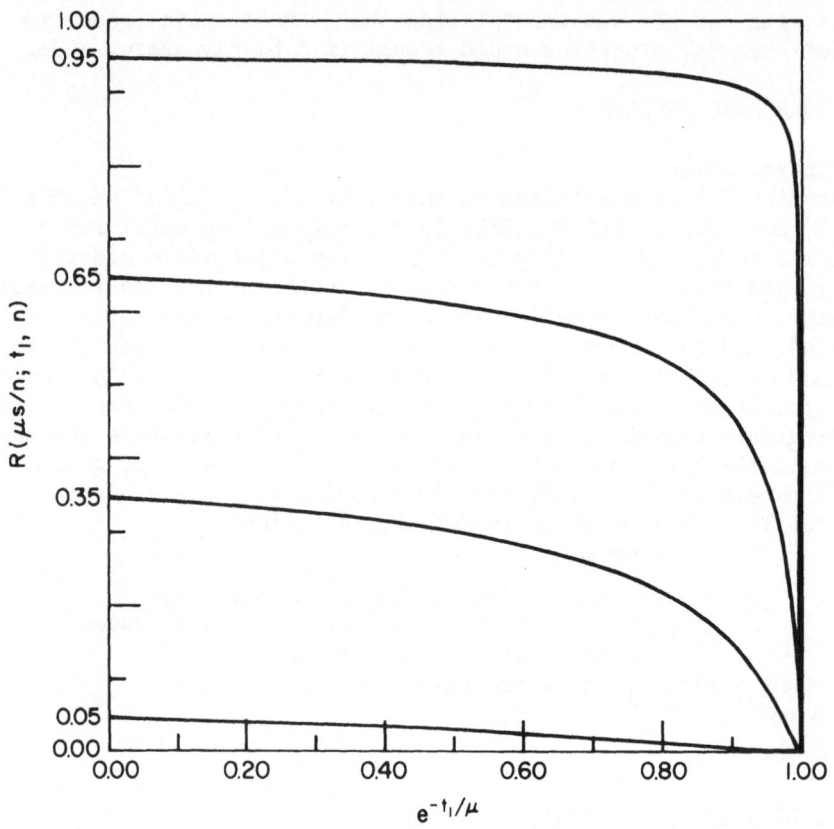

Figure 1. System reliability as a function of its age (t_1) and part MTTF (μ), for n parts with gamma failure law, shape parameter = 1/2.

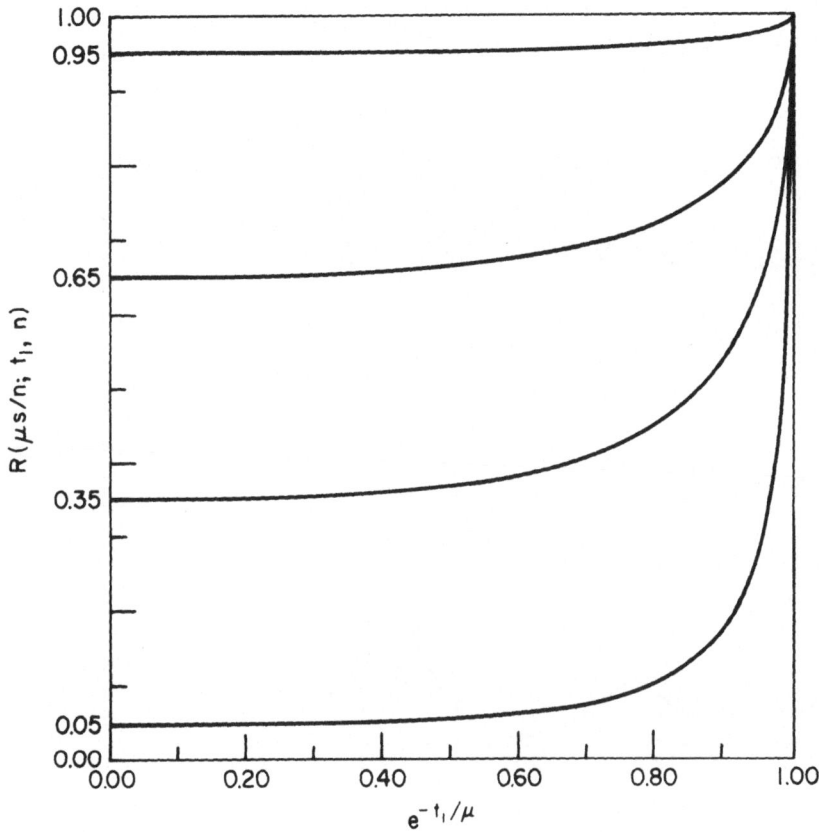

Figure 2. System reliability as a function of its age (t_1) and part MTTF (μ), for n parts with gamma failure law, shape parameter = 3/2.

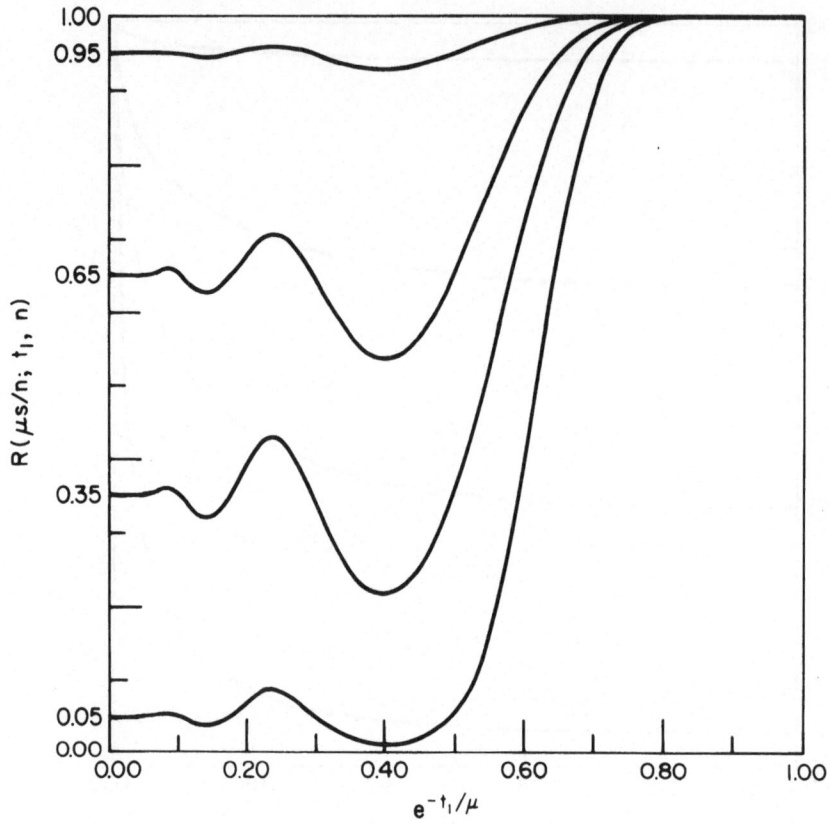

Figure 3. System reliability as a function of its age (t_1) and part MTTF (μ), for n parts with gamma failure law, shape parameter = 12.

right to left as system age increases. The four plots shown on each figure are for different values of s such that (3) takes on the values 0.05, 0.35, 0.65 and 0.95. For any given interval, w, for which R is to be calculated, w is set equal to $\mu s/n$. Then $s = nw/\mu$ and the curve which has its limiting value equal to exp (-s) is selected to calculate R as a function of system age. It should be noted that in the case where $\alpha = 1$ (exponential distribution) the four curves would be horizontal lines with ordinates 0.05, etc.

Clarification of how the plots are interpreted is best accomplished by means of an example, as presented in BGH (1973). Consider a system of n = 300 series parts each having the same gamma failure distribution with MTTF = 10,000 hours. The problem is to find the probability of surviving a mission time w of 100 hours when the system age is 15,000 hours and a failure has just been repaired. The pertinent abscissa value is exp (-1.5) = 0.22. Further, s = 3 and exp (-s) = 0.05. Therefore, this system has a time equilibrium ($t_1 = \infty$) reliability of 0.05 for a 100 hour period. Its reliability for a 100 hour period when it has an age of 15,000 hours is obtained from the graphs, looking at the curve which has an ordinate of 0.05 at $t_1 = \infty$. For $\alpha = 1/2$, 3/2, 12, R = 0.04, 0.055 and 0.085, respectively.

B. Application of graphical results to "electronic" and "mechanical" systems.

It should be realized that these plots are not exactly applicable to any real system. In addition to all the other assumptions required for their development (set B, less the requirements for infinite age and complexity) they assume parts with identical failure laws, a situation which seldom applies in the real world. Nevertheless, these results will be used to obtain a first order approximation for the age dependent reliability of "electronic" and "mechanical" systems.

(1) <u>Electronic systems</u>. The "failure rates" (actually, hazard functions as defined in this paper) quoted for most electronic parts in MIL-STD-721A are of the order of 0.1×10^{-6}/hr. to 1×10^{-6}/hr. If the constant hazard assumption used to obtain these figures is not exactly correct then the estimate of MTTF, obtained by taking the reciprocal of the hazard, will also be in error. However, to an approximation, μ can be taken to be of the order of 10^6 or 10^7 hrs. particularly if we can assume at least partial cancellation of error because of a mixture of parts with increasing and decreasing hazards. Then, even if $t_1 = 100,000$ hours (more than eleven years of continuous operation!) the value of the abscissa is 0.9 or more and the system cannot be assumed to be in time equilibrium even after this protracted period of operation. Hence, for electronic systems with parts with gamma failure laws, only the special case of parts with exponential

failure laws will result in equilibrium over operating periods of interest. Since no real part will have exactly constant hazard function it would appear that real electronic systems could not be considered to achieve equilibrium. The assumption of the gamma failure law is artificial, however, because of the sharp discontinuities in its hazard function for $\alpha \approx 1$ and $t = 0$. That is, for $\alpha = 1$ $h_T(0) = \lambda$, $0 < \lambda < \infty$, but for $\alpha < 1$, $h_T(0) = \infty$ and for $\alpha > 1$, $h_T(0) = 0$. The selection of a failure law, T^*, whose hazard had to following properties: $h_{T^*}(0) < \infty$ in the decreasing hazard case and $h_{T^*}(0) > 0$ in the increasing case, would result in more realistic alternatives to the constant hazard case. However, the choice of an appropriate failure law, and particularly, values of its parameters, would have to be based on data and adequate data on either the part or system level are woefully lacking.

Until such time when good data on the failure laws of parts become available, it would appear that the results derived from one viewpoint by Grigelionis (1964) and BGH (1973), and from another by Ascher and Feingold (1969), should be considered the appropriate model for electronic systems reliability whenever the HPP cannot explicitly be justified. That is, the NHPP should be adopted as a first order model until/unless evidence is available to refute its use. This step has already been taken by proponents of reliability growth, who, however, usually refer to the NHPP as a "Weibull Process". When clustering of failures is pronounced, as found in the case of three computers by Lewis (1964), either the branching HPP (Lewis 1964) or the nonhomogeneous branching Poisson process (Lewis 1967) must be substituted as an appropriate model.

(2) <u>Mechanical systems.</u> Most mechanical parts have wearout characteristics, implying $\alpha > 1$. Since the (truncated) normal distribution has often been used as a model for such parts and the gamma distribution approaches normality as $\alpha \to \infty$, Figure 3 which is based on $\alpha = 12$ will be used. Automobiles will be used as examples - they are not necessarily typical mechanical systems but they are the ones most familiar to both laymen and reliability analysts. Of course, the parts comprising an automobile are not physically identical but, to a first approximation, their mean lives may be comparable. That is, because of the highly competitive nature of the automobile industry, it is reasonable to assume that parts are not designed to greatly outlast the cars in which they are installed.

We will assume a total operating life of 5,000 hours based on say, 150,000 miles of operation at an average speed of 30 miles per hour. Then, if we assume that equilibrium is reached even at the end of the car's useful life, we must assume that $\exp(-5000/\mu) \approx 0.1$ or $\mu \approx 2000$ hours. Further, we will assume 10,000 parts in series based on an estimate of 12,000-15,000 parts in a car (Jacobs 1972). But then the equilibrium MTBF of the system is

$$\text{MTBF} = 1/(10,000/2,000) = 0.2 \text{ hours!}$$

Obviously, an MTBF of this order would be ample cause to scrap the car but a glance at Figure 3 indicates that reliability is <u>even poorer</u> in the region where the abscissa is approximately 0.4. Though this example is oversimplified, the extreme nature of the results implies that not only is equilibrium not reached, in addition, most cars are probably scrapped well before the first (local and global) minimum is reached. The existence of minima in Figure 3 is explained by the increased clustering of system failures near the MTTF's of the parts as the shape parameter of the gamma distribution increases. Hence, the local maxima are due to partial renewing of the system because of mass replacements in a short interval.

The conclusion to be drawn from the above analysis is that cars, as a rule, remain in the "early life" region for their entire useful lives. (Note that this implies that $\mu \gg 2,000$ hours.). Hence, the NHPP, which is the appropriate "early life" model may well be applicable over the entire life of a car. In other words, as the car's age approaches the MTTF of its parts, its progressively poorer reliability causes it to be scrapped.

The question remains, are there mechanical systems with large numbers of parts with small MTTF's, which nevertheless are operated for long periods of time? In other words, are there systems with combinations of n, μ and t_1 that would place them into the global minimum or equilibrium regions? There may be systems, particularly those with small numbers of parts, which have acceptably high equilibrium reliability. However, for large n, system overhaul may be a necessary substitute for equilibrium operation. Such overhaul may essentially renew the system, thus enabling it to deteriorate towards unacceptable reliability once again. The appropriate stochastic model then would be a regenerative process, see Ross (1970), or as noted earlier, BGH (1970).

5. STATISTICAL TECHNIQUES

Up to this point in the paper, the problem of determining appropriate system reliability models has been based on probabilistic and empirical viewpoints. Each of these approaches has serious drawbacks. The tractability of probabilistic models depends on invoking most of Assumption Set B. Since these assumptions are often questionable, the applicability of the results is equally conjectural. In a similar vein, the intuitive notions which exist about system reliability, particularly for automobiles, may also be in conflict with reality. The remaining approach, the statistical analysis of systems failure data, has great potential for improving our understanding of systems reliability. This is because the statistical approach is neither highly dependent on the assumptions required for a probabilistic analysis nor on intuitive considerations. It must be stressed, however, that it has its own limitations, for example, results are strongly dependent on test ground rules, failure criteria and the amount of data available.

In spite of their potential, statistical techniques have contributed very little to our overall knowledge of systems reliability. This is partly due to the scarcity of statistical analyses in the literature and partly a result of invalid techniques which have often been used in the few available studies. The prime problem is a lack of analysis of times-between-failures in their original chronological order. That is, instead of performing a test for trend on chronologically ordered data, analysts often consciously or unconsciously assume a renewal process as a model. The data are immediately reordered by magnitude and the scope of the analysis is restricted to determining an appropriate distribution and estimates of its parameters. The grossly misleading results which can arise from this approach are depicted in Bassin (1969) and Ascher and Feingold (1969). It is also noted that similar problems arise in fields other than reliability. For example, a series of papers and notes, Maguire, Pearson and Wynn (1952, 1953) and Barnard (1953) demonstrated the distortion caused by disregarding the chronological order of times between successive mine accidents in Great Britain.

6. CONCLUSIONS

It is certainly difficult, if not impossible, to make firm decisions about appropriate system models at the present time. While theory and intuition suggest various versions of the Poisson Process as suitable candidates, there may be large gaps between the assumptions needed for mathematical tractability and reality. Statistical techniques have the potential for assessing the magnitude of these gaps but very few valid statistical analyses are presently available. Therefore, the call put forth in the past (Krohn (1969), Ascher (1970)) for publication of systems failure data together with appropriate analyses remains as pertinent as ever. Until such analyses are forthcoming, our knowledge of systems reliability will remain rudimentary. Better implementation of existing techniques, particularly in the statistical area, is needed to provide a firm foundation for systems reliability modelling.

REFERENCES

Ascher, H. (1968). "Evaluation of Repairable System Reliability Using the 'Bad-As-Old' Concept," IEEE Rel. Trans., Vol. R-17 pp. 103-110.

Ascher, H. (1970). "Hazard Functions, Renewal Rates and Peril Rates," Ninth R&M Conf., Soc. Auto. Eng., New York.

Ascher, H. (1973). "Peril Rate/MTBF (t_1, t_2)," Annual R&M Symp., IEEE Cat. No. 73CH0714-6R.

Ascher, H. and Feingold, H. (1969). " 'Bad-As-Old' Analysis of System Failure Data," Eigth R&M Conf., Gordon and Breach, New York.

Barnard, G. (1953). "Time Intervals Between Accidents - a note on Maguire, Pearson, and Wynn's paper," Biometrika, Vol. 40, pp. 212-213.

Bassin, W. (1969). "Increasing Hazard Functions and Overhaul Policy," Annual Symp. on Rel., IEEE Cat. No. 69 C 8-R.

Blumenthal, S., Greenwood, J., Herbach, L. (1970). "Reliability and Availability Demonstration: The Renewal Model and Some Competitors," New York University Tech. Report No. 1559.03.

Blumenthal, S., Greenwood, J., and Herbach, L. (1973). "The Transient Reliability Behavior of Series Systems or Superimposed Renewal Processes," Technometrics, Vol. 15, pp. 255-269.

Downton, F. (1969). "An Integral Equation Approach to Equipment Failure," J. R. Statist. Soc. B, Vol. 31, pp. 335-349.

Drenick, R. (1960). "The Failure Law of Complex Equipment," J. Soc. Indust. Appl. Math., Vol. 8, pp. 680-690.

Evans, R. (1971). "Data We Will Never Get," IEEE Rel. Trans., Vol. R-20, p. 2.

Grigelionis, B. (1964). "Limit Theorems for Sums of Repair Processes," Cybernetics in the Service of Communism, Vol. 2, pp. 316-341. Clearinghouse for Fed. Sci. and Tech. Inf., Washington, D.C.

Jacobs, R. (1972). "Conference Report," Micro. and Rel., Vol. 11, pp. 257-261.

Krohn, C. (1969). "Hazard Versus Renewal Rate of Electronic Items," IEEE Rel. Trans., Vol. R-18, pp. 64-73.

Lewis, P. (1964). "A Branching Poisson Process Model for the Analysis of Computer Failure Patterns," J. R. Statist. Soc. B, Vol. 26, pp. 398-456.

Lewis, P. (1967). "Non-homogeneous Branching Poisson Processes," J. R. Statist. Soc. B, Vol. 29, pp. 343-354.

Maguire, B., Pearson, E., and Wynn, A. (1952). "The Time Intervals Between Industrial Accidents," Biometrika, Vol. 39, pp. 168-180.

Maguire, B., Pearson, E., and Wynn, A. (1953). "Further Notes on the Analysis of Accident Data," Biometrika, Vol. 40, pp. 213-216.

Ross, S. (1970). "Applied Probability Models with Optimization Applications," Holden-Day, San Francisco.

A TECHNIQUE FOR SELECTIVELY REDUCING SMALL SAMPLE ERROR IN
MAINTAINABILITY DEMONSTRATIONS WHEN APPORTIONMENT
IS BASED UPON PREDICTED RELIABILITY

Steven A. Austin

U.S. Army Electronics Command
Fort Monmouth, New Jersey

Developmental testing of U.S. Department of Defense equipment is a costly process. It is essential, therefore, that these tests be designed and conducted as efficiently as possible. Of special interest in this paper are two statistically related developmental tests--the reliability and maintainability demonstrations. The reliability demonstration involves operating the equipment under stressed conditions for a given period and recording the times to failure. On the other hand, failures are introduced or simulated for the maintainability demonstration, and the times to repair are recorded. Ideally, the failures to be simulated would be those that would occur naturally if the equipment were operated long enough, but getting accurate failure data to use for repair task apportionment is not easy. Often the data from the reliability demonstration is not yet available, or else it is too limited to be of much benefit. In the absence of any useful reliability test data, the apportionment of simulated failures is usually based upon the Reliability Mathematical Model (prediction) prepared by the equipment contractor. There are problems associated with this practice, however. The technique to be discussed here solves some of these problems.

PRESENT METHOD

It is standard practice for the maintainability demonstration tasks to be apportioned according to the technique outlined in MIL-STD-471A, Maintainability Verification/Demonstration/Evaluation. The purpose of this standard is the establishment of uniform procedures, test methods, and requirements for demonstrating the achievement of specified maintainability requirements.

Briefly, the technique is as follows:

1. The Reliability Mathematical Model is used as the basis for all task apportionment. The major units which comprise the system are identified. These now become the functional blocks of the system.

2. The failure rate of each of the functional blocks is calculated from the part failure rates.

3. The relative frequency of occurrence of the failures of each functional block is determined by Equation 1.

$$R_i = \lambda_i / \sum_{i=1}^{n} \lambda_i \qquad \text{(Equation 1)}$$

    R_i is the relative frequency of occurrence of the failures of the ith functional block, λ_i is the failure rate of the ith functional block, and n is the total number of functional blocks in the system.

4. R_i is then the fraction of the total maintainability sample that will be used as simulated failures for the ith functional block.

The purpose of this apportionment procedure is to create a pool of equipment failures that can be repaired in the maintainability demonstration. This procedure works well when certain conditions are satisfied: 1) The reliability prediction is accurate, and 2) the system reliability does not change during development. Realistically, these conditions seldom exist. The initial prediction is usually optimistic, and the equipment reliability almost invariably improves during the development phase. However, because of the nature of the apportionment technique in MIL-STD-471A, an optimistic reliability prediction alone does not adversely affect the maintainability demonstration. Equation 1 has the effect of normalizing the block failure rates, negating any effect of an overall high or low scaling factor. Once the relative frequencies of occurrence have been determined, only the total demonstration sample size is needed in order to determine the exact number of failures that will be introduced into each functional block of the system under test.

A simplified diagram of a Reliability Mathematical Model is shown in Figure 1. This is typical of the models used to divide a system into discrete functional blocks. The consolidation of Block 2a and Block 2b forms a series system. When the repair tasks are selected for the maintainability demonstration, it is only natural that more tasks be allocated to those blocks which

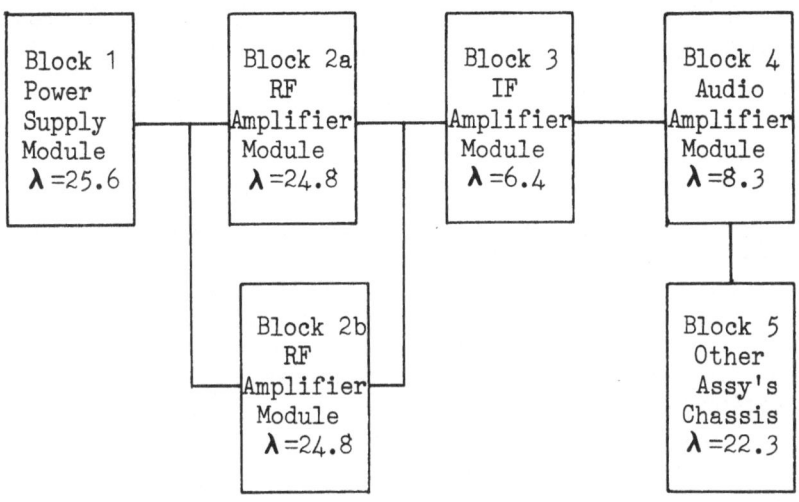

FIGURE 1. Simplified Reliability Math Model.
λ's in failures per million hours.

fail more often and vice versa. If the maintainability demonstration were conducted on the basis of this model alone, as it often is, it would be assumed that the failure rates are static. But, generally, the reliability will have changed somewhat by the end of development. Furthermore, a change in reliability results in a change in maintainability, since maintainability, or more specifically, mean time to repair, is a function of the individual block failure rates and of the failure repair times. The repair times will vary little with design changes, since the times are dependent primarily upon the equipment configuration. Most design changes for the improvement of reliability do not affect the overall equipment configuration. The block failure rates often change dramatically, however. For the sake of argument, let us assume that the reliability of a certain item was improved after the maintainability demonstration had been conducted. When we now attempt to recalculate the maintainability with the new failure rate data, we find that some blocks that were allocated very few simulated failures in the maintainability demonstration now require more tasks, perhaps an order of magnitude more. In essence, many of the smaller samples should have been larger, and the larger samples should have been smaller. When the original maintainability demonstration data is analyzed for the second time, we find an increase of extraneous information because of the small sample error introduced. Consider, for example, an equipment functional block to which only one failure was allocated.

When the new failure rate data is considered, the block may require five or even ten failures, causing the information obtained from the single repair to be weighted considerably higher.

THE DUANE MODEL

To improve our situation, let us examine a model for reliability growth proposed by J.T. Duane (Reference 1). Using a deterministic approach to reliability modeling, Mr. Duane revealed that, for the equipment he analyzed, the cumulative failure rate versus the cumulative operating hours fell close to a straight line when plotted on log-log paper. Mathematically, his failure rate equation may be expressed by

$$\lambda(t) = Ct^{-\alpha} \qquad \text{(Equation 2)}$$

where $C > 0$, α is between 0 and 1, and $\lambda(t)$ is the system cumulative failure rate at operating time t. For a constant failure rate system the mean time between failures (MTBF) of the system operating at time t is

$$MTBF(t) = ((1 - \alpha) C)^{-1} t \qquad \text{(Equation 3)}$$

(see Figure 2). A convenient feature of the Duane model is the use of α. This parameter may be compared to the level of effort applied to improve the reliability of a system. When $\alpha = 0$, it is assumed that the reliability exhibits no growth during development. When $\alpha = 1$, however, it is assumed that all possible resources are being devoted to improving the system reliability. A thorough discussion of this model may be found in Reference 2.

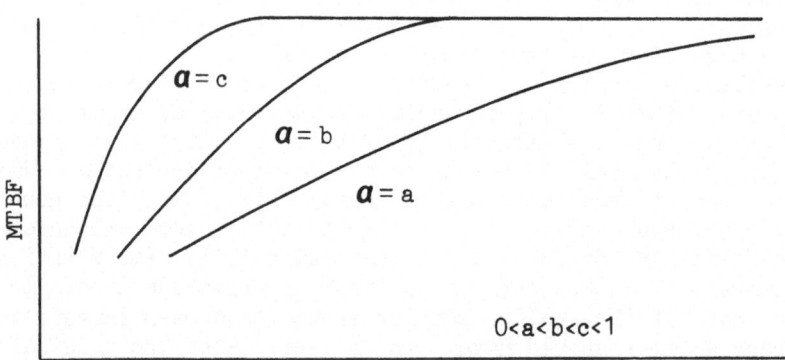

FIGURE 2. Reliability vs. Time of Duane's Model.

The technique to be developed in this paper uses the Duane model to calculate task apportionment for the maintainability demonstration. To implement this technique, we must make several assumptions:

1. The Reliability Mathematical Model can be reduced to a series system.

2. There are a finite number of functional blocks in the Reliability Mathematical Model.

3. On a single block basis, the greatest improvement in the system reliability is realized by reducing the failure rate of the worst functional block.

4. As the block failure rates are improved, one at a time, there is a tendency for the λ_i's to approach a common value.

As a limiting case the expression for the relative frequency of failure occurrence becomes

$$R_i = 1 / n \qquad \text{(Equation 4)}$$

where n is the number of functional blocks in the Reliability Mathematical Model. If the "level of effort" is considered, the above expression can be written

$$R_i = 1 / n \Big|_{\alpha = 1} \qquad \text{(Equation 5)}$$

In other words, all of the resources possible are being expended to improve the system reliability. Figures 3 and 4 illustrate.

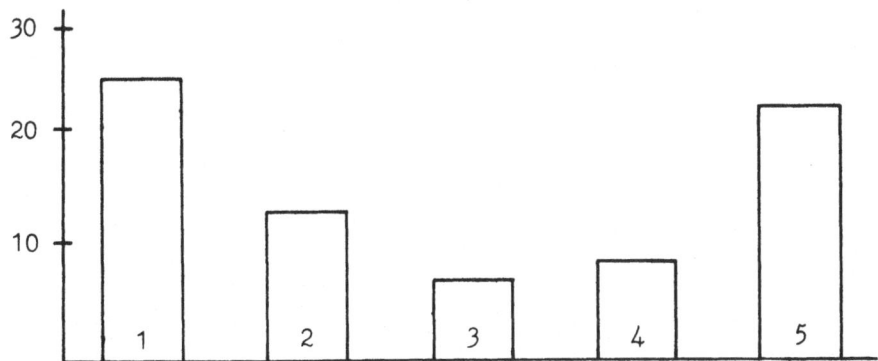

FIGURE 3. Block Failure Rates (λ_i's).

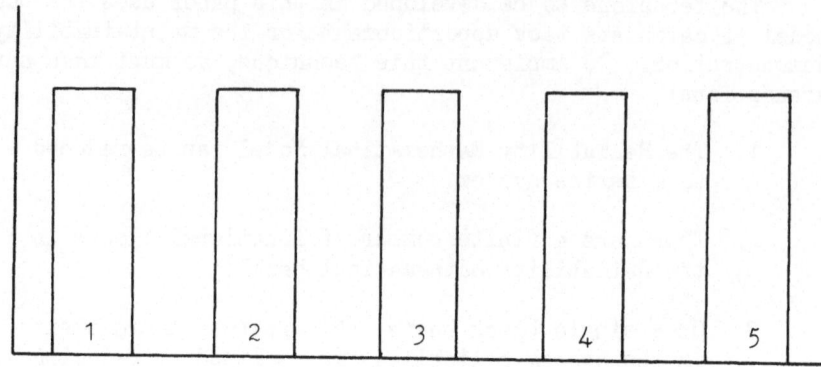

FIGURE 4. Relative Frequencies of Occurrence ($a = 1$).

At the other extreme of a, the "level of effort" is zero. This implies that the reliability will not be improved beyond its current value (no resources are being expended for improvement, i.e., the reliability is static). An expression for the relative frequency of failures may be written

$$R_i = \lambda_i / \sum_{i=1}^{n} \lambda_i \Big|_{a=0} \qquad \text{(Equation 6)}$$

This latter case is representative of the relative frequencies of failure calculated by the procedure of task apportionment in MIL-STD-471A, a method that relies upon the predicted reliability with no allowance for growth.

Theoretically, a would lie somewhere between 0 and 1. Experience has shown that a is between 0.3 and 0.7 for most equipment. The bounds on the relative frequencies of occurrence (R_i's) might be stated

$$\lambda_i / \sum_{i=1}^{n} \lambda_i \leftrightarrow R_i \leftrightarrow 1/n$$

which could be written as an equation

$$R_i = \lambda_i / \sum_{i=1}^{n} \lambda_i \ \times \ \text{Correction Term} \qquad \text{(Equation 7)}$$

To maintain the values of R_i previously stated at each extreme of a, the correction term must vary from 1, when a equals zero, to

$\sum_{i=1}^{n} \lambda_i / n \lambda_i$, when a equals one. Obviously a "power of a" expression is in order. We may write

$$\text{Correction Term} = (n \lambda_i / \sum_{i=1}^{n} \lambda_i)^{-a} \qquad \text{(Equation 8)}$$

When Equations 7 and 8 are combined, the expression for R_i becomes

$$R_i = (\lambda_i / \sum_{i=1}^{n} \lambda_i)^{1-a} / n^a \qquad \text{(Equation 9)}$$

A crucial examination of Equation 9 reveals that, when $a = 1$, the expression reduces to that of Equation 5; and, when $a = 0$, the expression becomes that of Equation 6. Consequently, when the level of reliability growth is zero, the maintainability demonstration tasks are apportioned exactly as in MIL-STD-471A. As reliability growth is applied, however, the apportionment changes appropriately. Figure 4 shows the graphical results of applying Equation 9 to the data of Figure 1. Observe that as a approaches the upper extreme, the maintainability demonstration sample size for each of the functional blocks of the Reliability Mathematical Model becomes

$$\lim_{a \to 1} N_i = 1 / n \qquad \text{(Equation 10)}$$

where N_i is the number of failures to be simulated for functional block i, n is the number of functional blocks, and a is the growth parameter of Duane's model.

IMPROVED PROCEDURE

Using Equation 9, it is now possible to modify our original maintainability apportionment procedure to consider reliability growth. The improved procedure is stated below.

1. The Reliability Mathematical Model is used as the basis for all task apportionment. The major units which comprise the system are identified. These now become the functional blocks of the system.

2. The failure rate of each of the functional blocks is calculated from the part failure rates.

3. A reliability growth parameter (a) is assumed (Duane's model), based upon previous data and system requirements (0.3 to 0.7 typically).

4. The relative frequency of occurrence of the failures of each functional block is determined as follows:

$$R_i = (\lambda_i / \sum_{i=1}^{n} \lambda_i)^{1-a} / n^a \qquad \text{(Equation 9)}$$

where R_i is the relative frequency of occurrence of the failures of the ith functional block, λ_i is the failure rate of the ith functional block, n is the total number of functional blocks in the system, and a is the reliability growth parameter from Duane's model.

5. The R's are then the fraction of the total demonstration sample that will be used for simulated failures for each of the blocks.

CONCLUSION

The technique described here provides a level of assurance for the maintainability demonstration data that varies little with subsequent design changes. This approach yields an increased level of cost effectiveness for the very expensive demonstrations required during developmental testing.

REFERENCES

1. Duane, J.T., "Learning Curve Approach to Reliability Monitoring," IEEE Transactions on Aerospace, Volume 2, Number 2, 1964.

2. Crow, Larry H., "Reliability Growth Monitoring," AMSAA Technical Report Number 35, U.S. Army Materiel Systems Analysis Agency, Aberdeen Proving Ground, Maryland.

3. Goldman, A.S. and T.B. Slattery, *Maintainability: A Major Element of System Effectiveness*, John Wiley & Sons, New York, 1967.

4. Military Standard, Maintainability Verification/Demonstration/Evaluation, MIL-STD-471A, 27 March 1973.

ON DEFINITIONS OF SOFTWARE RELIABILITY
AND COMPUTING RELIABILITY

Igor Bazovsky, Jr.

Igor Bazovsky and Associates, Inc.
Sherman Oaks, California

ABSTRACT

Extended definitions of both Software and Computing Reliabilities are given in terms of general items which need to be specifically considered when actual predictions are to be made. For Computing Reliability, not only the computer system and algorithm is considered, but also the user's operating procedures and the field conditions. The user procedures are treated as an algorithm which is an extension of the one in the computing system. The paper proposes that the given definitions be used to unify the many varying measures of algorithm complexity.

INTRODUCTION

There are many different ways of defining algorithm complexity, or equivalently algorithm efficiency, that the user may well be at a loss as to which algorithm is best for his situation. This paper combines many of these measures of complexity into a unified figure of merit by using the concepts of Software Reliability and Computing Reliability. This unifying approach is considered worthwhile because many of the measures of algorithm efficiency have mutually opposing effects, and hence the Dominated Ordering concept is not applicable.

The measures of algorithm complexity used today are (reference 1): (a) speed of computation, (b) error in computation, (c) storage space needed during computation, and (d) number of gates used.

From the viewpoint of the unified figure of merit given later these measures are just individual pieces of an algorithm's specification. As such they are analogous to the dimensions and delays of a black box. However, the user is interested in the transfer function so that he can calculate the advantages and disadvantages of choosing one particular black box over another. As far as the algorithm is concerned, the user wants the best assurance of adequate computation. But suppose that speed in computation is a factor in his situation. Perhaps he can make a faster algorithm at the expense of more error. Or perhaps he can speed up his algorithm by working in parallel instead of sequentially. But both these proposed improvements introduce a degradation factor. The first introduces more error, and the second requires more hardware, more energy consumption (because parallel computations in general require more steps of computation to get the same accuracy), and a lower hardware reliability. On the other hand keeping the hardware constant but introducing double precision arithmetic keeps the reliability high and reduces round-off errors, but it reduces the speed of calculation. What is the relative importance of these two tendencies? One step in answering this question is to determine whether the measures (a) through (d) above are for a worst case or for an average case. An algorithm is <u>minimax optimal</u> (or worstcase optimal) if no algorithm is more efficient in the worst case. An algorithm is <u>minimean optimal</u> (or average case optimal) if no algorithm is more efficient in the average case. However, Minimean and Worstcase specifications are in terms of all possible a priori inputs, whereas an individual user might have a unique input environment so that what is average in an a priori situation might be worst case for his, and what is worstcase in general might be impossible for a specific situation. These considerations motivate the definitions of Software and Computing Reliabilities given below. But first the question of program correctness will be considered.

Experience has shown that the probability of error, and especially of large, catastrophic errors is very much a function of the debugging effort that has accompanied the algorithm's implementation (reference 2). The following few paragraphs describe a method of proving program correctness which, because it uses the concepts of modern abstract system theory (reference 3), seems to hold a lot of promise. This is a method of proving correctness by imbedding the program in a previously proven program, or by unfolding it into a loop free version. These operations of imbedding and unfoldment are done via the notions of Category Theory, a very recently introduced concept.

<u>Definition 1</u>. A category C consists of: (a) two classes: one is denoted by $|C|$ and its elements are called <u>objects</u>. The other is denoted by C and its elements are called <u>morphisms</u>. Each morphism has two objects attached to it, one is called the <u>source</u>, the other is called the <u>target</u>. (b) Two functions ∂_i, $i = 0, 1$, called <u>source</u> and <u>target</u> respectively, which assign to each

morphism, f, its source and target objects. Thus $\partial_0(f)$ = (Source of f), and $\partial_1(f)$ = (target of f). (c) A function 1: $|C| \to C$ assigning to each object A in $|C|$ its <u>identity morphism</u>, 1_A. For 1_A we have $\partial_0(1_A) = \partial_1(1_A) = A$. (d) A <u>partial</u> binary operation, x, called <u>composition</u>, which is defined between two morphisms f and g if and only if $\partial_1(g) = \partial_0(f)$, and yielding a morphism fxg, also written fg. I.e., if the target object of g is the source object of f, then the two morphisms can be composed much like two functions can be composed to yield f(g(x)). (e) Composition is <u>associative</u> whenever defined; and $1_A f = f = f 1_A$ whenever defined.

<u>Definition 2.</u> A <u>Functor</u> from a category B to a category C consists of two parts: the <u>object part</u> is a function $|F|: |B| \to |C|$ and the <u>morphism part</u> is a function $F: B \to C$. These functions are connected by consistency requirements such that (a) $\partial_i(F(f)) = |F|(\partial_i(f))$, i = 0, 1, f in B; (b) F(fg) = F(f) F(g) whenever fg is defined in B, and (c) $F(1_A) = 1_{|F|(A)}$ for all A in $|B|$.

An example of a category is Pfn, the category with sets as objects and Partial functions as morphisms. Another example is Pa(G), the category of all paths from a directed graph G. Pa(G) is a category whose objects are the nodes of G, and whose paths are all finite possible compositions of edges in G.

<u>Definition 3.</u> A <u>Program</u> is a functor $\hat{P}: Pa(G) \to Pfn$, for some graph G.

Thus a program assigns sets to the nodes of a graph and partial functions to the edges of a graph. The sets correspond to the possible values of the program at each stage (node), and the partial functions correspond to the computations performed by the program in going from one stage ot the next.

<u>Definition 4.</u> A <u>Program Homomorphism</u> is a pair (F,E) between two programs $\hat{P}_0$ and $\hat{P}_1$, where $\hat{P}_0: Pa(G_0) \to Pfn$, $\hat{P}_1: Pa(G_1) \to Pfn$, and F: $Pa(G_0) \to Pa(G_1)$ is a Functor, and E: $\hat{P}_0 \Rightarrow \hat{P}_1 F$ is a <u>Natural Transformation</u>. A natural transformation E: $P_0 \Rightarrow \hat{P}_1 F$ from the functor $\hat{P}_0$ to the functor $\hat{P}_1 F$ is a <u>family</u> of morphisms, E_v indexed by the objects, v, of the source category $Pa(G_0)$ such that in the target category Pfn, the following relation is satisfied. When e is an edge in G_0, with nodes v, w for source and target respectively, then

$$\partial_0 \left(\hat{P}_1 F(e) \right) = E_v \left(|\hat{P}_0|(v) \right) = E_v \left[\partial_0 \left(\hat{P}_0(e) \right) \right]$$

$$\partial_1 \left(\hat{P}_1 F(e) \right) = E_w \left(|\hat{P}_0|(w) \right) = E_w \left[\partial_1 \left(\hat{P}_0(e) \right) \right].$$

Thus a program homomorphism provides a consistent mapping between programs such that program equivalence or imbedding can be proved much quicker and much more conveniently than by previous methods (reference 4).

SOFTWARE RELIABILITY

The essential idea of Software Reliability is that it gives the probability of adequate computation under a set of fixed conditions. In order to make precise the notions of "adequate" and of "fixed conditions," consider the following.

<u>Definition 5.</u> The Software Reliability of an algorithm, A, is the conditional probability, R(A), given the algorithm operates on input data, I, drawn by a specified method, M, from a specified class, D, of possible data sets, under specified means of calculation, C, under specified external conditions, E, and under a given amount of debugging effort, B, that the output, O, will be correct within specified tolerance limits, T(I,E). I.e.:

$$R(A) = P\left(O \text{ in } T(I,E) \mid D,C,E,I \text{ in } D, B\right)$$

(The computer hardware and organization characteristics are included via the parameter C). Denote by $\bar{x}$ the value of x, got via C.

<u>Example.</u> Consider an algorithm which computes the geometric series summation by $1 + x + \ldots x^N = S$. Let D be the set of numbers between zero and one, let I be a one element set, $\{x\}$, of the set D, and let M be characterized as selecting distribution function, $F(x) = P(I \leq x)$. If $T(I,E)$ is a given function $g(x)$ of I only, then $R(A) = P(S - \bar{S} \leq g(x))$. When there are no computer errors, i.e., perfect addition and multiplication, then $S - \bar{S}$ is given by $x^{N+1}(1-x)$. Since the x derivative $D_x(S - \bar{S}) > 0$ for all x in D, then with perfect computation, $S - \bar{S}$ is a monotonic increasing function of x. Thus when $g(x) = c$ (a constant), $R(A) = F(y)$, where y is the solution of $x^{N+1}/(1-x) = c$. However, the means of calculation are far from perfect, and round-off occurs with probability one. The amount of round-off error depends on the computer type. For complex quantities, such as when $x = e^{-a}$, the magnitude of the error depends also on the subroutine used for calculating x^n. The next section shows how the errors of $(e^{-a})^n$ and $e^{-(an)}$ are, in general, different even when calculated in the same computer. In even more complex situations the methods used for calculation is very critical in the sense that one method may work while others fail.

THE ALGEBRA OF COMPUTER ARITHMETIC

The characteristics of computer arithmetic are so complicated that there are about six or eight distinctly different algorithms for correctly solving the simple quadratic equation, the appropriate one being dependent on the numerical values of the coefficients. The reason for the difficulty is that the algebra of computer arithmetic is mathematically very pathological. For instance, the computer truncates the decimal expansion of numbers after a certain point.

Thus, when summing series it is better to sum the small terms first if one wants the best possible accuracy; (f) Furthermore, such pathologies as the following exist: $2x - \overline{2x} = 0.1 \not\Rightarrow x - \overline{x} = 0.05$, and $\overline{2x} = a \not\Rightarrow \overline{x} = a/2$.

Even more remarkable is the fact that the errors performed in assuming the implications, of point (f) above, depend on the magnitude of x.

The above remarks are of importance to purely numerical work, and apply to digital computers with the conventional accumulator registers (which store by counting pulses into a shift register). The problems illustrated will be multiplied when real time, hybrid processing is considered. For then the inertial response characteristics of the analog equipment will come into play in determining the limitations of the algorithm. Thus, it seems that the practitioner of Software Reliability would be interested in delineating conditions on the input data, I, so that by looking at the input one can tell whether the algorithm will give a good answer. This would involve first determining the limiting conditions on internal data (as in a to f above). Then given an algorithm, A, one would try to transform the internal data limitations to input data limitations. If this were done, then the user would be able to use the algorithm with confidence just be checking for some conditions on the input data. Ideally, one could like to say: "The SOFTWARE RELIABILITY is 1.0 whenever the input data is within the prescribed limiting conditions for the particular algorithm in question." (Note the similarity to the definition for HARDWARE RELIABILITY, where "limiting conditions" is replaced by "standard operating design limits."

COMPUTING RELIABILITY

When the external conditions, E, are dynamic in real time (e.g., number of aircraft being handled by an airport control tower), then definition 5 will give different values to R(A) as E changes. Different values of E will in general result in different values of the optimum data drawing method, M. So if E changes and M is optimized for each E value, then M will change also. Since algorithms have to be used even in dynamic conditions, and since under such conditions, the operator will be seeking to optimize other factors, such as M, C, or even A itself, according to some optimization procedure, the following definition is indicated:

Definition 6. Let S be a selection procedure for selecting (M, C, and A) according to observations of a stochastic process, F, and let P be the set of procedures used in controlling F. Then if the notation of definition 1 is used, the Computing Reliability is:

$$R_c(t_o) = P\left(O \text{ in } T(I, E) \text{ for all } t < t_o \middle| Z, P, S, R(t_o), B\right)$$

where R(t) is the combined hardware reliability of the computer, and of the data gathering process, and where E is a part of Z. (Note that both the software units - the computing algorithms and the operating procedures - are being used in an interactive mode.)

Example. Suppose there are n aircraft being handled by a control tower of an airport. Then Z is the 2 - tuple (operating procedures of control tower, operating procedures of pilots). The operating procedures enable one to predict the system behavior at different system states. Note that not all of Z need be controlled by P. For example, the weather may or may not be controlled, depending on the specific operating procedures (and the technology backing them up) being used by the control tower. If weather is assumed constant (for the purpose of simplified illustration), then E reduces to the set of data required to specify the configuration of n aircraft, and M is the way of sampling this set (including the radar resolution, and the processing delays of the equipment). The set I is thus a delayed and approximated version of E. Clearly the tolerance T(I, E) of the computations depends only on E, in this case. However, if A(I) denotes the result of massaging I with the algorithm A, then P depends on A(I). The output, O, of the system is the handling of the aircraft configuration (e.g., O in T(I, E) for all $t < t_o$, if all aircraft are handled and landed safely).

Some factors in the above example can be incorporated into $R_c(t_o)$ by the use of conditional probabilities and the theorem of total probability. For example, if there are no degraded operating models for the hardware (which can only be operating within specification, or completely down), and if p(i|Up) is the conditional probability of landing i aircraft safely, given that the hardware is up, then the probability of safe-handling of n aircraft, is:

$$p(n|W) = \sum_{i=0}^{n-1} P(n-i|Up) R_{i+1}$$

where R_i is the hardware's conditional reliability of surviving the time to the landing of the ith aircraft given that it has survived to the landing of the $(i-1)^{st}$. In general a Monte Carlo Simulation approach will be more accurate, not only in computing the Software and Computing Reliabilities, but also in getting the expected total cost of a hardware/software selection. In such an analysis, the data and calculations made in estimating the Software and Computing Reliabilities would be supplemented by the costs of the errors (i.e., errors of software, hardware, procedures, algorithmic approximation, etc.).

ACKNOWLEDGMENT

The author is thankful to professors Martin L. Shooman and J.A. Goguen, Jr. for fruitful discussions on the subject matter.

REFERENCES

1. Reingold, E.M., Establishing Lower Bounds on Algorithms - A Survey, Proceedings, Spring Joint Computer Conference, AFIPS, 1972, pages 471 - 481.

2. Shooman, Martin L., Probabilistic Models for Software Reliability Prediction, IEEE 1972 Fault Tolerant Computing Symposium, pages 211 - 213.

3. Goguen, J.A. Jr., System Theory Concepts in Computer Science, to appear in 6th Hawaii International Conference on System Sciences.

4. Goguen, J.A. Jr., On Homomorphisms Correctness and Subroutines for Programs and Program Schemes, Proceedings, 13th IEEE Symposium on Switching Theory and Automata Theory, 1972, pages 52 - 60.

ON SOME OBJECTIVE FUNCTIONS FOR RELIABILITY OPTIMIZATION

Igor Bazovsky, Jr.

Igor Bazovsky and Associates, Inc.
Sherman Oaks, California

ABSTRACT

This paper studies the relationship between work accomplishment by machines and the costs accruing therefrom, for the purpose of obtaining the Total Life Cycle Profit. This is proposed as the objective function of interest, as opposed to the currently popular total life cycle costs. Efficiency degradation in the form of increased operating costs, and decreased work rates is considered together with the Reliability and Maintainability characteristics of the machines. Efficiency degradation is one overall measure of wearout, and thus compliments the use of the increasing failure rate as a measure of wearout. In this paper both measures of wearout are incorporated into a renewal theoretic model which develops the expected work done, the expected costs, and the expected number of failures in two generic types of situations. The two situations are: Case 1: Machine renewals are carried on until a prespecified amount of work has been accomplished, and Case 2: Full utilization schedule in a calendar time T. The final section proposes to apply these concepts to the optimization of Redundancy, and references two papers where this has been done.

INTRODUCTION

There are two basic cases to consider in a "profit" oriented situation: <u>Case 1</u>. A fixed amount of work, W, is to be accomplished, and that is the only reason the machine is being considered (e.g. a production line tool is to be rented, built or bought for the purpose of manufacturing exactly n items).

Case 2. A full utilization schedule is to be followed for a calendar time T. The dichotomy in the two cases corresponds to that between different purposes for procuring the equipment. Machines meant for continuous production lines, would be treated via Case 2, whereas machines rented out for one specific work load would be treated via Case 1.

Since both cases can be considered simultaneously, the following NOTATIONAL CONVENTION will be used: When W, T, or w, t, appear as arguments of a function f, then f(W) and f(T) are two _different_ functions corresponding to Cases 1 and 2 respectively. E.g., if C(·) is the cost function, the C(t), C(w) denote: cost up to time t, and cost up to work-done w, respectively. (Note that the C denotes _different_ functions in each case). Z will be the generic notation for W or T, and z that for w or t. When a distinction has to be made then the usual suffix notation $f_w(x)$ and $f_t(x)$ will be used. Using the notational convention, the variables of the objective function will now be delineated.

Since Total Life Cycle Profits (TLCP) serve as a much more realistic objective function than do Total Life Cycle Costs, define, TLCP(Z) = -U(Z) - Procurement Costs, where, U(Z) = OC(Z) - GI(Z), and OC(Z) = Operating Costs incurred in (0,Z), GI(Z) = Gross Income accumulated in (0,Z).

$$U(Z) = \sum_{i=1}^{N(Z)} \left(c_i + b\, m_i + A(z_i) \right) + D(Z) \qquad (e.1)$$

where c_i = hardware cost of ith replacement, N(Z) = number of machines replaced in (0,Z); m_i = the replacement time taken to replace the ith machine; $A(z_i) = E_i + S_i + L_i - V_i$; V_i = gross income made by the ith machine, and E_i, S_i, L_i are the Energy Requirement costs, the Servicing costs (e.g. preventive maintenance), and Pollution costs, respectively, which are accumulated by the ith machine; D(Z) = the total additional costs incurred either by the $(N(Z) + 1)^{st}$ machine or by the $N(Z)^{th}$ maintenance action, whichever is occuring at Z; $b = c_3 + c_4$; c_3 = labor plus power per-unit cost for maintenance actions; c_4 = per-unit-time cost of delayed work delivery.

Remark. Note that pollution costs are being considered directly as part of the operating and/or manufacturing costs! This is the procedure advocated by the famous scientist, conservationist, and cinematographer Captain Jacques Cousteau (reference 1). Some analysis and classification of pollution effects (e.g. of the pollution efflux/pollution damage Transfer Function) was performed and combined with a Reliability Drift Failure Analysis in reference 2.

Lemma, 1. If the costs per replacement are independent, identically distributed random variables, and if $M(Z) = E[N(Z)]$ and $Q(z)$ is the unreliability, then using the cited notational convention, for both cases,

$$E[U(Z)] = [M(Z) + 1] \int_0^\infty A(z) dQ(z) - E[A(z_1 + z_2) - A(z_1)]$$

$$+ E[c_1 + b\, m_1] M(Z) + bE[z_3] \qquad (e.2)$$

where z_1 is the "age" of the machine in use at Z, z_2 is its remaining life, and z_3 the "age" of the maintenance time at Z.

Remarks. (1) For the idea of the proof, see, e.g., reference 3, Theorem 3.16, ff. (2) For Case 1;

$$M(W) = \sum_{n=1}^\infty \int_0^W f^n(w)\, dw,$$

where f^n denotes the n - fold convolution of $f(w) = dQ/dw$. For Case 2;

$$M(T) = \sum_{n=0}^\infty \int_0^T (f^{n+1} * g^n)(t)\, dt,$$

where * denotes convolution and $g(t)$ is the replacement time probability density function. (3) The relationship between $f(w)$, $f(t)$ is,

$$f_w(x) = f_t\bigl(a(x)\bigr) \frac{da}{dx},$$

where f_w, f_t correspond in the usual suffix notation to $f(w)$, $f(t)$ in our notational convention, and $a(x)$ = the (failure free) time taken to accumulate x work units by a machine, starting from age zero. $A_w(x) = A_t\bigl(a(x)\bigr)$. (4) Lemma 1 also applies when an age replacement policy is in effect (i.e. replace at failure or at Z_0, whichever comes first). Then $Q(x)$ has to be changed to have probability mass $R(Z_0) = 1 - Q(Z_0)$ at Z_0, and $E[c_1] = c_{11} Q(Z_0) + c_{12} R(Z_0)$, where the expected cost of unscheduled (scheduled) replacement is c_{11} (c_{12}). Similarly, $E[m_1] = m_{11} Q(Z_0) + m_{12} R(Z_0)$.

Corollary, 2. For Case 1, the expected time to complete W work units is of interest, and is given by (e.2) when $c_1 = 0$, $b = 1$ and $A(z)$ is replaced by $a(z)$, as defined above.

When Z is very large, an objective function which is easier to optimize is the long term, average, expected cost,

$$L_z = \lim_{Z \to \infty} \frac{E[U(Z)]}{Z}.$$

Lemma, 3. If $E[A(z)] < \infty$, then $L_z = (E[c_1] + E[bm] + E[A(z)])/u(z)$, where $u(z)$ is the mean time between renewals.

Proof. The lemma follows from the elementary renewal theorem and the fact that $E[A(z_1 + z_2)/z] \to 0$ (reference 3, Theorem 3.16, ff.).

Remark. When age replacement is being considered as in remark 4 above, then;

$$E[A(z)] = \int_0^{Z_o} A(z)dQ(z) + A(Z_o) R(Z_o), \text{ and}$$

$$u(z) = \int_0^{Z_o} R(z)dz + \left[m_{11} Q(Z_o) + m_{12} R(Z_o)\right]_t$$

Corollary, 4. L_w is not identically equal to L_t.

Proof. Consider the special case when the work rate is a constant, r. Then $a(w) = w/r$, $T = W/r$, $A_w(x) = A_t(x/r)$, and $d F_w(w) = d F_t(w/r)$. Thus $E[A(w)] = E[A(t)]$. However $u(w) = r$ MTBF, and $u(t) = $ MTBF $+ E[m_1]$.

Remark. The result of corollary 4 is anti-intuitive, and remarkably enough, this has resulted in the incorrect use of L_w instead of L_t as an objective function in many cases cited in the literature.

Corollary, 5. As an objective function, L_t is preferable to L_w.

Proof. L_w comes from Case 1, which essentially deals with operating time only (the maintenance times are added on as a random "cost" per replacement). However, real-life situations have to adhere to calendar time, and hence the maintenance times must be considered in their chronological positions, interspersed between the operating times. This is done in Case 2.

TRANSIENT STARTUP AND SHUTDOWN EFFECTS

The previous section obtained the TLCP objective function for machines with a deterministic work-done profile. Now consider that a machine is to be procured to perform an exactly known amount of work, W, and that the rate at which this work is performed is either r_1, r_2 or r_3 work units per operating time or periods: transient start up period (TSUP), normal period (NP), or transient shut down period (TSDP). The corresponding power consumption rates (used as a measure of instantaneous operating costs) are p_1, p_2, p_3. The corresponding servicing cost rates are s_1, s_2, s_3. The corresponding pollution cost rates are l_1, l_2, l_3.

Theorem, 6. If failures can only occur during NPs, and if the sojourn times during the TSUP, NP, and TSDP are mutually independent random variables, with pdf, f_i, $i = 1, 2, 3$, then

$$M(W) = \sum_{n=1}^{\infty} \int_0^W \left[\int_0^{W-v} P(\Sigma Y_i \leq W - v - u) dI_n(u) \right] dK_{n-1}(v)$$

where Y_i/r_2 = length of the NP of the ith machine, Σ = summation from $i = 1$ to n; $I_n(u) = P(\Sigma X_i \leq u)$; X_i/r_1 = length of the TSUP of the ith machine; $K_n(v) = P(\Sigma Z_i \leq v)$; Z_i/r_3 = length of TSDP of the ith machine. (Note the above assumed $r_j \neq 0$. If $r_j = 0$, then define the corresponding sojourn times to be zero.)

Proof.

$$P[N(W) \geq n] = P\left[\sum_{i=1}^n Y_i \leq W - \sum_{i=1}^n X_i - \sum_{i=1}^{n-1} Z_i\right]$$

$$= \int_0^W P\left[\sum_{i=1}^n Y_i \leq W - \sum_i^n X_i - v\right] dK_{n-1}(v).$$

Thus the nth term in the above summation is equal to $P\left(N(W) \geq n\right)$.

Example. Suppose that Z_i and X_i are deterministic and that the failure process during NPs is Nonhomogeneous Poisson, with failure rate function $\lambda(x)$, where x is NP time only. If the length of each TSUP (TSDP) is A(B), then from the proof of theorem 6,

$$P[N(W) \leq n] = \sum_{i=0}^{n} \frac{\left[m\left(\frac{W}{r_2} - T_n\right)\right]^i \exp\left(-m\left(\frac{W}{r_2} - T_n\right)\right)}{i!}$$

where $T_n = r_1 n_A + (n-1) r_3 B$, and $m(t) = \int_0^t \lambda(x) \, dx$.

Lemma 7. If work is only performed during NPs, if $Q_1(Q_2) = P$ (a failure in one TSUP (TSDP)), and if TSUP failures are followed by new TSUPs, then,

$$M(W) = S \sum_{n=1}^{\infty} P\left[\sum_{i=1}^{n} Y_i \leq W\right]$$

where

$$S = \left(\frac{1}{1-Q_1} + Q_2\right)$$

Proof. The number of TSUP failures occuring before a whole TSUP is traversed without failure has the geometric pdf with mean $Q_1/(1 - Q_1)$. Thus the expected total number of TSUP failures is $Q_1/(1 - Q_1) \, \Sigma P(\Sigma Y_i \leq W)$. Thus $M(W) = (1 + Q_{\bar{1}}/(1 - Q_1) + Q_2)\Sigma P(\Sigma Y_i \leq W)$.

Remark. Theorem 6 and Lemma 7 apply unchanged when work rates are non-constant, if I(u) denotes the distribution function of the work-done during a TSUP, and K(v) that during a TSDP.

Theorem 8. Let $I(t)$, $J(t)$, $K(t)$ be the distribution functions for X, Y, Z respectively, and let $I_n(t)$ etc., denote their n - fold convolutions. Then if $B_i(W,t)$, $i = 1, 2, 3$ denote the distribution functions of total calendar time spent in TSUPs, NPs, and TSDPs respectively, and if failures can only occur during NPs, then

$$B_1(W,t) = \sum_{n=0}^{\infty} I_n(tr_1) \left[(J*K)_n (W - tr_1) - (J*K)_{n+1} (W - tr_1) \right]$$

$$B_2(W,t) = \sum_{n=0}^{\infty} J_n(tr_2) \left[(I_n * K_{n-1})(W - tr_2) - (I_{n+1} * K_n)(W - tr_2) \right]$$

$$B_3(W,t) = \sum_{n=0}^{\infty} K_n(tr_3) \left[(I*J)_n (W - tr_3) - (I*J)_{n+1} (W - tr_3) \right]$$

Proof. Only the proof for B_2 is given. That for the others is similar. The following lemma is used in the proof.

Lemma. If x and y are non-negative random variables, then $P(y<t, \overline{x<W-t}) = P(y<t, x+y \le W) - P(x \ge W-t, x+y \le W)$. Let

$$y_n = \sum_{i=1}^{n} Y_i, \text{ and } x_n = \sum_{i=1}^{n-1} (X_i + Z_i) + X_n$$

Note $B_2(W,t) = P(r_2(\text{total NP time}) \le tr_2) = P(\Sigma Y_i \le tr_2)$. This latter event can occur in the following mutually exclusive ways: At the instant the W units of work are accomplished, the machine is in a NP, and n complete non-NPs have been worked through ($n = 1, 2, \ldots$), with total length $x_n > W - tr_2$; or the machine is in a non-NP, and n ($n = 1, 2, \ldots$), complete NPs have been worked through with total length $y_n \le tr_2$. The theorem of total probability thus gives:

$$B_2(W,t) = \sum_{n=0}^{\infty} P(x_n > W - tr_2, x_n + y_{n-1} \le tr_2 < x_n + y_n)$$

$$+ \sum_{n=0}^{\infty} P(y_n \le tr_2, x_n + y_n \le tr_2 < x_{n+1} + y_n).$$

But for any independent events A and B, $P(A \le tr < A + B) = P(A \le tr) - P(A + B \ge tr)$. Using this to break up each of the above summands and rearranging:

$$B_2(W,t) = \sum_{n=0}^{\infty} \left[P(y_n \le tr_2, x_n + y_n \le t) \right.$$

$$- P(x_n \ge W - tr_2, x_n + y_n \le t) \bigg]$$

$$- \sum_{n=0}^{\infty} \left[P(y_n \le tr_2, x_{n+1} + y_n \le W) \right.$$

$$- P(x_n \ge W - tr_2, x_{n+1} + y_n \le W) \bigg]$$

Now using the lemma and the independence between the x_n and Y_n gives:

$$B_2(W,t) = \sum_{n=0}^{\infty} P(y_n \le tr_2, x_n < W - tr_2) - P(y_n \le tr_2, m\ x_{n+1} < W - tr_2)$$

$$= \sum_{n=0}^{\infty} P(y_n \le tr_2) \left[P(x_n < W - tr_2) - P(x_{n+1} < W - tr_2) \right]$$

But $P(x_n < W - tr_2) = (I_n * K_{n-1})(W - tr_2)$.

Corollary, 9. If any arbitrary choice is made of one of the variables: total calendar time, total servicing costs, or of total power costs incurred in performing the required W work units, and if the corresponding consumption rates in the TSUPs, NPs, and TSDPs are denoted by q_i, ($i = 1, 2, 3$), then:

$$E[Q] = \sum_{i=1}^{3} q_i \int_{0}^{W/r_i} \left(1 - B_i(W,t)\right) dt.$$

where $q_i \equiv 1$ if Q = total operating time.

Proof. When $\lim_{x \to \infty} xR(x) = 0$, then $E[x] = \int_0^\infty R(x)dx$,
and similarly for $1-B_i(W,t)$. But at a rate of r_i work units per unit time, $B_i(W,W/r_i) = 1$, and so W/r_i for the upper limit of integration is sufficient.

Corollary, 10. Assume that a TSUP (TSDP) failure is always followed by a TSDP (TSUP), and that whenever a machine leaves a TSUP (TSDP) then there is a (constant) probability Q_1 (Q_2) that the TSUP (TSDP) is being left because of a failure. Then theorem 8 still applies if $J(t)$ is modified to have a singular probability mass Q_1 at $t = 0$, and if J_n is its n - fold convolution. Theorem 6 will apply if, furthermore $M(W)$ is multiplied by a factor of $(1 + Q_2)$.

Example 2. Assume that a TSUP (TSDP) failure is always followed by a TSDP (TSUP), and that TSUP (TSDP) durations are determined by $z_i = \min(x_i, y_i)$, ($i = 1$ for TSUPs, and $i = 2$ for TSDPs), where x_i has failure rate λ_i, and y_i has failure rate μ_i. If x_i denotes time to failure and y_i denotes time to normal transition from the TSUP (TSDP), then corollary 10 applies with $Q_i = P(\text{failure}) = \int \lambda_i \exp - (\lambda_i + \mu_i) x_{dx} = \lambda_i/(\lambda_i + \mu_i)$, and the TSUP (TSDP) duration pdf being exponential with failure rate $(\lambda_i + \mu_i)$. Thus $E[U(W)]$ can be calculated via theorems 6 and 8, modified by Corollary 10.

SOFT CONSTRAINTS

The standard redundancy optimization problem is to maximize Reliability under "Linear Constraints," where the nth part adds a cost (weight, volume) of c units. In keeping with the spirit of TLCP optimization as above, the optimization criteria can be formulated in terms of TLCP. Thus the addition of each unit of redundancy affects factors such as power consumption, pollution efflux, maintenance costs, fuel consumption rates, and profit losses due to lowered cargo capacity. Denote by $c_1(n)$ the additional total cost (including the above factors) of adding the nth unit, when the first $(n - 1)$ have been already added. $c_1(n)$ is not in general a constant. Some factors will tend to make it an increasing function (weight, space and power saturation effects) of n, and others a decreasing function (combined maintenance activity, lower procurement costs, proration of user training, etc.).

The expected total cost, given $(n - 1)$ back up units, is:

$$E[C|n] = \sum_{i=1}^{n} c_1(i) + E[N|n] E[C_2]$$

where $E[N|n]$ expected number of failures occuring within the situation of interest, and $E[C_2]$ expected cost per failure. This objective function is optimized in references 4 and 5, for some cases of interest.

ACKNOWLEDGMENT

Thanks to Professor Richard L. Scheaffer for interesting the author in optimum age replacement with increasing cost factor, (reference 6).

REFERENCES

1. Personal Communication, Made at U.S. Navy Deep Sea Oceanic Conference, San Diego, USA, 1970.

2. Bazovsky, Igor Jr., Optimum Design of Anti-Pollution Devices, Proceedings, 1972 International Conference on Cybernetics and Society, pages 612-614, IEEE.

3. Ross, Sheldon M., <u>Applied Probability Models with Optimization Applications</u>, Holden-Day, San Francisco, 1970.

4. Bazovsky, Igor Jr., An Algorithm for Optimizing the Safety Effectiveness of Parallel Alarm Systems, Proceedings, 1973 Annual Reliability-Maintainability Symposium (ASME), pages 610-613.

5. Bazovsky, Igor Jr., Safety Effectiveness of Sequential Contingency Systems, The System Safety Society Symposium, (Denver, July 1973).

6. Scheaffer, Richard L., Optimum Age Replacement and Increasing Cost Factor, Technometrics, February 1971.

A COMMENT ON RELIABILITY EVALUATION OF SOFTWARE

R.G. Bennetts

Department of Electronics,
University of Southampton, England.

'... it is not only the programmers task to
produce a correct program, but also to demonstrate
its correctness in a convincing manner.'
                                          E.W. Dijkstra

INTRODUCTION

   The qualitative assessment of the reliability of computer
system software has recently become a major topic for research
workers - it has always been a problem! This short comment
is mainly concerned with the problem of software testing and
it is suggested that a meaningful assessment of software
reliability cannot be made until this problem has been solved.
In particular, the paper discusses two recent developments -
one in hardware synthesis procedures and the other in software
synthesis - and indicates a possible link between them that may
allow some limited algorithmic testing of software.

THE TESTING OF DIGITAL HARDWARE

   To fully appreciate the difficulties of testing software, it
is instructive to consider first the philosophy behind the
testing of hardware. Obviously, exhaustive application of all
input sequences is out of the question for all but very small
logic networks and early research workers turned their attention
to the type of faults that occurred most frequently. It was
discovered that the _effect_ of most of these faults could be

modelled as one or more logic connections being stuck-at-logical 1
or stuck-at-logical 0 (s-a-1, s-a-0) and nearly all the techniques
for generating testing sequences for logic networks are now based
on this 'stuck-at' fault model [1] i.e. this common _effect_ is
treated as the root cause.

These techniques, which may be based on the network topology
itself (D-algorithm), the equivalent Boolean algebraic description
(Boolean difference) or on a computer simulation (Partitioning),
all consider the effect of a single (or multiple) s-a-1 or s-a-0
fault on the behaviour of the circuit and the aim is to provide
a sufficient number of tests covering the occurrence of all such
faults. There are two important points to be made here:
a) a standard fault model is used as a basis for generating the
tests, and
b) the actual tests required are dependent on the _structure_ of
the network i.e. the actual gates, bistables, etc. and their
interconnection.

There are obvious reservations about the use of this fault
model since the tests will diagnose only those faults whose
effect falls within the s-a-1, s-a-0 category (e.g. most solder
splashes, open- and short circuits, etc.) and not those whose
effect falls outside this category (some wiring errors,
intermittent crosstalk, etc.). Within this limitation however,
the model has proved very useful and much work has been and still
is being conducted into efficient computer algorithms for
generating the tests.

CAUSES OF SOFTWARE FAILURE

Turning now to software, let us examine the causes of
failure and their effects on reliability.

During the preparation and running of a program, a heavy
reliance is placed on the correctness of the flowchart describing
the initial algorithm* together with the correctness of the
various translatory and service programs (compilers, assemblers,
loaders, operating system, etc.). In its simplest form, the
reliability of a computer program is a function of the product
of the reliabilities of a number of translation and shifting
stages together with the programmer reliability and hardware

---

* Here, algorithm is used to mean a decision tree that is directed,
bounded and terminated and gives correct answers for the specified
range and type of data input.

reliability. Research workers are, quite rightly, considering improvements to these individual components but the major problem of a priori testing still remains.

It is not difficult to see why this is so when one considers the many sources of error and their possible effects. There are far too many to list here but the following sample will serve as an illustration:

- Choice of language : the programmer who is more familiar with FORTRAN than ALGOL will (hopefully) write a more reliable program in FORTRAN.
- Machine architecture : here, all sorts of hardware limitations will occur e.g. round-off, finite word length (implications on absolute magnitude of fixed and floating point numbers), etc.
- Choice of compiler : in some machines, optimising compilers now exist alongside the standard one.
- Boundary constraints of data : apart from the hardware limitations above, most algorithms are only valid for certain types of data e.g. integer variables, Boolean variables, real (rather than imaginary) numbers, etc.

There are many more causes of error and the unfortunate fact about them is that, in a large number of cases, their effect on <u>all</u> possible sets of data input to the program is just not known and furthermore, cannot practically be predicted.

This problem is similar to that encountered in hardware today. In order to exhaustively test an n-input, single output combinational logic network, 2^n input combinations (tests) must be applied and the outputs monitored and compared with a truth table. For n = 50 assuming a testing rate of 10^6 per second, it would take 32 years to test the network! It is because of this that the 'stuck-at' fault model is used and it is a sobering thought when one considers that, for such a combinational network, the number of tests to detect all single s-a-1 and s-a-0 faults is of the order of 2n i.e. a testing time of 100 μs for n = 50.

The question naturally arises therefore - does such a fault model exist for software? The sad truth at this point in time is that the answer is 'no' and, an even sadder truth is that there does not appear to be much research effort directed towards seeking such a model. This is either because a model capable of embracing a large percentage of software error effects does not exist, or because most research workers are more concerned with improving the individual component reliabilities, or both. The author does not know whether a suitable model exists but would like to offer the following observations:

1. In the end, the effect of a software error will manifest itself, when the program is actually executed by the computer with real data. (This disregards the detection and correction of syntax errors and electro-mechanical peripheral read-in errors, but these are generally relatively easy to diagnose.)
2. One of the inherent features of hardware testing is that knowledge of the <u>structure</u> of the digital network is fundamental to fault diagnosis.
3. The actual binary information residing in store and defining the algorithm and its data represents a finite state machine that has been created especially to execute the initial algorithm and, as such, should be equivalent to it.

Combining these three observations, we see that the key point is that, if we can show that the actual coded binary information in store is equivalent to the original algorithm, then we may conclude that, in implementation at least, we have successfully conducted the various stages of writing, preparing and compiling the program.

The question now arises - can this be done and if so, what is its practical value? Considering the first part of the question, we see that some form of representation common to both the initial algorithm and the coded binary implementation is required. Also, if error detection <u>and</u> correction is required some knowledge of the algorithm structure and implementation structure would appear to be essential (a point also made by Dijkstra [2, page 5]). Recent developments in structured programming [2,3] and digital system specification [4,5] may be instrumental here since both methods of representation seek to retain the structure of the process. In particular, Doran and Tate [3] define the algorithm by a structure diagram that contains not only the actual sequence of events (the flowchart), but also the hierarchical structure of the algorithm itself, whereas Patil and Dennis [4] describe the digital hardware process by means of two directed graphs - one of which relates to the data-flow structure and the other, known as a Petri-net, to the control and sequencing of events. Also, the data-flow/Petri-net system mirrors almost exactly the register and microprogram structure of a conventional digital computer and should therefore be relatively easy to determine.

It appears that, since both these representations retain algorithm structure, there potentially exists a common 'bridging' representation. If this is so, behavioural identicality or otherwise may be demonstrated.

This then raises the second part of the question - what purpose will it serve? The point here is that if a common representation exists and techniques for demonstrating identicality or otherwise are known, then the effect of almost any fault can be projected through to the implementation representation and assessed. This may lead to more insight into the final effect of certain software errors and may also suggest some form of fault model to be used as a basis for test generation at the initial structure diagram/flowchart stage. Furthermore, once this projected effect has been assessed, steps may be taken to either avoid the initial faults' occurrence or mask out its effect, i.e. fault-tolerant software.

CONCLUDING REMARKS

The observations and suggestions in this paper have raised more questions than answers. The main comment really is that algorithmic testing of software cannot be achieved until a useful fault model has been established (similar in concept to the 'stuck-at' fault model in digital hardware testing). Until such algorithmic testing can be carried out no realistic a priori assessment of software reliability can be made.

Finally, I would end the paper in the same way as I started - with a quotation from Dijkstra's monograph [2]. The starting quotation is what the paper is all about whereas the final quotation is what the research suggested should attempt to disprove!

'Program testing can be used to show the presence of bugs, but never to show their absence!'
                                                E.W. Dijkstra

REFERENCES

1. BENNETTS, R.G. and LEWIN, D.W. Computer Journal, 14, 2, 1971, 109-206.
2. DAHL, O-J, DIJKSTRA, E.W. and HOARE, C.A.R. 'Structured Programming' Academic Press, 1972.
3. DORAN, R. and TATE, G. Massey University Computer Unit (New Zealand), June 1972.
4. PATIL, S.S. and DENNIS, J.B. 6th Annual IEEE Computer Conference 1972, 223-226.
5. GLASER, E.L. Ibid, 1972, 175-192 (5 papers).

POWER SYSTEM RELIABILITY INDICES

Roy Billinton

Power System Research Group
University of Saskatchewan
Saskatoon, Canada

ABSTRACT

The application of probability techniques in the quantitative evaluation of power system reliability has increased considerably within the past decade. With this increase in application and particularly within the past few years there has been several developments in theoretical concepts which are proving extremely useful in practical application. It also appears increasingly evident that in certain areas, system adequacy cannot be suitably described by a single index such as the probability of satisfying the system requirement. This paper illustrates some fundamental concepts of power system reliability evaluation together with the development and practical application of additional indices in several basic system areas.

INTRODUCTION

The need for quantitative reliability indices has long been recognized by power system engineers and some of the first recorded applications in the available literature were published almost forty years ago[1]. Since that time there has been numerous theoretical developments and practical applications documented. Many of the basic fundamentals are illustrated in the texts, "Power System Reliability Evaluation"[2] and "Power System Reliability Calculations"[3].

One aspect of the basic power system problem which makes it somewhat different from most reliability applications documented in the available literature is that a power system is a continuously operable maintainable system. The basic definition of reliability as the probability of a device or system performing its purpose adequately for the period of time intended and under the operating conditions encountered is quite useful but it is not sufficient in many areas of power system application. As in all maintainable systems, the steady state or limiting probabilities associated with equipment or system availability or unavailability have become basic parameters in long term evaluation. This is particularly true in the field of generating capacity planning, where generating unit statistics such as the forced outage rate have been collected for many years*. Additional system indices such as the probability of being able to satisfy the system load, the expected load curtailed or the expected deficient days per year have been developed using the generating unit availability and unavailability statistics. The necessary mathematical manipulsations can become quite complex in a large system.

Transient or time dependent probability values are required in the field of operating capacity assessment. The inclusion of corrective maintenance effort in the analysis again makes this somewhat different from much of the generally published material which deals with mission orientated systems.

Additional indices in the form of more physically relatable parameters were first recognized in the generating capacity planning field in 1958[12]. The basic concept of the frequency of encountering a particular generating capacity condition was presented as an additional index. The initially available technique was quite restrictive and the indices difficult to develop in a general system. The basic need for more physically oriented indices was, however, first generally accepted in the field of distribution system engineering. It was easily seen that an overall average annual outage time of ten hours at a customer load point could have quite a different impact on the customer if it occurred ten times per year with an average outage time of one hour or once per year with an average outage time of ten hours. At least three indices appear necessary to suitably describe the load point reliability. They are, the frequency of failure, the average duration of an outage and the average total hours of outage in a year. The last index is the expected outage time and can be obtained by multiplying the unavailability probability by the number of hours in a year. It can also be obtained as the product of the outage frequency and the average outage duration. In distribution system analysis, the number of outage occurrences is usually represented by the failure rate while in generating capacity applications, the actual frequency of a capacity condition is calculated. The two quantities frequency and failure rate are fundamentally quite different but are often numerically quite similar, particularly in the distribution field.

---

* See the Section on Equipment Outage Data in the "Bibliography on the Application of Probability Methods in Power System Reliability Evaluation", R. Billinton, IEEE Transactions, PAS-91, No. 2, March/April 1972, pp. 649-660.

This can be seen using the diagram in Figure 1 where the random behaviour of a single component or system is represented by the average Up and Down times.

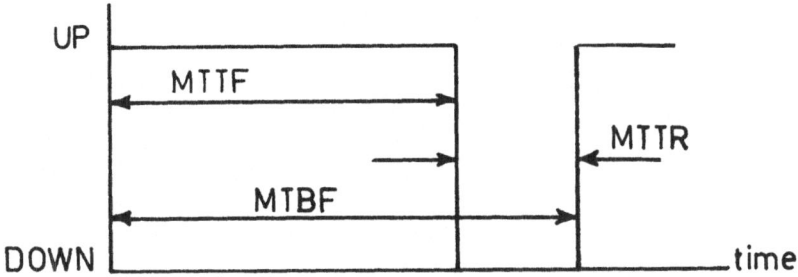

Figure 1. Average System Performance Indices.

In Figure 1,

MTTF = Mean Time To Failure = m
MTTR = Mean Time To Repair = r
MTBF = Mean Time Between Failures = T = Cycle Time

In the reciprocal form

$\lambda = \dfrac{1}{m}$ = Failure Rate

$\mu = \dfrac{1}{r}$ = Repair Rate

$f = \dfrac{1}{T}$ = Cycle Frequency

If the load point failure rate is 2.0 failures/year and the average repair time is 4 hours, the cycle frequency will be 1.9982 occurrences/year. As the repair time increases, the difference between the two indices becomes more noticeable. This numerical similarity has caused many power system engineers to consider the expressions, failure rate and failure frequency to be quite interchangeable. Given one value and the average downtime, it is a simple matter to calculate the other value. The basic techniques used to calculate failure rate in distribution system applications and the frequency of encountering a given condition in a more general system are quite different. The impetus for the development of the general approach was created in the generation field. The conventional index of loss of load probability or expectation is not compatible with accepted transmission and distribution indices and therefore not completely suitable as an index of overall or composite system reliability[4]. Unfortunately many power system engineers consider the frequency approach as a generation concept and therefore have not recognized the complete generality of the technique.

This paper illustrates the basic development of failure rate and failure frequency indices in power system application together with the related indices of average outage duration and average annual outage duration.

THEORETICAL CONCEPTS AND APPLICATIONS

As noted previously, the basic indices of failure rate, average outage time and average annual outage time were first recognized in the distribution field. The first major publication on the subject appeared in 1964[5]. This paper presented a series of equations for series and parallel system reduction which could be used to obtain a single equivalent component characterized by a failure rate and an average repair or downtime. The basic equations in this approach are as follows[1].

Consider a system with two components designated 1 and 2 in series,

$$\lambda_s = \lambda_1 + \lambda_2 \qquad r_s = \frac{\lambda_1 r_1 + \lambda_2 r_2}{\lambda_1 + \lambda_2} = \frac{\Sigma \lambda_i r_i}{\Sigma \lambda_i} \qquad (1)$$

If the two components are in parallel

$$\lambda_p \simeq \lambda_1 \lambda_2 (r_1 + r_2) \qquad r_p = \frac{r_1 r_2}{r_1 + r_2} \qquad (2)$$

In the parallel system case, the two components are assumed to be completely redundant and continuity is the sole criterion. The application of these equations can be shown using a relatively simple example.

Consider the following system.

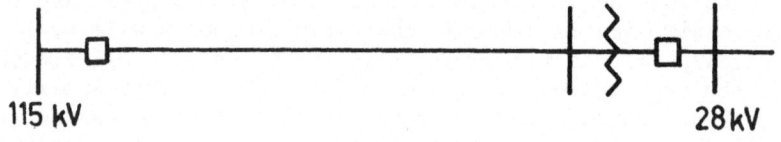

Component	Failure Rate	Average Repair Time
115 kV line	0.519 f/year	9 hours
115 kV breaker	0.0066	3 x 24
115/28 kV transformer	0.0126	14 x 24
28 kV breaker	0.0050	2 x 24

λ = 0.519 + 0.0066 + 0.0126 + 0.0050 = <u>0.5432 f/year</u>

$r = \dfrac{(0.0066 \times 3 \times 24) + (0.0126 \times 14 \times 24) + (0.0050 \times 2 \times 24) + (0.519 \times 9)}{0.5432}$

= <u>17.72 hours</u>

Average total annual downtime = $U \simeq \lambda r$
= (0.5432)(17.72) = <u>9.62 hours</u>

In the above analysis it has been assumed that the 28 kV bus is completely reliable.

Assume that the bus has a failure rate of 0.0113 f/year with an average downtime of 4 hours.

The load point reliability indices are now.

$\lambda = 0.5432 + 0.0113 = 0.5545$ f/year

$r = \dfrac{(0.5432)(17.72) + (0.113)(4)}{0.5545} = \underline{17.4 \text{ hours}}$

$U = \underline{9.66 \text{ hours/year}}$

Consider the following system.

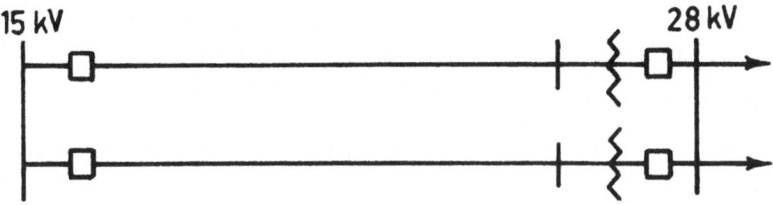

Use the same data as the previous example and assume 100% redundancy.

For one series path

$\lambda = 0.5432$ f/y
$r = 17.72$ hours

Both paths in parallel

$\lambda = (0.5432)(0.5432)(\dfrac{17.72 + 17.72}{8760})$

$= \underline{0.00119 \text{ f/y}}$

$r = \dfrac{17.72}{2} = \underline{8.86 \text{ hours}}$

Add on the 28 kV bus

$\lambda = 0.00119 + 0.0113 = \underline{0.01249 \text{ f/year}}$

$r = \dfrac{(0.00119)(8.86) + (0.0113)(4)}{0.01249} = \underline{4.46 \text{ hours}}$

$U = (0.01249)(4.46) = \underline{0.0556 \text{ hours/year}}$

It can clearly be seen that the bus value dominates the load point failure rate. This is because of the series aspect. In many systems an awareness of series element dominance can save considerable computational effort.

MAINTENANCE EFFECTS

If preventive maintenance is not performed at the correct time, the component failure rate will increase with time and the reliability will decrease. In the case of a series system, if a component is removed from service, the load point or points suffer a maintenance outage. In the case of a two unit redundant system, no outage occurs but the system is in a series state while the maintenance outage exists.

Consider a two identical unit redundant system, the failure rate due to maintenance is given by

λ_m = 2 (Maintenance rate)(Probability that the other unit fails during maintenance)

Example:

Consider the system previously studied.

Scheduled outage data

Component	Outages/year	Average Outage Time
115 kV line	4.0	8.2 hours
115 kV breakers	1.5	8.0
115/28 kV transformer	2.0	8.0
28 kV breaker	1.5	4.0
	9.0	

Assume no simultaneous maintenance outages.

Average outage time = $\frac{(4.0 \times 8.2)(1.5 \times 8.0) + (2.0 \times 8.0)(1.5 \times 4.0)}{9.0}$

= 7.42 hours

$\lambda_m = 2[9.0(0.5545 \times \frac{7.42}{8760})] = \underline{0.00845}$ f/year

The expected value of the system downtime due to a maintenance outage overlapping a forced outage in a two unit identical component system is given by

$\frac{r_f \cdot r_m}{r_f \cdot r_m} = \frac{(17.72)(7.42)}{25.14} = \underline{5.23 \text{ hours}}$

The overall rate including permanent outages alone and those overlapping maintenance outages is obtained by summing the two values previously obtained. The permanent outage contribution should be modified by an exposure factor. This value is usually very close to unity and therefore neglected.

0.01249 + 0.00845 = $\underline{0.02094}$ f/year

Expected outage duration

$$= \frac{(0.01249 \times 4.46) + (0.00845 \times 5.23)}{0.02094} = \underline{4.76 \text{ hours}}$$

Average total outage time = $(0.02094)(4.76)$
= $\underline{0.0997 \text{ hours}}$

In many parallel redundant systems, the incidence of short or temporary failures can be quite significant. Similar indices to those calculated above can be obtained for failures which fall into this category. It is obvious at this point that some rather specific definitions are required to facilitate data collection to permit the prediction of load point indices. One example of such a list of definitions has been prepared by the IEEE Committee on Performance Records for Optimizing System Design[6]. In addition to the independent failures in parallel facilities, the data should also provide statistics which permit the calculation of failure rates due to failure bunching under adverse weather conditions. This is an important point particularly in low voltage networks[1]. The basic equations given on page 3 can be modified using a conditional probability approach to recognize possible overload conditions. They can also be used to analyze more complicated configurations which do not lend themselves to direct series parallel reduction by using a cut-set approach to reduce the network. The available equations provide a very useful and and flexible tool for distribution system reliability prediction which does not require extensive mathematical manipulation or knowledge of stochastic process or probability theory. They do require, however, consistent statistics on equipment failure, repair and maintenance parameters and this at the moment is the biggest stumbling block to their applications. Recent surveys in the IEEE and the CEA have found that although most utilities have comprehensive outage reporting schemes, these schemes are customer orientated and do not provide consistent component outage data. In many cases data is recorded only if the outage results in an actual loss of customer load.

The system depicted in Figure 1 can be represented by a state transition diagram as shown in Figure 2.

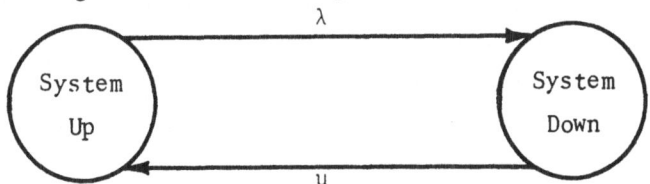

Figure 2. Two State Model.

If λ and μ are constant, the failure and repair density functions are exponential and the system can be modelled as a Markov process.

Given that the system is in the UP state at time $t = 0$,

$$P_{UP}(t) = \frac{\mu}{\lambda+\mu} + \frac{\lambda}{\lambda+\mu} e^{-(\lambda+\mu)t}$$

$$P_{DOWN}(t) = \frac{\lambda}{\lambda+\mu} - \frac{\lambda}{\lambda+\mu} e^{-(\lambda+\mu)t}$$

As illustrated later in the paper, time dependent probabilities of this form are used in operating capacity assessment.

The limiting or steady state probabilities are obtained by letting $t \to \infty$.

$$P_{UP} = \frac{\mu}{\lambda+\mu} = \frac{m}{m+r} = \text{Availability} = A$$

$$P_{DOWN} = \frac{\lambda}{\lambda+\mu} = \frac{r}{m+r} = \text{Unavailability} = \overline{A}$$

These equations can be manipulated to give an expression for the frequency of encountering a state

$$A = \frac{m}{m+r} = \frac{1}{\lambda.T} = \frac{f}{\lambda}$$

$$\therefore \ f = \lambda.A = \mu\overline{A} \tag{3}$$

In more general terms, the frequency of encountering a particular state is the probability of being in the state multiplied by the rate of departure from it or the probability of not being in the state multiplied by the rate of entry[1].

This simple formula can be extended in terms of overall utilization by considering the frequency as the expected number of transitions across a particular boundary. Consider a general system in which the state transition rates are again time invariant[7]. Under steady state or equilibrium conditions, the expected number of transitions per unit time into and out of a system state is the same, i.e.,

$$E(OTR) = E(ITR) \tag{4}$$

where,

OTR = The transition rate out of the state
ITR = The transition rate into the state

and

E = The expected value.

Using equation (4), the steady state equations can be written in the following form for the ith state,

$$P_i \sum_{\substack{k \in X^- \\ i \in X^+}} \lambda_{ik} = \sum_{\substack{k \in X^- \\ i \in X^+}} P_k \lambda_{ki} \tag{5}$$

where,

P_i = The steady state availability (probability) of state i

λ_{ik} = The transition rate from state i to k

X^+, X^- = The disjoint subsets of the sample description space X such that

X = {All the states of the system or all the nodes of the state transition diagram}

X^+ = {State i}

X^- = {All states except i}

Arranged in matrix form, equation (5) becomes

$$AP = B$$

where

P = The column vector of steady state probabilities

B = The column vector of all zeros

and

A = The transpose of the matrix of transition rates

The steady state availabilities can be obtained solving (n-1) out of n equations from (5) and the equation

$$\sum_{i=1}^{n} P_i = 1$$

After the availabilities have been evaluated, the frequencies of the individual or cumulative states may be obtained using the following relationship

$$f^+ = \sum_{i \in X^+} \sum_{j \in X^-} P_i \lambda_{ij}$$
$$= \sum_{j \in X^-} \sum_{i \in X^+} P_j \lambda_{ji} = f^- \qquad (6)$$

where,

f^+ = The cumulative frequency of the states in the subset X^+

and

f^- = The cumulative frequency of the states in the disjoint subset X^- which contains all the states not in X^+.

The cycle times and the mean durations can be obtained using the following relations,

Cycle Time = 1/Frequency   (7)
Mean Duration = Availability/Frequency   (8)

This is a very powerful technique and is now being used in a wide range of power system applications. In the case of generating capacity planning, the generating unit outages are considered to be independent events and therefore the probabilities associated with single or multiple outage conditions can be found by straight multiplication. The frequency of individual or cumulative outage states can be obtained using the basic formula given by equation (6). Some very quick computational algorithms have been developed for this purpose. In some systems, the state probabilities cannot be found by simple multiplication. This is illustrated in the state transition diagram shown in Figure 3.

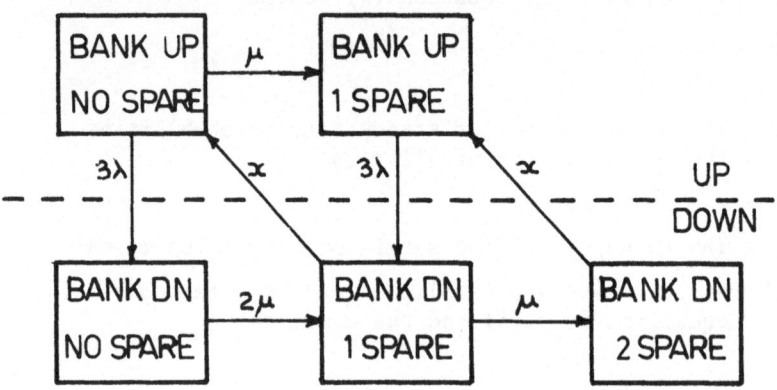

Figure 3. State Transition Model of a System with Dependent Events.

The physical configuration[8] represented in Figure 3 is that of a three phase transformer bank containing three single phase transformers and one spare unit. The failure and repair rates for a single transformer are λ and μ respectively. The installation rate for a spare is represented by x. The resulting system equations in this case are as follows.

$$P_{UP} = \frac{2x^2\mu^2 + 2x\mu^2(3\lambda+\mu) + 6\lambda x^2\mu}{Z}$$

$$P_{DN} = \frac{9\lambda^2 x^2 + 6\lambda x\mu(3\lambda+\mu) + 6\lambda\mu^2(3\lambda+\mu)}{Z}$$

$$Z = 9\lambda^2 x^2 + 2\mu(x+\mu)(3\lambda+x)(3\lambda+\mu)$$

The frequency of encountering the outage condition is:

$$f_D = \frac{6\lambda x\mu(3\lambda+\mu)(3\lambda+x)}{9\lambda^2 x^2 + 2\mu(x+\mu)(3\lambda+x)(3\lambda+\mu)}$$

Average duration of the UP state = $\frac{1}{3\lambda}$

Average duration of DOWN state = $\frac{1}{x} + \frac{1}{\mu}\left[\frac{3\lambda x}{2(3\lambda+\mu)(x+\mu)}\right]$

It quickly becomes impractical to attempt to develop general equations for more complicated configurations. Numerical results for specific failure, repair and installation rates can, however, be easily obtained using a digital computer. The following table shows the change in bank unavailability as the number of spares increases for a system with λ = 0.25 f/year, r = 30 days and an expected installation time of 2 days.

The basic difference between failure rate and failure frequency can be easily seen from the results in Table I. The system failure rate is 0.75 failures/year. The upper limit for the system frequency is 0.747 occurrences/year. The difference is due to the two day separation between the MTTF and the MTBF. If frequency is used as the sole index of system adequacy, the system appears to become less reliable with the addition of the first spare. The increase in failure frequency of course arises from the fact that the system cannot fail again until it is put back in operation. All three indices, unavailability, frequency and average outage duration are required to give a complete picture.

TABLE I

Number of Spares	Unavailability	Frequency	Average Outage Duration
0	0.061697	0.704 o/yr	768.0 hours
1	0.005754	0.746	67.6
2	0.004123	0.747	48.4
3	0.004093	0.747	48.0

The incremental reliability benefits associated with spare components can be readily determined using the approach illustrated[9]. The technique can be applied to any practical configuration for which the success and failure states can be defined. In many systems, the failure and repair rates cannot be considered as being time invariant. This is particularly true in regard to the repair process. Many power components, however, do operate within their useful life period and receive adequate preventive maintenance. The difficulty in this case arises in finding the individual state probabilities not in the actual calculation of the state frequencies. Techniques in addition to basic simulation are available for this purpose and the application of the methods of supplementary variables, semi-Markov processes and the device of stages has been documented in the literature[10].

As noted in the Introduction to this paper, quantitative reliability indices have been utilized in the power field for close to forty years. The bulk of application activity has been in the field of generating capacity planning for single and interconnected systems. The basic technique which has been used for many years and probably will continue to be used is the standard loss of load probability or expectation. The index in this case is the probability that the load will exceed the capacity available within a designated period. The most common index is in terms of a mathematical expectation and is the expected number of days in a year that the load will exceed the available capacity. A basic risk level which has been quoted widely in North America is that of 0.1 days/year. The reciprocal value is often used, i.e., 1 day in 10 years and given a frequency connotation which does not entirely exist. It is not valid to present the 0.1 days/year value as a standard index unless the basic calculation procedure is also standardized. There are many possible variations in the procedure with regard to the load model used, the inclusion of large unit derating levels, maintenance schedules, load forecast uncertainty and interconnected system effects.

The recursive approach to the problem of calculating frequency indices as represented by equations (6), (7) and (8) has introduced a new set of dimensions into the capacity planning area. The basic approach in either the loss of load method or the recursive technique is to develop a model of the system generating capacity and to combine this with a suitable load model to obtain the appropriate indices for the margin between the load and capacity. Generating unit outages are assumed to occur independently and therefore the state probabilities can be found by simple multiplication. There are some excellent algorithms available for computing the cumulative probabilities and frequencies of the negative margin state directly without first[11] finding the individual state values.

The available indices in the static capacity area can be most easily illustrated using a simple system example. Consider the system shown in Table II.

Table II

No. of Identical Units	Unit Size MW	Mean Down Time (r) Years	Mean Up Time (m) Years
1	250	0.06	2.94
3	150	0.06	2.94
2	100	0.06	2.94
4	75	0.06	2.94
9	50	0.06	2.94
3	25	0.06	2.94

Total installed capacity = 1725 MW

Load Level MW	(L_i) %	No. of Occurrences
1450	100	8
1255	85.6	4
1155	79.6	4
1080	74.5	4
		20 days

The basic load model used in the frequency and duration approach is assumed to consist of a random sequence of daily peak load levels each with a mean duration designated as "e" the exposure factor[1]. In between every two peak loads is a load period with a mean duration of (1-e). The load level is assumed to have a constant magnitude. The exposure factor was assumed to be 0.5 in the studies shown in this paper. The cumulative probability of the first negative margin for the 20 day period expressed on an annual basis is 0.8988137×10^{-4} which corresponds to a loss of load expectation of 0.0656134 days/year. The cumulative frequency is 0.1879385×10^{-3} encounters/day which when expressed in the reciprocal form as cycle time becomes 0.5320887×10^{4} days. The variation in loss of load expectation and cycle time as a function of the peak load is shown in Figure 4.

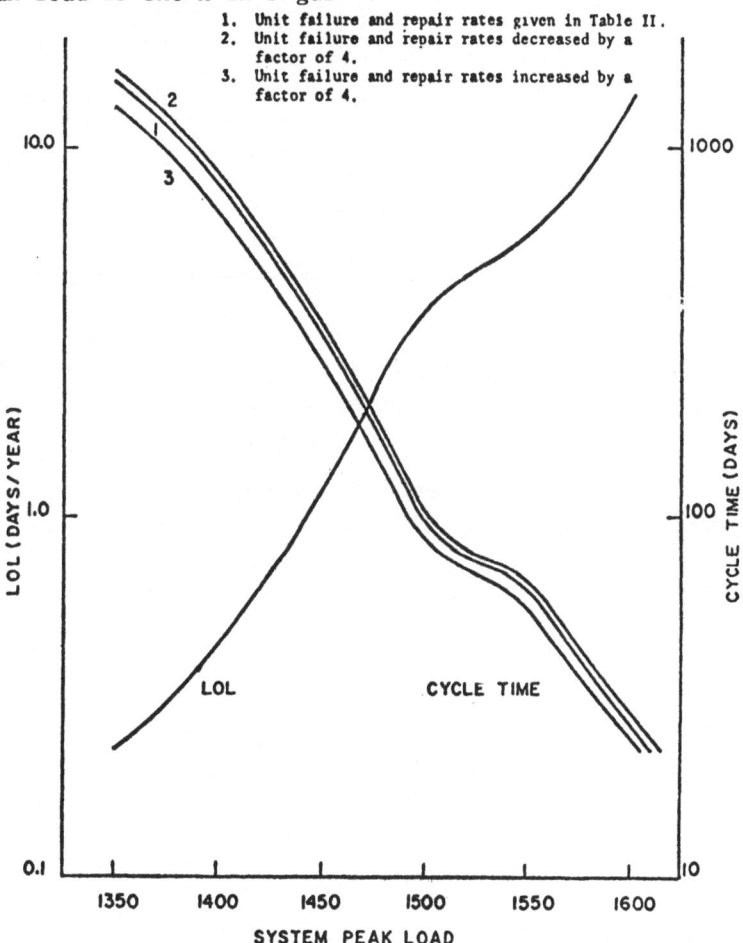

Figure 4. Loss of Load Expectation and Cycle Time as a Function of the System Peak Load.

In computing the reliability indices shown in Figure 4, it has been assumed that a year is composed of a series of 20 day periods with the load model shown in Table II. In actual practice, the year is divided into intervals during which the generating capacity out for maintenance can be considered to be constant. The non-stationery effects of seasonal load changes are then incorporated by using a valid load model for each interval. The annual LOL expectancy and outage frequency are obtained by summing the interval values. The load can be represented by a more continuous load model rather than by a two step approximation. This model can also be used to determine the expected energy curtailment due to a capacity outage[13].

The unit unavailability is given by,

$$\overline{A} = \frac{\lambda}{\lambda+\mu}$$

and if both λ and μ are varied in the same proportion, the unit unavailability and availability will remain unchanged. The loss of load expectation will not respond to a change in unit cycle time if the unavailability is unchanged. This effect is shown in Figure 4 where the failure and repair rates have been varied by factors of 0.25 and 4. The frequency and duration approach is, however, responsive to these variations and therefore provides a means of examining the sensitivity of the indices which is not available in the loss of load approach.

Studies to determine the unit additions required to meet subsequent load growth can be performed using either the loss of load method or the frequency and duration approach. In most cases, the computed increase in load carrying capability due to a given unit addition is virtually the same by either method.

Figure 5 shows the effects of adding a group of 250 MW units to the previously described 22 unit system. It was assumed that each year was composed of a series of 20 day periods with the basic load model of Table II and a 10 percent load growth each year. The peak load for Year 1 is therefore 1450 MW and for Year 2, 1595 MW etc. The risk for Year 1 is 1.2138 days/year (0.0656134 x $\frac{365}{20}$). This value is considerably higher than normal due to the assumption that the year is composed of a group of equal peak load periods. The contribution to the annual expectancy at other periods of the year would normally be less than the contribution due to the peak load period. Figure 6 shows a similar expansion pattern with the risk expressed in terms of cycle time. The dotted lines in Figures 5 and 6 show the changes in annual risk as units are added in a given year. The effective load carrying capability of a unit is the increase in load carrying capability of the system at a given risk level due to the unit addition. In the system shown in Figures 5 and 6, the 250 MW unit additions increase the load carrying

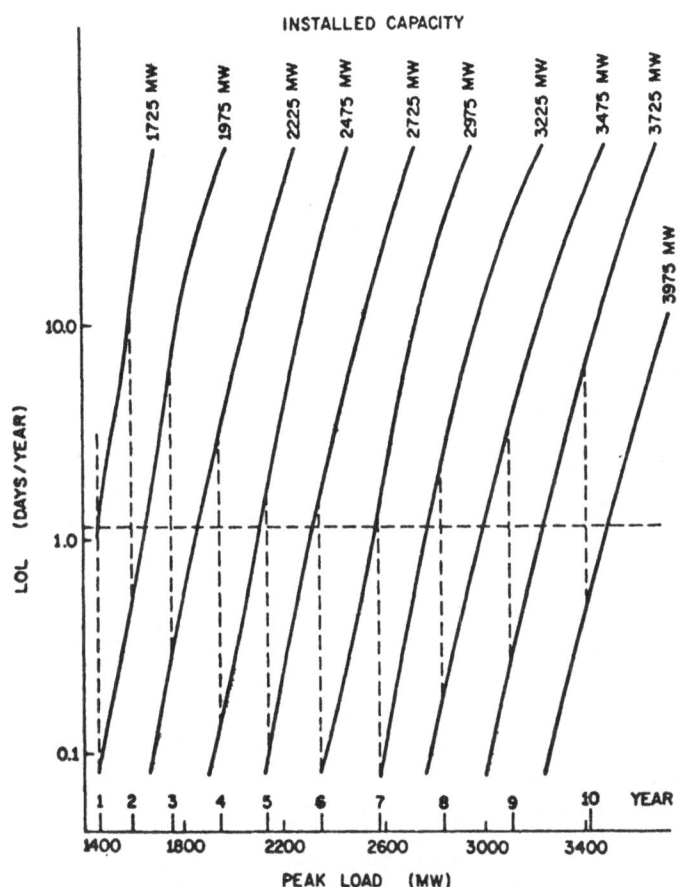

Figure 5. System Expansion Using Loss of Load Expectation as the Criterion.

capability by 225 MW in the first few years. The increase approaches 250 MW in the last year after 9 - 250 MW units have been added. The addition of a large unit to a system which contains only relatively small units can result in a substantial penalty in load carrying capability assessed against the large unit. This penalty will diminish as additional large units are added in future years. The economic appraisal of the expansion pattern must therefore be spread over a sufficiently long period to permit the true benefit of each unit addition to be assessed.

The same basic concepts can be utilized in the analysis of interconnected systems[14]. Figures 7 and 8 show the variation in indices as a function of the capacity of the interconnection. The two interconnected systems are assumed to be identical with the capacity and load data shown in Table II. Figure 7 shows the availability of the cumulative negative margin as a function of

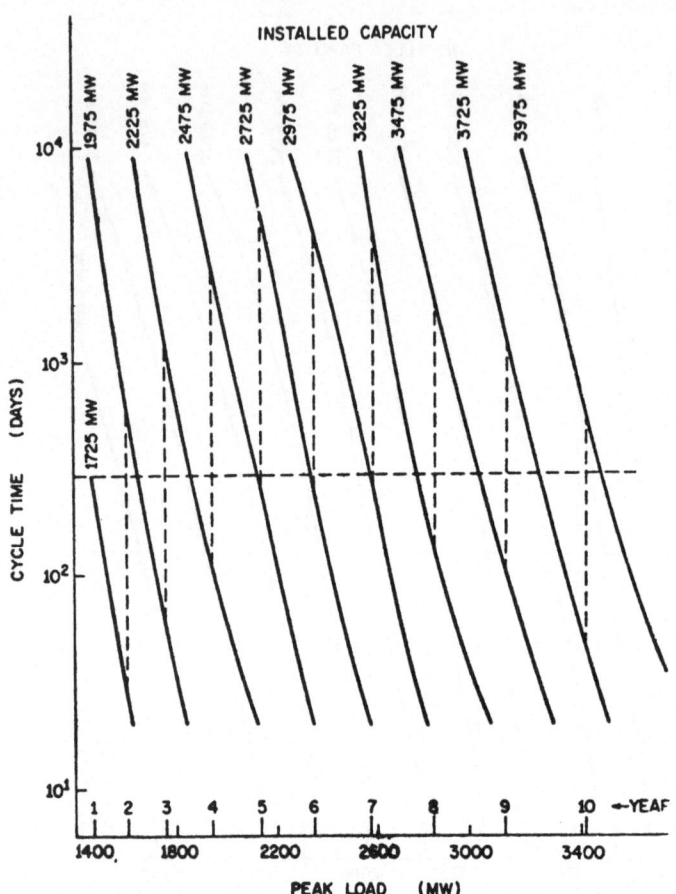

Figure 6. System Expansion Using Cycle Time as the Criterion.

the tie capacity.

The system risk indices are also sensitive to the reliability of the interconnecting tie line. This can be seen in Figures 9 and 10. The peak load in both systems was held at 1450 MW and the tie capacity at 200 MW. The availability and the frequency of the failure state were computed under the following conditions.

1. Keeping μ_{ab} (the mean repair rate of the interconnection between Systems A and B) fixed at 2.5 per day (i.e. 1.0 p.u.) and varying λ_{ab} (the mean failure rate of the interconnection between Systems A and B) from 0.005 per day (i.e. 0.5 p.u. to 0.03 per day (i.e. 3 p.u.)).

2. Keeping λ_{ab} fixed at 1.0 p.u. and varying μ_{ab} from 0.5 p.u. to 3.0 p.u. and

3. Varying both λ_{ab} and μ_{ab} from 0.5 p.u. to 3.0 p.u.

77

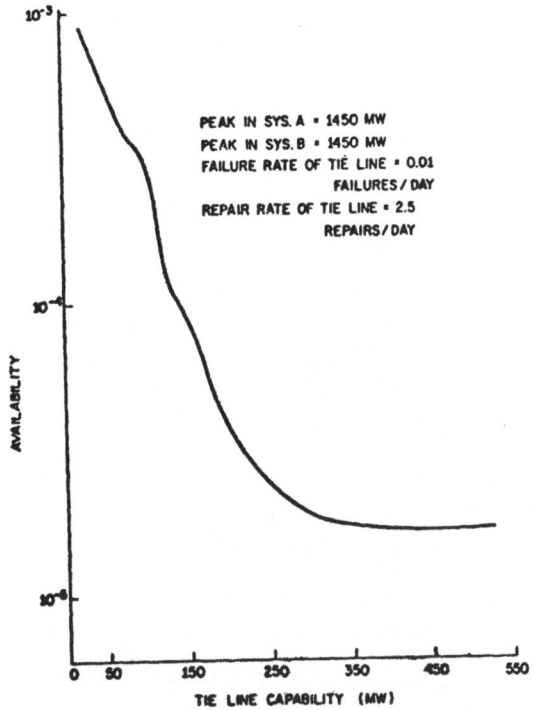

Figure 7. Variation of Risk Level (Availability) in SYS.A with the Variation of Tie Line Capability.

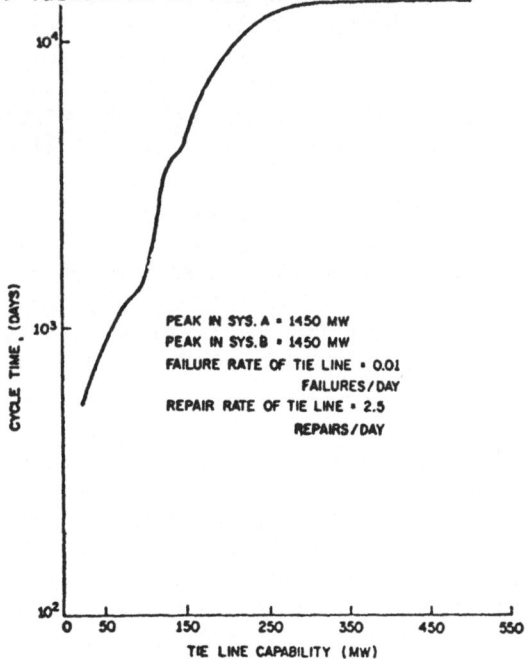

Figure 8. Variation of Risk Level (Cycle Time) in SYS.A with the Variation of Tie Line Capacity.

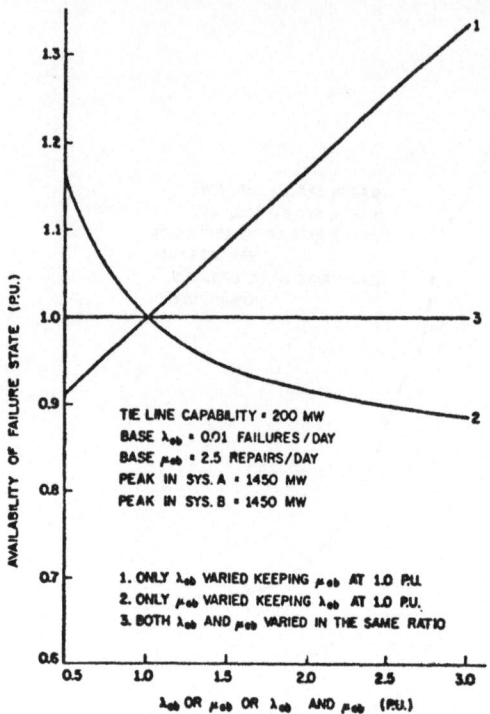

Figure 9. Variation in Risk Level (Availability) of SYS.A with Variations in Tie Line Mean Failure and Mean Repair Rates.

The results of these studies are shown as the ratios of the base availability and the base cycle time in Figure 9 and 10. The base availability and the base cycle time were computed with λ_{ab} = 0.01 per day and μ_{ab} = 2.5 per day. As in the results shown in Figure 4, the frequency and duration approach responds to changes in the failure and repair parameters of the tie line even if these changes are in the same proportion.

The frequency or cycle time of the failure state is more sensitive to the λ_{ab} variation than μ_{ab} variation. This is shown by the characteristics of Figure 10. These studies are important in that they show how the reliability may be improved by controlling the failure rate of the tie line or improving its repair rate. These parameters depend both upon the design and the operation of the line.

As previously noted, the bulk of the available literature in the power system field deals with static generating capacity reliability evaluation. Within the past few years there has been considerable development work done on operating capacity reliability assessment[15]. The operating problem is conceptually quite different from the planning application[1]. The generating unit probabilities are time dependent, i.e. given a unit is operating at time 0 what is the probability of it operating at t hours from time 0. The basic index in this case is probability and if the repair aspect

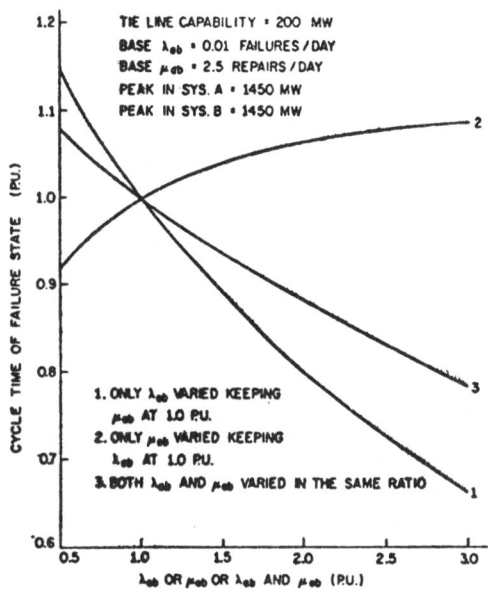

Figure 10. Variation in Risk Level (Cycle Time) of SYS.A with Variation in Tie Line Mean Failure and Mean Repair Rates.

is omitted, the probability of a unit failing during the time interval t is $1 - e^{-\lambda t} \simeq \lambda t$. The units in operation can be combined to form a generation model which can then be merged with the load model to provide a risk index in terms of the probability of satisfying the load demand. The approach itself can become quite complicated if rapid start units, load forecast uncertainty, interconnection effects and unit derating levels are included in the analysis. The frequency and duration approach has recently been extended to the area of short term reliability evaluation[16]. In addition to the time specific probability of an operating capacity deficiency, additional indices of interval frequency and fractional duration can be computed which are analagous to the indices of frequency and mean duration in long term steady state analysis.

CONCLUSIONS

This paper has attempted to illustrate by application the basic indices used in power system reliability evaluation. It appears increasingly obvious that more than one single index is required in order to make a valid assessment of the effects due to parameter and system changes. The three indices of failure rate or failure frequency, average outage duration and system unavailability or a directly related parameter appear to provide the required information. Techniques are available for computing these indices in the distribution, transmission and generation areas.

The availability of consistent data will slow the application of these techniques in some sectors but this difficulty is being overcome as more utilities recognize the potential benefits of consistent data gathering and subsequent reliability analysis.

BIBLIOGRAPHY

1. Billinton, R., "Bibliography on the Application of Probability Methods in Power System Reliability Evaluation", IEEE Trans., PAS-91, No. 2, March/April 1972, pp. 649-660.

2. Billinton, R., "Power System Reliability Evaluation", (book), Gordon and Breach Science Publishers, New York, N.Y., 1970.

3. Billinton, R., Ringlee, R.J. and Wood, A.J., "Power System Reliability Calculations", (book), M.I.T. Press, Massachusetts, 1973.

4. Billinton, R., Bhavaraju, M.P., "Transmission Planning Using a Reliability Criterion - Part I - A Reliability Criterion", IEEE Transactions, PAS-89, No. 8, January 1970, pp. 28-34.

5. Gaver, D.P., Montment, F.E., Patton, A.D., "Power System Reliability - I - Methods of Calculation", IEEE Transactions, PAS, July 1964, pp. 727-737.

6. IEEE Committee Report on Proposed Definitions of Terms for Reporting and Analyzing Outages of Electrical Transmission and Distribution Facilities and Interruptions, IEEE Transactions, PAS-87, May 1968, pp. 1318-1323.

7. Billinton, R., Singh, C., "Reliability Evaluation in Large Transmission Systems", IEEE 1972 Summer Power Meeting, Paper No. C 72 475-2, San Francisco, California.

8. Billinton, R., Ringlee, R.J., "Models and Techniques for Evaluating Power Plant Auxiliary Equipment", Joint Power Generation Conference, September 1971, St. Louis, Mo.

9. Billinton, R., Prasad, V., "Quantitative Reliability Analysis of HVDC Transmission Systems, Part I - Spare Valve Assessment in Mercury Arc Bridge Configurations", IEEE Transactions PAS-90, No. 3, May/June 1971.

10. Billinton, R., Singh, C., "Reliability Evaluation in Systems with Non-Exponential Down Times", Proceedings 1972 NATO Conference, Testing and Reliability Evaluation, The Hague, Holland, September 1972.

11. Billinton, R., Singh, C., "The Frequency and Duration Method of Generating Capacity Reliability Evaluation", Engineering Journal, March 1972, Vol. 55/3.

12. Halperin, H., Adler, H.A., "Determination of Reserve Generating Capability", AIEE Transactions, PAS Vol. 77, August 1958.

13. Billinton, R., Singh, C., "System Load Representation in Generating Capacity Reliability Studies, Part I - Model Formulation and Analysis, Part II - Applications and Extensions", IEEE Transactions, PAS-91, September/October 1972.

14. Billinton, R., Singh, C., "Generating Capacity Reliability Evaluation in Interconnected Systems Using a Frequency and Duration Approach, Part I - Mathematical Analysis, Part II - System Applications", IEEE Transactions, PAS-90, No. 4, July/August 1971.

15. Billinton, R., Jain, A.V., "Unit Commitment Reliability in a Hydro Thermal System", IEEE 1973 Winter Power Meeting, Paper No. C 73 096-5.

16. Billinton, R., Singh, C., "A Frequency and Duration Approach to Short Term Reliability Evaluation", 1973 IEEE Winter Power Meeting, Paper No. T 73 094-0.

RELIABILITY TECHNIQUES IN ASSESSMENT OF NUCLEAR
REACTOR SAFETY

Annick Carnino

Departement de Surete Nucleaire, Commissariat
a l'Energie Atomique

FOREWORD

Reliability methods allow us to perform a functional and logical analysis of various reactor systems; then follows a deep investigation about the equipments and their failure risks. With appropriate reliability data, it is possible to assess with a sufficient accuracy the probability of failure of the systems and then the probability of an accident.

1. ACCIDENT ANALYSIS

When we consider a given type of accident, we know the presumable sequence of its occurrence and we can imagine its consequences. In order to protect from these consequences and especially the radioactive risks, we have to prevent, survey and detect all possible failures on the different barriers existing between the radioactive fuel and the public. For instance, in P.W.R. reactors, we have three barriers : fuel cladding, primary circuit, containment. If any of them is ruptured during the accident, we must make sure that two or at least one will remain intact during the accident sequence.

Let us take for instance the loss of coolant accident on a P.W.R. reactor : the major accident would be the rupture of a primary pipe. The pressure will drop almost instantaneously, the coolant will get out of the primary circuit through the rupture; the core will not be cooled any more and will probably fuse (partially or entirely). But the heat produced in it will not be removed and it is necessary to inject cold borated water very quickly in the core in order to cool it again. As there has been a partial or entire fusion of the core and expansion of the primary coolant, different fission products are present in the primary containment; it is then necessary to confine them inside the containment and to filter and trap them : a special system has been devised for the purpose of cooling and confining this type of accident: the Emergency Core Cooling System (E.C.C.S). The E.C.C.S. is used for cooling the core by injection of borated cold water and for reducing the pressure in the primary containment and removal of the fission products.

In the occurrence of such an accident, we need the following systems to be operational : electric power, E.C.C.S., containment spray.

We have then to examine separately these different systems and to assess how reliable they are in order to minimize the possible consequences.

2. FUNCTIONAL AND LOGICAL ANALYSIS OF SYSTEMS

After the accident analysis is performed, we begin the stage of systems study. For each system, it is necessary to define very precisely the conditions, the environment and the functions in which it is to operate. In the example we took, let us consider the overall system of emergency :we know that it has to operate under accidental circumstances. During normal operation, this circuit will not be called upon; its only solicitation will be made by an emergency signal, but anyway the system must be operational at any time. To illustrate how the functional analysis can be done, we will develop it only on one part of the emergency system, that is the containment spray system.

Figure 1 shows the principles of the operation of the system : two independent lines come from the tank and lead to separate spread lines; there is a special line for testing a part of each of the lines (one after the other); a test circuit is used in the spread line to insure that the ejectors are not plugged. For the second part of the function of the spray system, we have lines of recirculation permitting us to reinject the water injected at the first step and cool it in the heat exchanger. From Figure 1, we can deduce the diagram of Figure 2 showing what failures are necessary to lose the function of the spray system. This diagram is presented as an event tree, using the classical symbols of tree analysis methods. Here to lose the spray system, the two lines must fail simultaneously; on each line, the function will be lost either by failed components or by the recirculation line still open or even by the test line.

3. FAILURE MODE AND EFFECT ANALYSIS

When an event tree diagram has been established, we have to look at the different systems and components to determine their failures and the subsequent effects.

The different modes of failure of a component have not the same effect : some will decrease the availability or even stop the system, others will reduce the safety of the systems or will cause both availability and safety to be reduced.

At this stage of the analysis, we take each of the events defined in the event tree and assess the modes of failures of the subsystems or of the components that could lead to the same undesired event.

For instance, let us take the motor valve : the different modes of failure are:
failed closed, failed open, external leakage, internal leakage.
If we know how a definite valve operates in our system - let us say valve N° 01VB in our spray system - we determine easily the failure mode which induces the function to be lost, here failed closed and external leakage.

For the whole system, the failure mode and effects analyses lead to a table with typical columns:
- component identification, function, possible failure modes, effects on system, methods of detection, values of failure rates.

The last column will be used to assess the overall reliability of the system and is not necessary if a quantitative analysis is not planned.

4. FAULT TREE CONSTRUCTION

From the previous failure analysis and from the plans of the system, it is easy to derive the fault tree. For the system of containment spray, we deduce the fault tree of Figure 3 by taking into account the failure modes and effect analysis.

The study we present on Figure 3 concerns only the hydraulic part of the containment spray system in order to simplify the tree. The structure of the fault tree is exactly similar to the event tree.

5. FAULT TREE USED FOR CALCULATION

The fault tree we obtained can be expressed in a similar form in terms of the various probabilities of failure of each component (Figure 4).The gates OR and AND of the first tree are now Boolean Operators for the calculation of the final probability.

We are not going to show all calculation steps; we will just indicate the results of the study: the failure probability of the hydraulic part of the spray system would be 7.10^{-3} for a period of mission of 5000 hours without taking into account the test and maintenance program.In the next section we will come back to the influence of the test program.

The result of this first calculation lights up the "critical path" which shows the most important component contributions to the final unreliability of the system.These components must then be chosen with great care, using a quality assurance program.

6. INFLUENCE OF THE TEST PROGRAM

In a reliability study, a very important part is played by the maintenance and test program. We are going to illustrate this by adjoining to the previous study a test program with different periodicities for testing : the first part of the circuit being tested is located between the exit on the tank and isolation valves (see Figure 1) and the second part to be tested is the spread heads.We are then faced with a problem of partially tested lines.We assumed that the failed components were repairable.

With a period between tests of 200 hours, the failure probability is thus decreased to 6.10^{-4} for the same mission period of 5000 hours.

The critical path thus obtained is modified and the major contributors to the system unreliability are represented on a similar tree.Another advantage of this study with tested components is that it is not necessary to decrease too much the period between tests as the probability failure reaches an asymptotic value.

7. CONCLUSIONS

The benefit of failure analysis and fault tree is very important to quantify the safety problems: the logical and deep analysis of the possible accidents, the knowledge of components under normal operation with their failure capabilities, the determination of major contributions to system unreliability, and the assessment of the overall failure of a system involved in a safety action are various tools now essential to safety analysts. A very important part in the safety analysis should be given to the influence of the test and maintenance programme and to the human reliability either in direct operation or in implementing the maintenance programme.

REFERENCES

- B.V. KOEN
  Methodes nouvelles pour l'evaluation de la fiabilite.
  Reconnaissance des formes
  Rapport CEA N° 4368

- J.P. SIGNORET
  Probabilite de defaillance du systeme d'aspersion de l'enceinte de confinement de la Centrale de Fessenheim
  Rapport interne SETS N° 17 (unpublished)

- A. CARNINO
  Terminolgie et methodologie utilisees en analyses de fiabilite
  Rapport interne SETS N° 11 (unpublished)

- A. CARNINO - R. QUENEE
  La fiabilite dans une methode d'analyse de surete
  IAEA-SM/169-13 (Symposium on principles and standards of nuclear safety, February 1973)

- A. CARNINO
  Memento pour la preparation d'analyses de surete par arbres de defaillances
  Rapport interne SETS N° 14 (unpublished)

- A.E. GREEN - A.J. BOURNE
  Reliability Technology
  Wiley Interscience, London

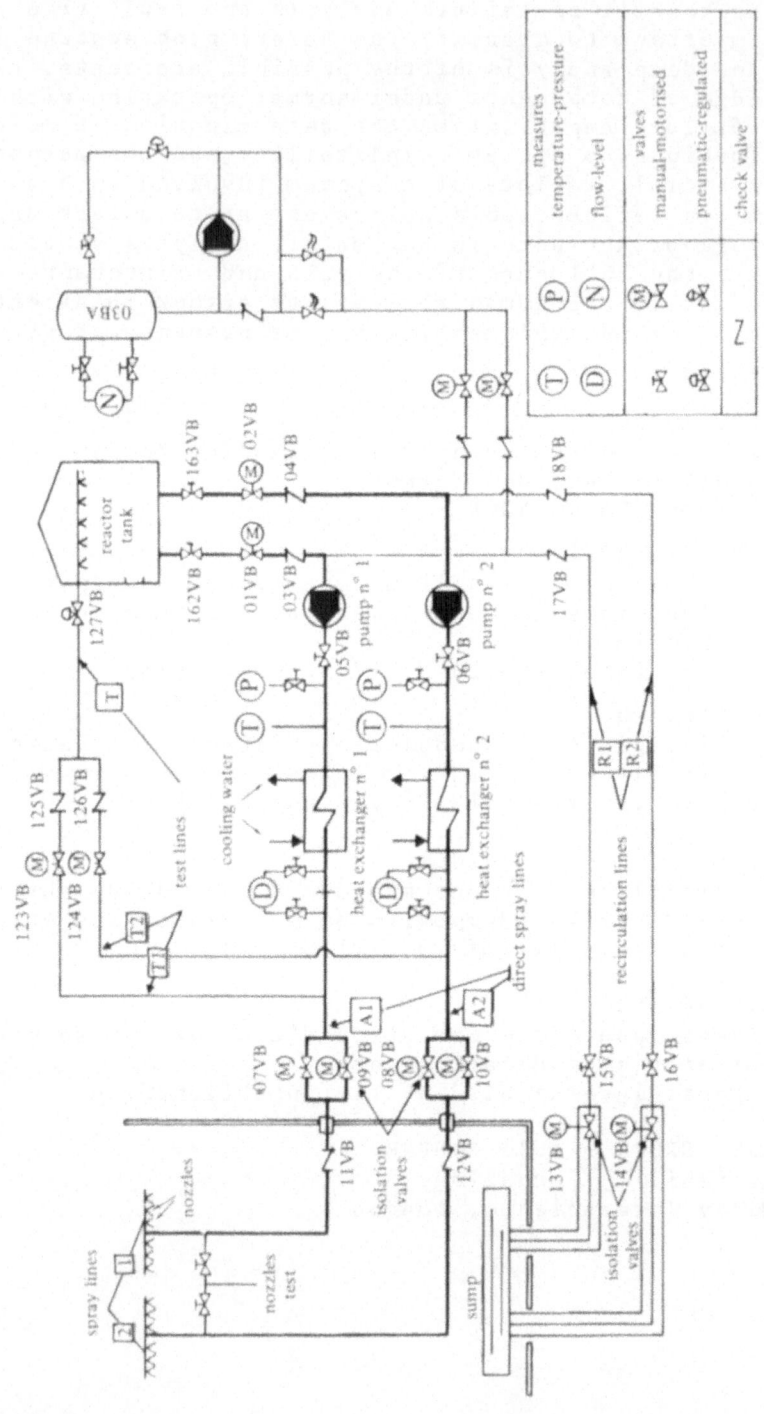

Figure 1 - Complete diagram of the hydraulic spray system

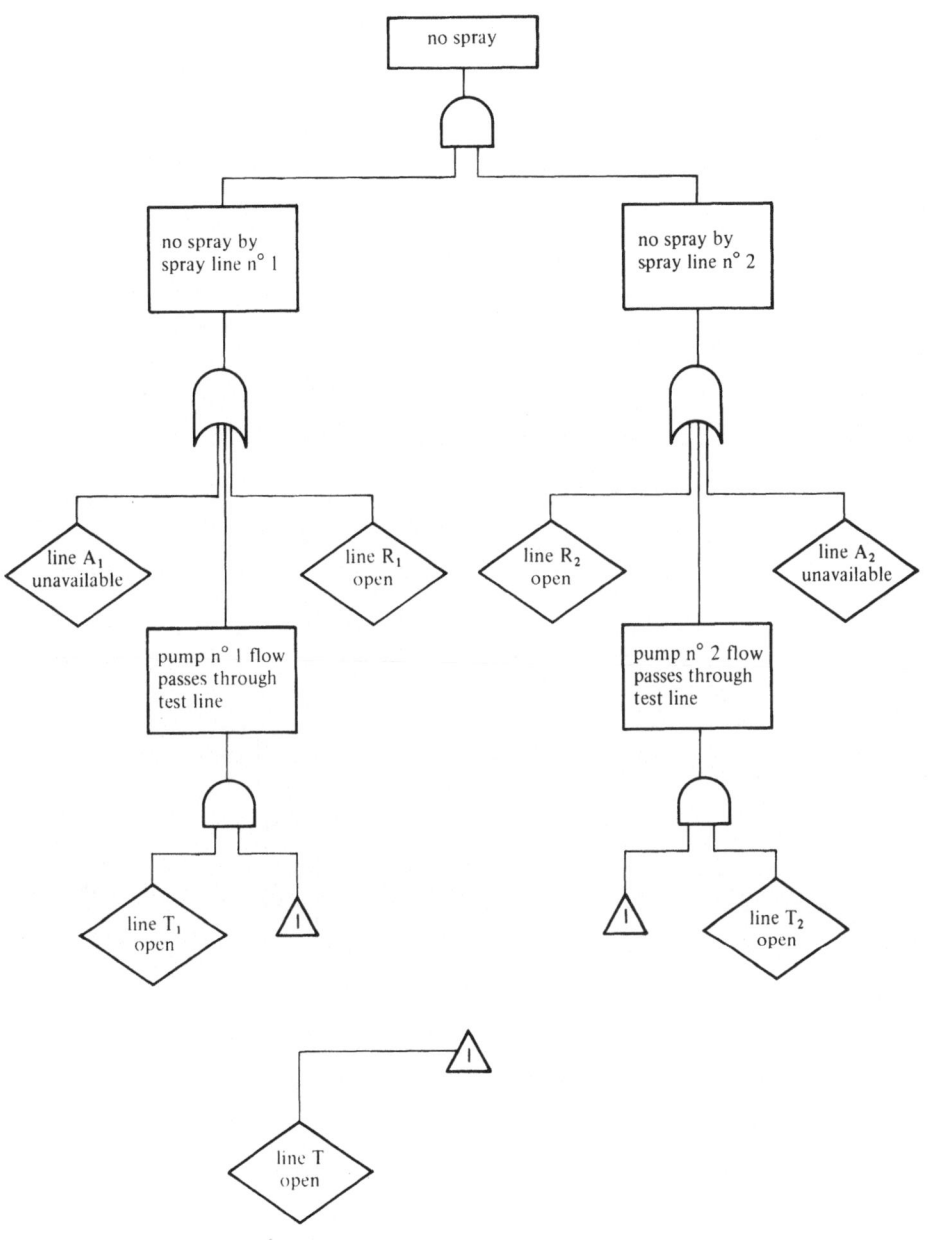

Figure 2 - Event tree of loss of spray system

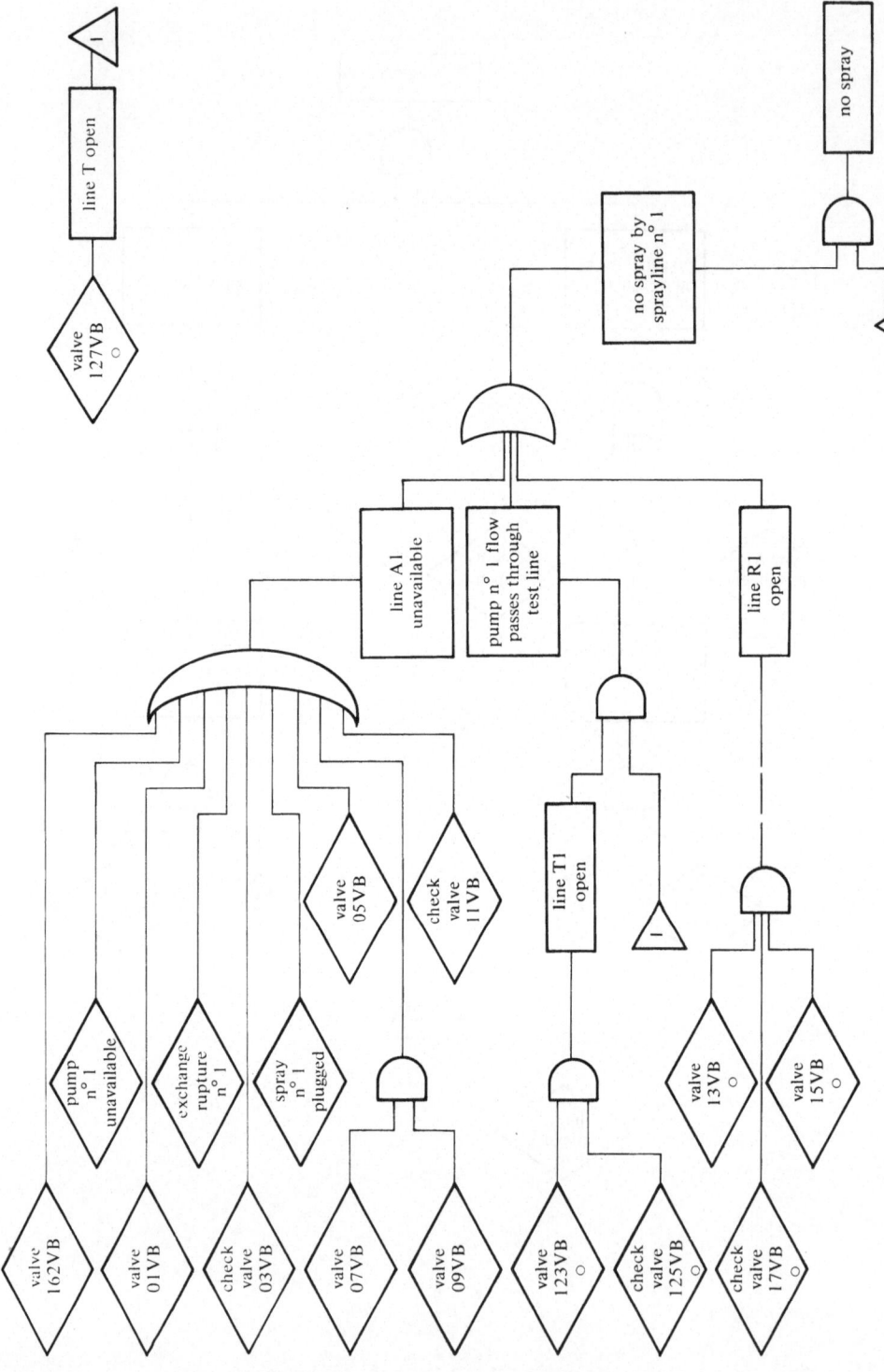

Figure 3 - Fault tree of the spray system failure

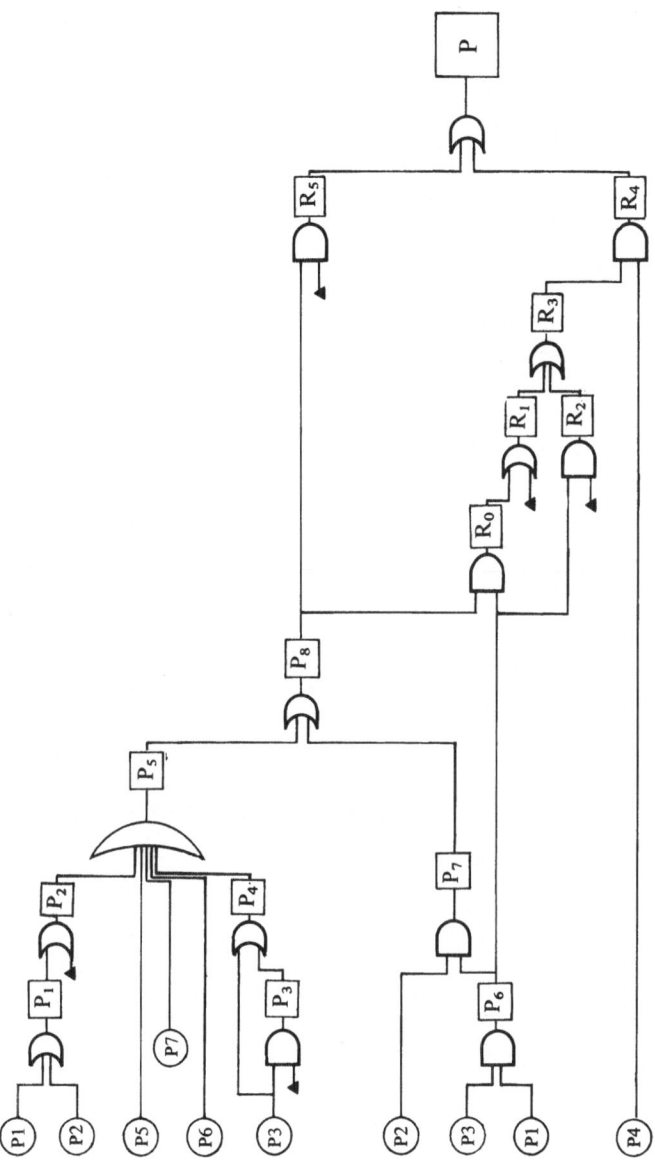

Figure 4 - Equivalent probabilities tree used for calculation

DIGITAL ANALYSIS OF AN ELECTRONIC AMPLIFIER FOR RELIABILITY STUDIES.

G.J. Dasani, J. M. Kontoleon and J. W. Lynn

Department of Electrical Engineering, University of Liverpool, Liverpool.

INTRODUCTION

Recent rapid growth in the size of electronic systems used in automatic process control, telecommunication and transport industries, has made it necessary to ensure a high degree of reliability. Improvement in reliability of such systems has mainly been achieved by use of redundancies. However, consideration of reliability at the design stage requires a detailed analysis of the electronic system. The work presented in this paper is the development of a computer program for analysing larger size electronic networks and overcoming the problems of storage and accuracy of solution. There are of course computer programs in existence which can solve such circuits but most of these can only be used on large computers and some suffer from slow convergence which is ensured only on the prior knowledge that the initial guess is sufficiently close to the solution. The techniques used here ensure faster convergence, keep track of round-off errors and are suitable for larger networks. The program is capable of performing

      (i)   steady state d.c. analysis

      (ii)   small signal frequency analysis

and   (iii)   transient analysis

NON-LINEAR NETWORK ANALYSIS

The analysis uses the nodal admittance approach. The non-linear elements present in the networks are replaced by their equivalent circuit models, based on the quantities measured across the terminals. The elements used in the models are limited to resistors, capacitors, current- or voltage-dependent sources and independent sources. Typical values of the known parameters of the model as supplied in the manufacturer's data sheets are initially used for obtaining the solution. The bias voltages determined by the solution are then utilized to search for the region of operation of the non-linear device. The new values of the parameters of the model at the next iterative step are therefore calculated from each solution together with the device characteristics which are already stored in the computer. Since the parameters of the model are inter-dependent it is not necessary to compare all the parameters to determine the error. Since the current gain of the transistor ($\beta = I_c/I_b$) is a single valued function of the current it is used to determine the error (E_n) in the solution at the nth iteration,

$$E_n = \beta_n - \beta_{n-1}$$

The parameters of the model are iteratively modified until a solution within a specified limit of accuracy is obtained.

In direct optimization the value of the parameter is increased or decreased by the error value. However, to give stable solution and faster convergence the criterion used is

$$\beta_{n+1} = \beta_n \pm aE_n$$

where a takes the value 1.0 or 0.5 depending on E_n and E_{n-1}. When the specified error criterion is reached the iterations are terminated after an additional iteration during which the error is further reduced and the states of the non-linear devices checked. An accuracy of about 1% is achieved in 3 to 5 iterations.

ERRORS

Due to finite word length of the computer there are round-off errors in the numbers stored. These are particularly important when the circuit admittance values are very far apart. (i.e. the difference in the powers of the values is greater than the word-length). In such cases using gaussian elimination the round-off errors can significantly affect the solution. In the computer program a check is maintained on this type of error and when it is significant this is indicated by a diagnostic print-out. The presence of this type of error is illustrated by considering the solution of the circuit shown in fig. 3. The direct solution of the circuit is given in Table 2(a). Simulation of a short circuit by an admittance 10^8 mhos leads to significant error (Table 2b) whereas an admittance 10 mhos while eliminating this error problem still gives reasonable results (Table 2c). Thus round-off errors can be avoided by proper choice of admittance values which evaluate extreme conditions. Use of double precision arithmetic can help in reducing such errors but this increases the storage and computation time considerably.

DIAKOPTICS

In the present case Kron's diakoptic technique is used for the analysis of large networks. The network is torn across the branches as shown in fig. 3 and the cut branches removed. The subdivisions of the networks are analysed separately and then combined by the connection matrix to obtain the solution. The admittance matrix of each subdivision (Y), the current source vector (I), the impedance matrix (Z_r) of the cut branches and the connection matrix (C) relating the currents in the subdivisions to the currents in the cut branches are formed from the circuit data. The steps in the solution of the node voltage vector (E) are

$$
\left.\begin{array}{rcl}
Z &=& Y^{-1} \\
ZC &=& Z*C \\
ZO &=& C_t*ZC \\
VA &=& Z*I \\
VB &=& C_t*VA
\end{array}\right\} \text{for each sub division separately}
$$

$$
\begin{array}{rcl}
ZST &=& Z_r + ZO_1 + \ldots \ldots \\
YST &=& (ZST)^{-1} \\
VBT &=& VB_1 + VB_2 + \ldots \\
IC &=& YST*VBT
\end{array}
$$

$$
\left.\begin{array}{rcl}
VN &=& ZC*IC \\
E &=& VA-VN
\end{array}\right\} \text{for each subdivision}
$$

From the above steps in solution the following points emerge (i)　　　Only the admittance matrices of the subdivisions have to be formed

(ii)　　An inversion of one extra matrix, 'the intersection matrix' (ZST) is required the order of which is equal to the number of cut branches.

(iii)　　The matrix multiplication operations are with C the elements of which are ± 1 or O. The other multiplication operations are with vectors only.

Use of this technique in the solution of large networks results in saving of storage and computation time. The advantage in time obtained even when the network is divided into only 2 parts is illustrated by fig.1. Further advantages in storage and time can be obtained by subdivision of large networks into more parts. However the maximum number of subdivisions to gain advantage in time and storage is limited by the size of the 'intersection matrix' which should not normally exceed the size of any of the subdivisions. This technique also prevents the accumulation of round-off error and even in extreme cases by tearing across the branches with large admittance values the round-off errors can be kept small.

The computer program developed here is used for the study of amplifiers such as that shown in fig. 2. The d.c. steady state solution is shown in Table 1. These results were also in agreement with those obtained using the diakoptic technique.

CONCLUSION

Diakoptic analysis can be a useful technique in circuit design, since the designer can assess the performance of a number of alternatives very quickly. The program is used for analysis of transistor networks under various fault conditions and, together with data about failure rates, for the prediction of reliability of amplifiers at the design stage.

Table 1

BRANCH	ELEMENT	VALUE	CURRENT	VOLTAGE
1	CA-CITOR	0.470 MF	-0.0000 MA	-1.1123 V
2	TRA-B-EM	2.599 K	-0.0097 MA	-0.5253 V
3	TRA-B-EM	26.297 K	-0.4978 MA	-1.8439 V
4	RESISTOR	1.00 K	-0.5870 MA	-0.5870 V
5	RESISTOR	15.000 K	0.5046 MA	7.5681 V
6	RESISTOR	82.000 K	0.0097 MA	0.7974 V
7	RESISTOR	68.000 K	0.0795 MA	5.4055 V
8	TRA-B-EM	3.108 K	-0.0068 MA	-0.5212 V
9	TRA-C-EM	37.304 K	-0.4092 MA	-4.0828 V

Solution of network shown in fig. 2

Table 2

	(a)	(b)	(c)
(1)	1.11304 V	1.21374 V	1.11304 V
(2)	0.58765	0.68247	0.58765
(3)	2.42878	2.71387	2.42878
(4)		2.71387	2.42878
(5)	1.90940	2.18782	1.90940
(6)	5.98892	5.56461	5.98892

Solution of network shown in fig. 3
(a) Considering it as a 5 node network
(b) Simulating the S/C between A & B by admittance of 10^8 mhos
(c) Simulating the S/C between A & B by admittance of 10 mhos

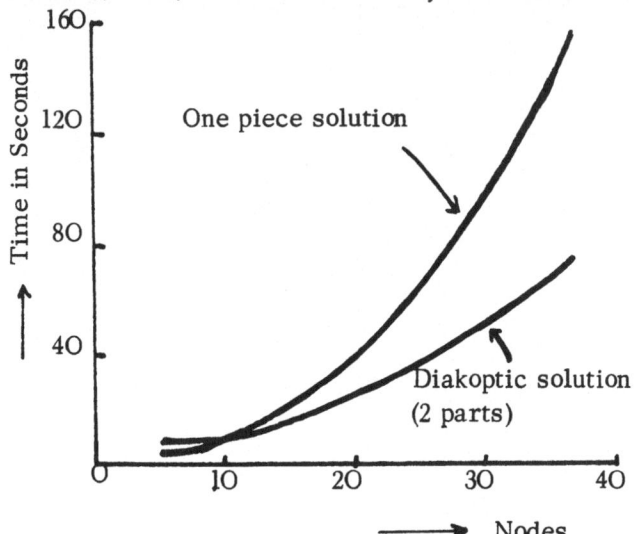

Fig. 1. Solution time

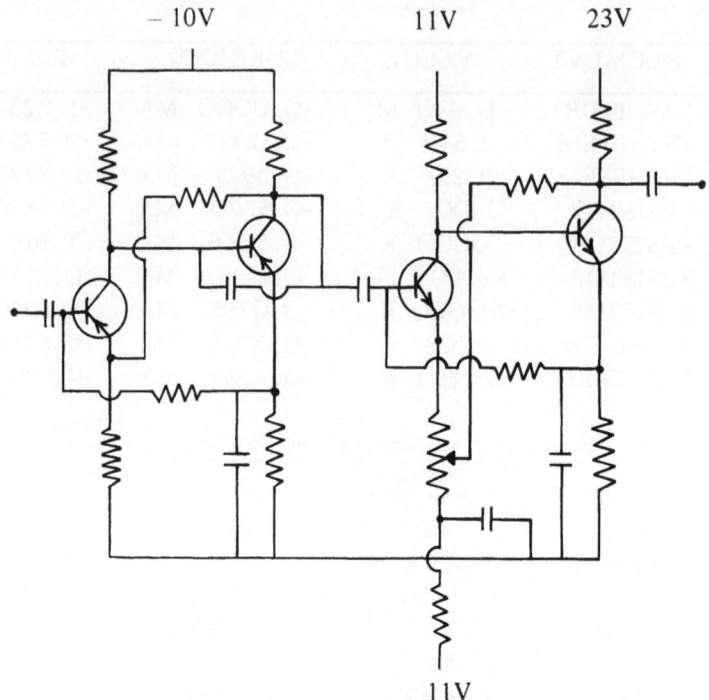

Fig. 2. Circuit diagram of the Amplifier

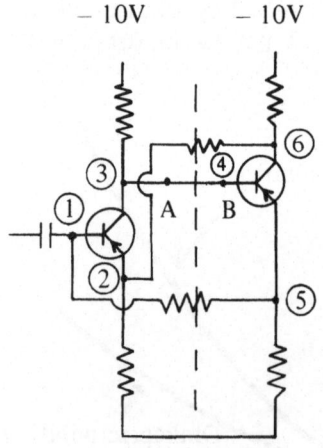

Fig. 3.

OPERATION OF A RELIABILITY DATA BANK

by A. R. Eames, C.D.H. Fothergill, R. Roughley, E. R. Woodcock

U.K. Atomic Energy Authority, Risley.

As an example of the operation of a reliability data bank the bank set up by the U.K.A.E.A. Systems Reliability Service (SRS) is outlined. This bank is designed to collect raw data on the behaviour of a variety of items of equipment in an industrial environment; to store this data in a computer file; to collate and analyse it so as to give statistical reliability data; and to store this reliability data together with that derived from other sources so that it can be readily retrieved as desired.

As a byproduct of this procedure the raw data can be analysed so as to feed back to the donors useful information on the operation of their equipment and many of the analyses available have been designed for this purpose.

The SRS data bank has now been operating for several years and has developed into a fully practical system.

THE COLLECTION AND RECORDING OF EVENT DATA

Data Bank Operation

The Reliability Data Bank, Fig. 1, comprises an Event Data Store and a Reliability Data Store. The former is supplied with detailed information from the operation and maintenance experience of industrial plants; this data is processed into the computer files of the Event Data Store which can provide both a feedback of information for the guidance of plant management and derived generic reliability data as one of several inputs to the Reliability Data Store.

Collection and Recording of Data from Plant

The first 3 stages in the collection and recording of data from the plant are shown in Fig. 2; these stages can be described as identification (area of reporting and inventory), event information and operating history.

Coding

Any viable data collection system must maintain consistency in terminology and flexibility, with economy, in computer processing. The basic system to achieve these aims is included in the Coding Manual which contains the codes for recording information related to each of the 3 stages (see Fig. 2), at the installation (plant), system and item levels, onto computer load sheets.

Stage I - Identification (Inventory Data)

The first stage of a project is to define the area of reporting, make an inventory of the systems and items involved and store in coded form the basic design data on each inventoried item. It is recommended that, initially, the area of reporting and the number of reportable items be kept small to promote a smooth reporting procedure, increasing the scale of the project as experience allows. Fig. 3 shows an extract from an item inventory and the nature of the basic information is shown at Stage I, Fig. 2. The provision of inventory data constitutes a once-off exercise and will only become invalid if there is a change in the basic data or the item is removed from the exercise for some reason such as 'end of life'.

Stage II - Event Information

An 'event' may be defined as any occurrence which affects the normal operation of an installation (plant), system or item. Some typical event types and other codes are shown on Fig. 4.

The Event Report is the vehicle used by the reporter of an event to record the details of the occurrence, as defined by the 'type of event' and it is essential that consistency in reporting is maintained within each reporting area. In order to achieve this the format and content of the Event Report should be agreed between SRS and the participating Site before it is used on the plant.

This form can be tailored to suit local practice and can usually be developed from an existing job card or works order form. Fig. 5 shows one example of an Event Report which includes the information needed to obtain reliability data as required for a particular project. The information required varies from project to project and SRS insists only on a minimum of information to derive **item failure rates** for processing to the Reliability Data Store, but will handle as much as is necessary to enable desired analyses to be fed back to the project.

The format of an event report may vary from the total descriptive type, i.e., written format, through the partial descriptive and partial 'tick box' type, as shown of Fig. 5, to the total 'tick box' type in which the reporter selects the 'tick boxes' appropriate to the conditions of the event.

Stage III - Operating History

The operating history of a system or item related to the operation of its parent installation is sometimes required to obtain an estimate of system/item operation between any two consecutive events. A procedure for recording the operation of the installation from which the system/item operation may be estimated, has been developed and a computer program has been written to utilise this information. This procedure is only used as an alternative to the preferred direct recording of actual operating history.

REDUCTION OF DATA FOR COMPUTER OPERATION

Reasons for Coding

The raw data as received in the form of Event Reports or Fault Reports is unsuitable for direct computer operation and storage. This data is first subjected to manual coding and transfer to the format of 80 column computer punched cards. Coding is necessary to reduce as much of the data as possible to numerical form, or alpha-numeric form, in order to achieve the maximum concentration of information in the minimum storage area, with uniform format. At the same time the facility for retrieval and decoding into literal form, where necessary, must be maintained, during data analysis. Two retrieval processes are used in the SRS Event Data Store Analysis program; numerical identification and character recognition.

Analysis Selection Procedure

Event data is divided into three levels or categories, according to the specification of the particular Event Reporting Scheme. These three categories can be employed in any configuration to suit individual requirements; only those categories needed are used, the others omitted - they can be duplicated if necessary. The SRS Analysis system is most flexible.

The three categories have been given the names (KEYWORDS) INSTALLATION, SYSTEM and ITEM to enable the analyst to select them uniquely for computer operations. Fig. 6 lists these three keywords together with three more, all used for the first selection of the event data for analysis; they restrict program interrogation to the required category of data for analysis. These keywords are associated with second selection keywords listed in Fig. 6. These correspond to the items of data which can be stored in the corresponding category of each event record.

The keyword +ITEM broadens the area of selection to include the Installation and System category entries corresponding to the Item category data selected when this keyword is used (see Fig. 6). Blank Card (i.e., on 80 column computer card with no holes punched) selects all records corresponding to the second selection keywords listed under this heading in Fig. 6. The selections mentioned so far will transfer the event data entries from tape to magnetic disc for analysis. However, the keyword NO SELECTION enables analysis direct from the data tape.

Analysis Procedures

Eleven analysis routines have been developed so far in the SRS Event Data Store Computer Program. They are listed in Fig. 7,

(i) PRINT - The entire contents of each of the selected event records is printed together with the relevant additional particulars transferred from the Item Inventory List when the Event data is added to tape from punched cards.

(ii) COUNT - A classified count of the selected item or group of items of interest is given. Two examples of this analysis are shown in Fig. 8.

(iii) SYSTEM FAILURES - This provides a list of those main systems of a plant which the customer wishes to be interrogated. Against each system is printed the total outage time (or repair time) average outage time and the number of events contributing to these times.

(iv) LONGEST OUTAGE - This analysis identifies those events for which the longest outages have been reported. The date and time of the events is given together with any descriptive detail stored with the data entry.

(v) DESCRIPTION - This analysis provides a condensed version of the data entries. The program prints the time and date of each event together with any descriptive information and up to three extra items of interest tabulated under their separate headings.

(vi) SERIAL - This is similar to Description but the Item Serial Number is printed at the start of each descriptive entry.

(vii) MANAGEMENT - This provides a concise summary of the behaviour of the plant over a limited interval of time. It provides two tables per entry, the first makes use of all events which have been recorded and the second only maintenance events. The tables show Outage time, Repair time, Skilled Labour Manhours, Total Labour Manhours, Materials and Parts Cost. The total and mean values are listed together with the number of events.

(viii) SUMMARY - This is similar to Management but makes use of all events. The separate table of maintenance information is not given.

(ix) LIFE - This analysis is designed to obtain information on failure rates and times to failure of items of equipment recorded in the bank, whether they be in the Item, System or Installation category, from the 'hours since last event' recorded in each entry. In order to carry out this analysis it is first necessary to pick out those entries dealing with one particular type of equipment so as to compare like with like. The sum of the 'hours since last event' recorded divided by the number of breakdowns will then give the mean failure rates to which confidence limits can be attached assuming the familiar Poisson distribution for number of breakdowns.

The information stored, however, enables statistical information to be obtained about times to failure where this term is taken to mean the time the equipment was in working order after being repaired before suffering a breakdown. By gathering together the stored data it is possible to tabulate the lengths of time each individual item operated between breakdowns. Most of these intervals will be complete, i.e., starting with a repair and terminating with a breakdown but some will not be complete. Because of these it is not possible to obtain directly a complete histogram of time between failure. However points on a cumulative distribution may be obtained by calculating estimates of $P(t)$, the probability that the time between failures is no greater than 't' as

$$P(t) = \frac{\text{No. of units failed in time t or less}}{(\text{No. of units failed in time t or less}) + (\text{No. of units operating for time t without failure})}$$

Having calculated $P(t)$ in this way a search is then made to find the 2-parameter unimodal distribution from which this sample is most likely to have been drawn according to the Kolmogarov criterion (Reference 1). The distributions examined are the Normal, Lognormal, Weibull, Exponential and Special Erlangian.

From the fitted distribution the failure rate as a function of time since last repair is calculated.

(x)&(xi) OUTAGE & COST - The data stored also contains information of the outage time caused by each event, the repair time, labour required to repair and the cost. This information can also be analysed and, if desired, a statistical distribution obtained by forming a cumulative distribution and analysing it in a similar way to that described for the Life analysis.

The SRS Event Data Store permits, as outlined above a variety of analyses, some of which are of value in the day to day operation of the plant and others which give reliability parameters that can be used over a wider field.

When sufficient data of the latter sort has been gathered it is stored in the Reliability Data Store from which it can be readily retrieved and used in reliability assessment.

ASSESSMENT OF EVENT DATA PREPARATORY TO THE GENERATION OF RELIABILITY INFORMATION

Ideal Case

Any items under event surveillance which have at least:-

(a) A population of 100.

(b) 100 recorded events.

(c) All events properly reported from field operation over a reasonable period of time.

(d) A complete engineering description.

is, without doubt, a suitable aspirant for the production of good quality reliability information.

Guide Lines

However, few items conform to this ideal specification and so a measure of the quality of event data (and reliability information) must be made. In order to do this all data should conform to the following guide lines:-

1. A sufficient number of failures, and equally significant occurrences, must be accumulated to make a positive contribution to the sample size for analysis.

2. As large a number of similar items as possible should be undergoing, or have already undergone, a suitably long period of event surveillance in order to provide the largest sample size possible.

3. The project rendering the raw data must have been set up on a suitable basis; its disposition must be to a satisfactory standard.

4. The subject of the event reports must be given a recognised engineering description.

Auxiliary Guide Lines

Data Bank Operators should implement the following:-

(a) Regular interrogation and test of the accumulated data to assess the quality.

(b) A periodic list of all items under surveillance, containing the associated number of fault reports and surveillance period, should be assessed independently by experts.

(c) Outage time, Repair time and Manhours expended are satisfactory adjuncts to Reliability Information.

THE USE OF ANALYSED DATA

In the established failure rate data banks, data for electronic components are available on a fairly generous scale and much has also been published in open literature. An increasing amount of data is now also being published on the reliability of heavy items of plant. Unfortunately much of this published information is inadequate because many originators publish the figures for average failure rate alone. They omit all reference to sample size, number of faults, types of failure, etc.

During the performance of safety assessments the need to supplement fault data, relating to the type of plant under consideration, with published data, has always been necessary. Although the results so far have proved this practice to be justified, gaps in the available data have led to difficulties, delays and additional work for the assessors. This indicates a need for much more work to be done in the data bank field so that additional information becomes available to the reliability assessors. Experience in assessing the reliability of equipments and systems has shown how certain basic items of fault data are required before successful analyses can be accomplished. Examples of such items of information are:-

(i) Mean failure rate.

(ii) Mean repair time.

(iii) Number of items involved.

(iv) Number of faults recorded.

(v) Length of surveillance period.

(vi) Length of operating time.

(vii) Environmental conditions.
etc.,

and all this data is needed for a wide range of items in the electrical, electronic, mechanical, pneumatic, hydraulic, and other fields.

It is often presupposed that this data can be adequately classified under such headings as "components", "equipments" and "systems" but experience has shown that the best way of classifying reliability data is by "item description". This involves drawing up an explicit engineering description for each item of interest.

In reliability assessment work there is a need to store and retrieve data and to process it as required; any processing must retain the original raw data intact. It is also necessary to be able to extract information based on parameters other than the item description; e.g., identity, manufacturer, design date, environmental conditions, etc. In order to store reliability data of this quality and to retrieve and analyse it, to the Reliability Analyst's and Safety Assessor's requirements, computer techniques must be resorted to and for this purpose a partly coded storage system has been developed by the Systems Reliability Service and it is known as the SRS Reliability Data Store.

The figures available from the SRS Reliability Data Store range from those relating to the failure rates of minute components to those obtained from assessments of presently operating plants, systems and assemblies. The most up to date information of this kind and quality is that transferred from the SRS Event Data Store. But much reliability data is being obtained from many other sources; for example:-

(i) Published figures in open and classified literature, if the quality, source and author are known to be dependable.

(ii) Figures from event data already accumulated in private recording systems and data banks in industry.

(iii) Figures obtained during proving trials and life tests.

(iv) The results of laboratory tests, climatic and durability tests and similar exercises.

The computer techniques developed for the Reliability Data Store enable further "processing" of the data; failure rates and other figures can be "pooled" statistically for identical items of, say, different manufacturers, etc. Also, different types of computer print can be made available to users of the Reliability Data Store. These are:-

FLARD: This allots one line to each item description. The whole contents of the Data Store can be printed out or alternatively a specified selection of the stored items. It is largely in code but the most significant part of the literal description is included. The principal parameters listed are - item description code, fault mode, source reference, number of items, sample size, number of faults, mean failure rate and several minor entries indicating that additional information is available in store for that item. This print is intended as a first interrogation of the Reliability Data Store to identify the items of interest.

MAGPIE: This provides a detailed account of the information stored, one item per page. All entries which have been put on the data tape, for each item selected, are printed out including the full literal comments on the data.

<u>AVER</u>: It is often necessary to "pool" data statistically to increase the sample size and so improve the estimates of population parameters. Analysts usually need to do this manually but to save their time the program AVER does it automatically. The print out lists the items, each described by a short plain language title, in alphabetical order together with the derived failure rate and it indicates whether surveillance was over operating time or history time.

CONCLUSION

In order to confirm that the reliability of a plant or system continues to meet the specified safety and economic criteria then prediction techniques and an up to date store of reliability data are two necessary requirements. The SRS Data Bank is a live and working system for the storage of this data allowing continual up dating by the analysis of new information collected from the observation of operating plant.

Reference 1. N. Smirnov, Bull. Math. d l'Univst. de Moscow, Vol 2 (1939), fasc 2.

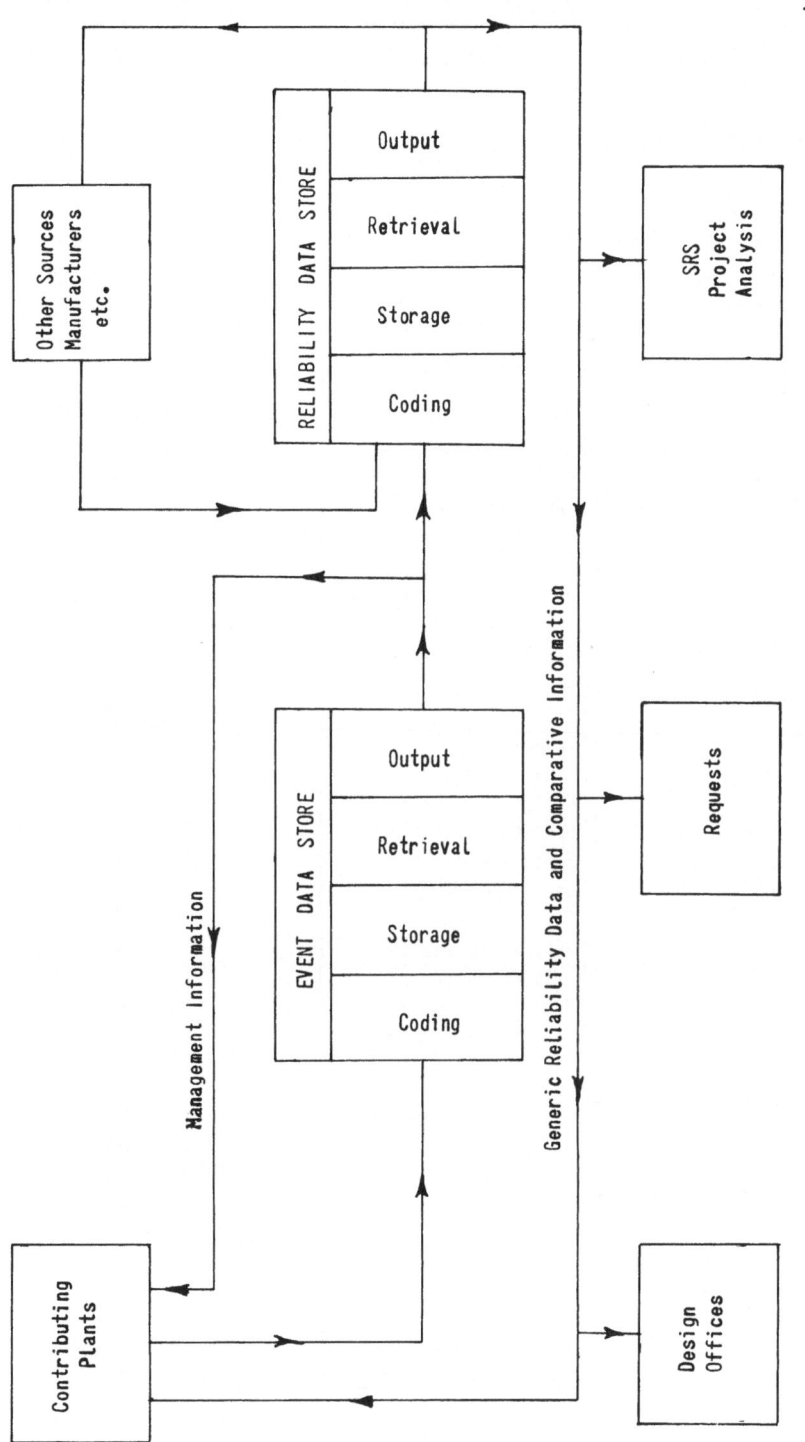

FIG. 1. Reliability Data Bank Information Flow

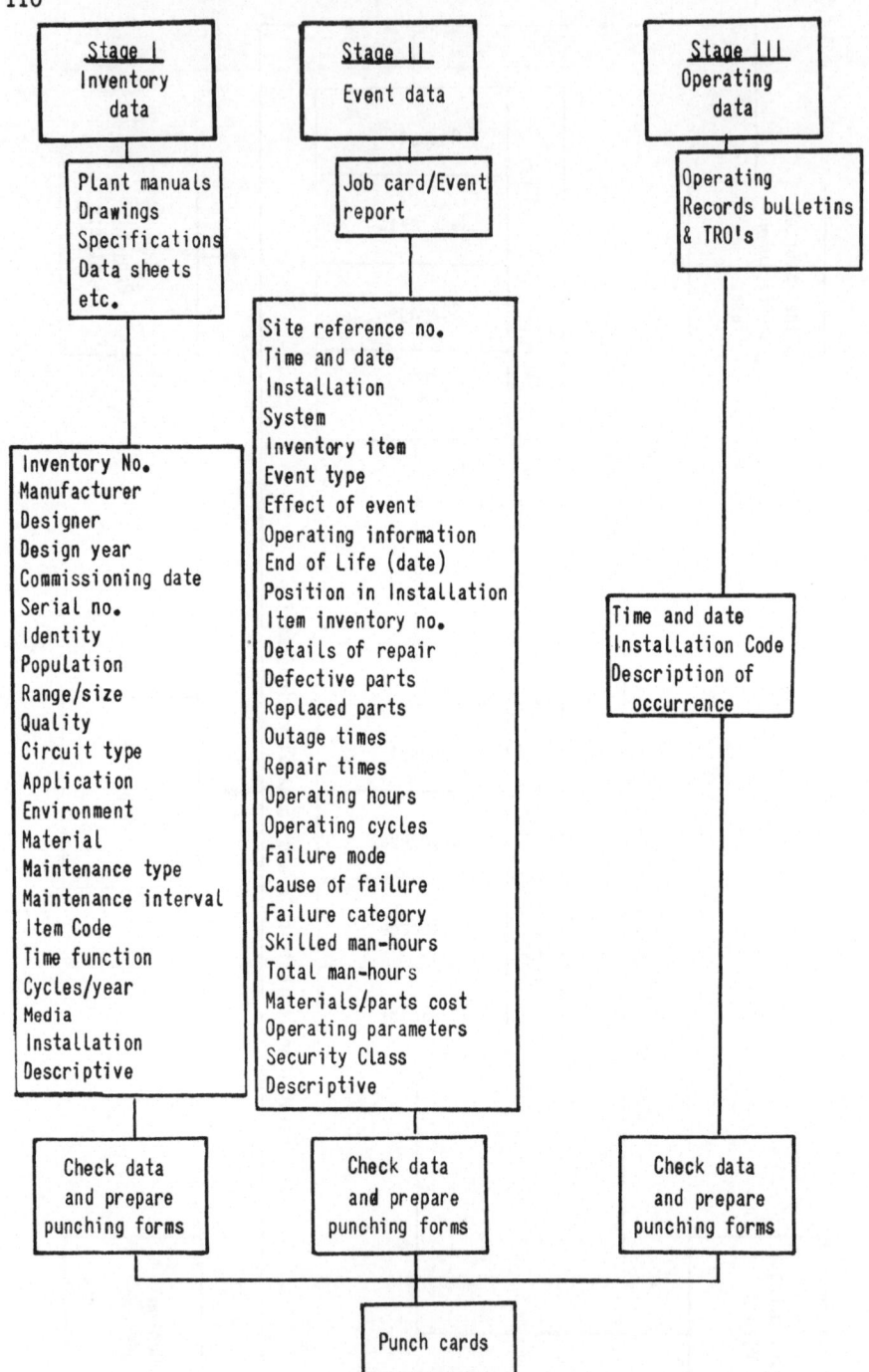

FIG. 2. The Three Stages of Data Collection

REFERENCE NUMBER	ITEM DESCRIPTION	SRS INFORMATION		
		SYSTEM CODE	ITEM CODE	INVENTORY NUMBER
A-01	Main turbine - High pressure	2430 131	1936 211	1010 P11
A-02	Main turbine - Low pressure	2430 131	1936 211	1020 P11
A-03	Reduction Gearing	2430 134	0608 200	1030 P11
A-04	Thrust Bearing	2430 134	0109 030	1040 P11
B-01	Main Boiler No. 1	2430 122	1840 100	2010 P11
B-02	Main Boiler No. 2	2430 122	1840 100	2020 P11
B-03	Air Preheater No. 1	2430 121	0259 300	2030 P11
B-04	Air Preheater No. 2	2430 121	0259 300	2040 P11
B-05/1	Sootblower - Main bank No. 1	2430 122	1831 000	2051 P11
B-05/2	Sootblower - Main bank No. 2	2430 122	1831 000	2052 P11
B-06/1	Sootblower - Air preheater No. 1	2430 121	1831 000	2061 P11
B-06/2	Sootblower - Air preheater No. 2	2430 121	1831 000	2062 P11
B-07/1	Forced draft fan unit No. 1	2430 121	0250 101	2071 P11
B-07/2	Forced draft fan unit No. 2	2430 121	0250 101	2072 P11
B-08	LP Steam Generator	2430 171	0701 420	2080 P11

FIG. 3. Item Inventory Coding List

SELECTIONS FROM EVENT TYPE CODE

10000	Breakdown
20000	Modification
41000	In Situ Maintenance
62300	Item Change, Item Renewed.

SELECTIONS FROM ITEM CODE

0007212	Amplifier, Main, Linear, Power
0319120	Diecasting Machine, Hydraulic, Cold Chamber
0708200	Hoses, Metal Reinforced
1509161	Pumps, Axial, Vane, Single Stage.

SELECTIONS FROM FAILURE MODE CODE

1410	Electrical, Overheated, Burnt Out
2510	Mechanical, Intermittent Operation
3330	Failed In Open Position
8000	Failure Mode Undiagnosed

FIG. 4. A Selection from the SRS Coding Manual

INSTALLATION:	REPORT NO.
SYSTEM:	DATE OF EVENT:
	TIME OF EVENT:
ITEM:	EVENT TYPE: 　　BREAKDOWN
ITEM DETAILS: DATA BANK INVENTORY NO: OP. HRS. SINCE LAST EVENT: NO. OF CYCLES SINCE LAST EVENT[1]: OUTAGE TIME: REPAIR TIME:	MODIFICATION 　　TEST/CALIBRATION 　　IN SITU MAINTENANCE 　　WORKSHOP MAINTENANCE 　　REQUEST 　　CENSUS
DEFECT/WORK REQUEST:	EFFECT ON SYSTEM: 　　NO LOSS 　　PARTIAL LOSS 　　TOTAL LOSS OF PERFORMANCE
REASON FOR ATTENTION	STATE OF INSTALLATION WHEN EVENT OCCURRED: 　　OPERATING 　　SHUTDOWN
DETAILS OF REPAIR:	EFFECT ON INSTALLATION:
DEFECTIVE PARTS	NONE 　　AUTOMATIC SHUTDOWN 　　MANUAL SHUTDOWN
FAILURE MODE	OUTAGE TIME:
RENEWED ITEMS	TOTAL MANHOURS: SKILLED: UNSKILLED: COST OF MATERIALS AND PARTS:

FIG. 5. Typical Event Report

FIRST SELECTION (keywords)

    INSTALLATION
    SYSTEM
    ITEM
    +ITEM
    BLANK CARD
    NO SELECTION

SECOND SELECTION (keywords)

INSTALLATION	SYSTEM	BLANK CARD
		(i.e. a computer card with no holes punched in it)
SITE REFERENCE NUMBER	SITE REFERENCE NUMBER	SITE REFERENCE NUMBER
INSTALLATION	INSTALLATION	INSTALLATION
TYPE OF EVENT	TYPE OF EVENT	TYPE OF EVENT
EVENT NUMBER	EVENT NUMBER	EVENT NUMBER
DATE OF EVENT	DATE OF EVENT	DATE OF EVENT
TIME OF EVENT	TIME OF EVENT	TIME OF EVENT
OUTAGE TIME	OUTAGE TIME	MULTIPLE SYSTEM EVENTS
HOURS SINCE LAST EVENT	REPAIR TIME	ENTRY NUMBER
CYCLES SINCE LAST EVENT	HOURS SINCE LAST EVENT	
SKILLED LABOUR	CYCLES SINCE LAST EVENT	
TOTAL LABOUR	EFFECT	
MATERIALS/PARTS COST	ITEMS FAULTY	
EFFECT	SYSTEM CODE	
OPERATING STATE	ENTRY NUMBER	
SHUT DOWN METHOD		
POISONED OUT		
FAULT IMPORTANCE		
ENTRY NUMBER		

ITEM AND +ITEM

SITE REFERENCE NUMBER	MATERIALS/PARTS COST	POSITION IN INSTALLATION
INSTALLATION	IDENTITY	INVENTORY NUMBER (uses
TYPE OF EVENT	COMMISSIONING DATE	numerical part only)
EVENT NUMBER	YEAR OF DESIGN	CIRCUIT TYPE
DATE OF EVENT	FAILURE MODE	PARAMETER - TEMPERATURE
TIME OF EVENT	SYSTEM CODE	PARAMETER - SPEED OF ROTATION
OUTAGE TIME	CATEGORY OF FAULT	PARAMETER - PRESSURE
REPAIR TIME	CAUSE OF FAILURE	PARAMETER - FLOW
HOURS SINCE LAST EVENT	CODE FOR ITEM	PARAMETER - MEDIA HANDLED*
CYCLES SINCE LAST EVENT	MANUFACTURER	PARAMETER - X
SKILLED LABOUR	DESIGNER	PARAMETER - Y
TOTAL LABOUR	APPLICATION	ENVIRONMENT
		ENTRY NUMBER

FIG. 6. Selection Of Data For Analysis

       (i) PRINT

      (ii) COUNT

     (iii) SYSTEM FAILURES

      (iv) LONGEST OUTAGE

       (v) DESCRIPTION

      (vi) SERIAL

     (vii) MANAGEMENT

    (viii) SUMMARY

      (ix) LIFE

       (x) OUTAGE (with REPAIR TIME)

      (xi) COST   (with TOTAL LABOUR)

FIG. 7. The Eleven Analysis Keywords

ANALYSIS BY TYPE OF EVENT

NAME (CODE)	NUMBER OF OCCASIONS
BREAKDOWN	234
MAINTENANCE	28
IN SITU SCHED. MAINTENANCE	53
IN SITU UNSCHED. MAINTENANCE	3
REQUEST	1
ITEM REMOVED	2
ITEM INSTALLED	3
TOTAL NUMBER	324

ANALYSIS OF ITEMS REPORTED AS FAULTY

	NAME (CODE)	NUMBER OF OCCASIONS
3	ALARMS	26
222	COMPRESSORS	1
1112	LOGIC UNITS	2
1715	RELAYS	7
1718	REFRIGERATION UNITS	7
1907	THERMOCOUPLES	1
1921	THERMOSTATS	1
2103	VALVES, MECHANICAL	4
	TOTAL NUMBER =	49

FIG. 8. An Analysis from the Event Data Store

RELIABILITY OPTIMIZATION

Ralph A. Evans, PhD

Product Assurance Consultant
804 Vickers Avenue
Durham, North Carolina   27701   U.S.A.

ABSTRACT

The reliability of a system can be improved (up through the design stage) by one or more of the following methods. 1) Change the specifications; 2) Change the configuration of the system and/or some subsystems; 3) Get more information and use it well; 4) Pay more attention to detail; 5) Use more-reliable components; 6) Use redundancy/maintenance. Still more things can be done during the manufacture and use of the system, including using feedback from the field to modify decisions already made. There are many constraints on what can be done, e.g., time, money, and system specifications such as size, weight, and performance. In practice some of the constraints will be vague, ill defined, and/ or qualitative. Optimization is the process of finding the set of alternatives that maximizes Worth without seriously violating any of the constraints. Techniques for the analyst range from direct evaluation of all possible points, through schemes such as Lagrange multipliers and steepest ascent, to linear and dynamic programming. Accurate reproduction of the real constraints and goals is often sacrificed for tractability of the analysis. The very nature of optimization requires that all benefits and costs be measured on a single scale of Worth. Since this is so difficult, and since it is so easy to become enamoured of mathematical/ computational techniques, the analyst must always work within the shadow of "Garbage In, Garbage Out."

INTRODUCTION

It is much easier to talk about optimizing reliability and to analyze ways of doing it than it is to get a physical system which is optimized--it is easier said than done.

Before we can optimize reliability, we need to look at ways reliability can be changed and the kinds of constraints that can be imposed upon our efforts. These classifications are for our convenience in discussion, they do not in themselves limit our activities. Not all changes which are made with the intention of improving reliability actually do improve it--especially when there is insufficient information about the physical situation.

Reliability can be modified by changing--

1) The overall approach to the problem (e.g., leased phone lines or private microwave link for a communications system);
2) The configuration of the system (e.g., an aircraft can have propellor or jet engines, wings over or under the fuselage, and, the mounting and number of engines are variable);
3) Some of the modules or subsystems (e.g., motor functions can be performed electrically, hydraulically, or by mechanical levers and gears);
4) Some components (e.g., use high reliability parts or commercial ones);
5) Details of manufacture (e.g., holes in steel can be punched, drilled, reamed, and/or burned);
6) Materials (e.g., wood, plastics, metal alloys);
7) Method of operation (e.g., the operator of a radio receiver can be required to tune each rf stage separately or it can be done all with one switch);
8) Definition of required performance (e.g., range and resolution of a radar);
9) Definition of reliability (e.g., use 'fraction of time system is available for profitable use' rather than 'probability of completing a mission');
10) Amount of attention to detail (e.g., an alloy can simply be selected from a handbook table, or, many tests can be run on many alloys to find the one which holds up best in service).

Efforts to improve reliability are constrained by--

1) Cost of design effort;
2) Cost of parts or manufacture;
3) Calendar time schedules;
4) Manpower available to do the job;
5) Availability of purchased components or materials;
6) Volume or weight of finished product;
7) Operator training limitations;
8) Uncertainty about actual use conditions;
9) Maintenance philosophy and logistics;
10) Logical consequences of various company policies;
11) Consumer resistance to some configurations;

12) Management refusal to effect administrative changes;
13) Lack of knowledge about material or component properties, or about the way a part will be made.

Other techniques and constraints are likely to be important in any particular job. Some of the changes and constraints are not easily quantifiable; and the ones listed are certainly not mutually exclusive. This makes a complete mathematical analysis virtually impossible. The remainder of the paper deals with feasible mathematical analyses and shows how their incompleteness should be taken into account. The topics are--

<u>Conceptual Models</u> - we analyze our picture of the situation, not the situation itself.
<u>FOM's for Reliability</u> - there are many possible figures-of-merit for good performance.
<u>Generating the Mathematical Model</u> - abstracting the essence of a real problem into a solvable form is difficult.
<u>Kinds of Optima</u> - what we are looking for.
<u>Region of Interest</u> - the regions around the maximum points are important.
<u>Mathematical Methods</u> - how the mathematical part of the problem is actually solved.
<u>Perspective</u> - what to do when faced with your own problem.

CONCEPTUAL MODELS

The idea of a conceptual model is adapted from the idea of a physical model such as a model car. In a physical model, the characteristics of importance to us are reproduced quite well. In a model car these might be proportions, shape, and color. The characteristics of little or no importance are not usually reproduced at all; e.g., there may be no motive power and the tires may not be pneumatic. The "inbetweens" receive indifferent treatment. The physical model is an abstracting of something important from the physical world; it is an imitation.

A conceptual model is analogous to a physical model. Since everything in the universe affects everything else to some degree, however slightly, any exact treatment would be hopelessly complicated. Therefore we decide how to look at the situation and we make a set of assumptions (both explicit and implicit) about what we will ignore and what we will include in our conceptual model. By its very nature a conceptual model is incomplete; it ignores some things and describes others in an approximate fashion.

After having made a set of assumptions for a conceptual model, we then operate on those assumptions with mathematics and logic;

we analyze them by any means at our disposal. While developing the logical implications of a set of assumptions, we often don't like the results: they don't seem to fit, they appear to be inconsistent with our beliefs, etc. Then we have two rational choices:

1) Change our beliefs about the way the world actually is, if we are convinced that the set of assumptions is very realistic; and/or
2) Go back and modify the assumptions, so that their logical implications do in fact fit our beliefs about the world.

The creation of a conceptual model is a circular, often haphazard, process wherein ideas come from everywhere and get analyzed, tested, compared, junked, and accepted. Some good ideas usually filter through the process.

A conceptual model is often mathematical in nature and the same formalism will describe several different situations. It is important to keep the distinction between the mathematics itself (which is quite general, completely impersonal, and always "true") and what we have it represent in an engineering sense.

FOM'S FOR RELIABILITY

Choosing a good figure-of-merit (FOM) for a system is more easily done arbitrarily than accurately. Two of the more popular concepts are 1) Reliability - probability of successful performance of a mission, and 2) Availability - probability the system is available for use when needed. ("Reliability" is used in two senses in this paper: as a proper noun it has the by-now standard definition of successful mission completion; otherwise it has the usual dictionary definition of dependability, etc.) Just choosing good FOM's, to maximize and to use as constraints, is a challenging problem. The procedure is a circular iterative one. It makes little difference where we begin, because we must go through the steps several times before we will be satisfied with the results.

1) Define the levels of performance, but no finer than necessary. Many people just choose a dividing line between good and bad. Others prefer to list several degraded levels of performance. The kinds of failures may need to be segregated, e.g.,
A) the system does something when it shouldn't (prematures) and
B) the system doesn't do something when it should (duds).

In many systems it is very worthwhile to have many of the failure modes such that the system fails gracefully; viz, there is a very degraded mode of operation which is still feasible after the major failure. For example, suppose a train fails, especially in a tunnel or on a bridge; if it could proceed at a very slow speed to a safe, convenient place, that is much preferable to being completely broken down.

2) Assign a Worth to each performance level. There may be many different kinds of performance and several levels for each. Most often, although not always wisely, 1 is assigned to satisfactory performance, and 0 is assigned to unsatisfactory performance.

3) Distinguish between a subsystem failure and a system failure. For example, if a radio receiver is not down for more than 5 minutes, it may not damage the mission. In that case, there would have been a subsystem (hardware) failure, but the system itself would not have failed (by definition).

4) Prepare the first few levels of a fault-tree analysis. Be sure to be reasonably complete and general at this step. For example, some failures might not have severe consequences; some may affect operator safety; some may cause property damage. Include operators and users as subsystems; and include maintenance and repair activities.

5) The FOM's should respond to the question--what are the more important things about the behavior of the system. Typical FOM's are System Worth, Safety, various kinds of Performance, Availability, Reliability, and Cost. Decide whether any FOM's are to be combined, either as one grand FOM to be optimized, or as a means of simplifying the constraints. Assign each failure in step #4 to the appropriate FOM's which have been chosen. Many of the failures will appear in several FOM's.

It is usually wise to avoid much simplification at this point; wait until the mathematical model is being prepared. That is, try to be as accurately descriptive as feasible in this step; don't oversimplify for tractability yet. Many examples in the literature have been grossly oversimplified for pedagogic purposes; so much so that it is easy to get the incorrect impression that most problems are solved in a very straightforward manner. This step is very complicated and time-consuming except for the smallest, simplest systems.

6) The repair philosophy must be stated explicitly. Standby redundancy can often be considered a special case of repair--it is just a question of how the changeover is effected in case of failure. In making the calculations for an FOM, the kind of book-keeping to be done on failures is important, viz, to what level of detail will the kinds of failure be recorded. At one extreme merely the general fact of a failure is recorded; at the other extreme one records the component that originally failed, the circumstances surrounding the failure, and all secondary failures

caused by the original one. It pays to be realistic about the
mechanism for filling out forms and reports of failures. Accurate
field reporting of failures is notoriously difficult and expensive.
It is too easy to expect more in the way of field reporting than
will actually happen.

In what state will a repair leave the system? Is the entire
system to be restored to a like-new condition after each failure?
Will only a subsystem be restored to like-new? Many if not most
of the analyses available in the literature make one of those two
assumptions. Often, one can only infer the correct assumption by
looking at the mathematics. Another assumption which is sometimes
made is that the system will be restored to the statistical state
it was in just before the failure occurred. These are engineering
considerations, not statistical ones.

7) Go back over the previous steps; refine the work until
it is reasonably self-consistent and accurate. This is not something
that will be done in a few days; it will take man-weeks to
man-months. Very large systems will consume man-years of effort.
(This is a self-regulating admonition: sooner or later we will
run out of time, money, or patience. Besides, we can always come
back to this part of the problem after we have more insight into
the whole activity.)

Much of the engineering judgement and expertise for the project
will have been exercised in this section. The work has been
outlined so that it requires a minimum of statistical and mathematical
knowledge up to this point. Getting an optimum system is
largely an engineering/administrative effort. Statisticians,
mathematicians, and computer programmers will be necessary at
some stage, but they are not the ones who create reliable hardware
and systems.

GENERATING THE MATHEMATICAL MODEL

The results of the previous section are the raw data for
this section. The engineering opinions have been recorded and
the analysis is about to begin. It will usually be necessary to
go back and revise those engineering opinions and decisions after
some of this mathematical analysis has been done. Optimizing a
system is not a matter of following a set of procedures step-by-step;
no one knows enough about our physical world or about our
desires for it, to do that.

1) Prepare the algorithms for computing the FOM's as a
function of the reasonable design/production choices. If it
hasn't been done already, the system elements will have to be
classified into subsystems, and each subsystem subclassified still

further. We stop when a subsystem is of manageable size, when its behavior can be visualized, and when the number of choices and parameters within the subsystem is tractable. Usually much of this has been done during the overall system layout--for the same reasons as we do it here: if it's too complicated, we can't handle it.

2) Some simplifications have to be introduced, so that the analysis is not hopelessly bogged down in details. We must keep a list of these simplifications; so we don't forget them. Qualitative restrictions may be eliminated, but they too must not be forgotten. As the time and cost for calculations go down, we can afford to do more of it, and simplify less. As a rule of thumb, I suggest that the cost for the analytic effort for this section and beyond should not exceed the engineering effort expended in the previous sections.

3) It is easy for the system to be so large that only the most simple failure and repair distributions can be used and still have the equations analytically tractable. If simulation and Monte Carlo trials are used, the distributions can be more complicated and still be manageable. Even if the system equations are solved analytically, some simulation is wise as a check.

4) Go back to step 1, as necessary, until the mathematical model and computational algorithms for the system and all subsystems are complete and reasonably simple. If the response surfaces for each will be too difficult to visualize, there is the severe danger that "Garbage In," whether in the form of bad input data or bad assumptions, will not be recognized; nor will "Garbage Out" be obvious should it occur.

Only in the rarest, most oversimplified cases will the optimization problem be solvable analytically. Thus computer programs need to be generated. The usual tradeoffs between computational efficiency and programming costs have to be considered. Choose a programming language and techniques which allow changes to be most easily made. After some trial runs, many parts of the program will undoubtedly have to be redone. It may even be necessary to go back and change some of the details of an FOM.

Now, all of the models are "ready to go," except for an occasional "back to the drawing board" as the implications of the models become more obvious.

A TEST METHODOLOGY ALLOWING TRADE-OFF OF RELIABILITY/MAINTAIN-
ABILITY REQUIREMENTS

David N. Farnan

US Army Electronics Command
Fort Monmouth, New Jersey

Testing to demonstrate reliability and maintainability is becoming increasingly difficult. As new and more reliable components and designs are discovered, the mean-time-between-failures (MTBF) of electronics equipment are becoming higher and higher. As the equipment performs more functions and uses higher levels of technology, it becomes more complex, making the mean-time-to-repair (MTTR) more difficult to achieve. Therefore, it becomes increasingly more important to develop new and better ways of stating our Reliability Availability Maintainability (RAM) requirements and demonstrating that these requirements have been achieved.

At the US Army Electronics Command, our mission is to develop and procure tactical electronic equipment for Army users in the field. Until recently, the user has stated separate, specific reliability and maintainability needs. No availability index was stated. Our procurement specifications have been written to satisfy these specific user requirements. So, firm requirements for reliability and maintainability were stated with a separate demonstration for each. This method was very rigid on the contractor, allowing him no room for trade-off analyses.

LIFE CYCLE COST METHOD

We tried to improve that method of stating requirements. The objective was to design equipment which would be reliable enough to reduce maintenance and other logistics costs to a minimum, to provide a high degree of cost effectiveness and to lower significantly the total lifetime cost of ownership. The primary goal was to design for the least total life cycle cost. To accomplish this, we simply stated reliability and maintainability goals. Reliability and maintainability index determination tests were conducted to determine the extent the goals were achieved. It was found that this allowed the contractor too much latitude in making his trade-off studies and did not necessarily meet the user's operational availability needs.

NEW POLICIES

Recently the user has begun to state his needs in terms of bands of performance. He now states reliability and maintainability as an upper and lower bound. This allows us, as the procuring agency, a certain amount of freedom in writing the equipment specification requirements and tests. The Army is constantly seeking new and improved methods of procuring the best possible equipment at the lowest possible price.

A procurement method is being tried on an experimental basis. The method is called "price limited electronics prototypes development". The primary feature of this method is a not-to-exceed unit production price which is established before awarding a development contract. The equipment specification contains flexible requirements which may be traded-off to design a development model which can be produced for the not-to-exceed unit production price. A few minimum acceptable technical performance requirements are necessary, but the emphasis in this process is to specify performance by adequately describing both the equipment's functions and the mission in which it will be used.

This policy gives the developer much more information to guide him in determining his design. Previously, the contractor was given only the equipment specification requirements as design criteria. With this procedure the contractor bases his design on the essential requirements and the not-to-exceed unit production price. He has the flexibility to trade-off non-essential requirements with the knowledge of the mission and the function the equipment must perform.

Competition between two development contractors is the mechanism which assures that we achieve the most performance which can be delivered at the specified unit production price and lowest total life cycle cost. At the completion of the development phase, the competing designs will be tested. The contractor who has produced the best performance at the unit production price and who has achieved the lowest projected life cycle cost will be awarded the initial production contract.

RAM TRADE-OFF METHOD

The methodology that this paper is proposing attempts to overcome some of the weaknesses of the old procurement methods and utilizes the flexibility allowed in the experimental method described above. The objective of procuring the best design at the stated price with the lowest life cycle cost is best achieved by allowing the contractor maximum freedom to perform trade-offs within bounds acceptable to the procuring agency.

The user's statement of a band of performance for reliability is translated into a minimum specified MTBF. The user's statement of a band of performance for maintainability is translated into a maximum MTTR for each level of maintenance. Although the user does not state a specific availability requirement, we specify an inherent availability requirement in the specification. This requirement is based on several factors. The first factor is the user's stated reliability and maintainability needs. The other factors considered are the current state-of-the-art capability, the development time available, and funding constraints.

The maintainability requirement is made up of a mean corrective time (Mct) and a maximum corrective maintenance downtime (Mmaxct) at each level of maintenance. The first level of maintenance encompasses cleaning, inspection and replacement of the lowest replaceable units, cables and making minor adjustments. The second level includes identification and replacement of failed subassemblies, modules, printed circuit cards, and chassis mounted parts. The highest level maintenance includes identification and replacement of failed piece parts.

The inherent availability (Ai) is determined by calculation based on the demonstrated reliability and maintainability

$$A = \frac{MTBF}{MTBF + MTTR} \quad \text{where MTTR} = M_{ct_1} + M_{ct_2}$$

Mct_1 and Mct_2 are the mean corrective times for the first and second levels of maintenance, respectively. Only the first two levels of maintenance are used in calculating the inherent availability. This is done since only the first two levels directly affect the inherent availability as far as the user in the field is concerned.

The RAM requirements are shown on the availability nomograph in Figure 1. The shaded area is the allowable trade-off region. Naturally, the MTBF may be above the shaded area, but the MTTR must stay below the maximum MTTR. The only constraint on designing to an MTBF above the shaded area or an MTTR below the shaded area is achieving the not-to-exceed unit production price.

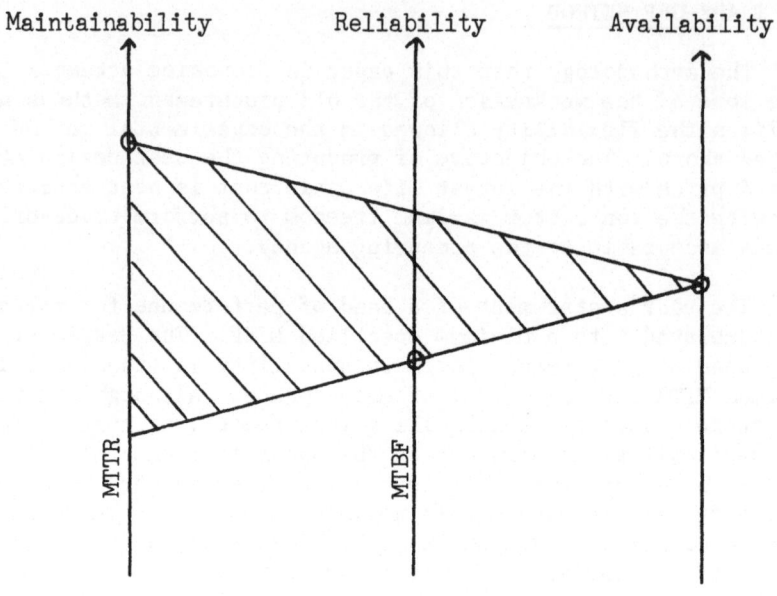

FIGURE 1. Availability Nomograph

Competitive development contracts are awarded with one contractor to be selected for an initial production contract. The contractor will be selected on the basis of overall equipment performance, achievement of the not-to-exceed unit production price, and lowest life cycle cost. Achievement of the RAM requirements are primary factors in all three criteria.

In order to give the competing contractors maximum flexibility to perform trade-off analyses, only a few minimum acceptable technical performance requirements are specified. The RAM requirements are included in these firm requirements. The remainder of the specification contains goals which can be traded-off to achieve acceptable performance at the not-to-exceed unit production price. The contractor is allowed freedom as to total design, provided the equipment performs the functions and meets the few minimum acceptable requirements. This freedom in trade-offs allows the contractor to use any type of components. He may trade-off cost versus reliability and may use commercial grade components rather than military standard or established reliability components. Through trade-off analyses, the contractor should maximize such things as design simplicity, ease of assembly, process repeatability, product inspectability, and interchangeability. Unit cost, critical materials, critical processes, and use of special test equipment should be minimized. Through these types of trade-off analyses, the contractor maintains maximum control over the system design and is allowed to make appropriate trade-offs among the design goals, which allow him to meet the essential reliability, maintainability, and availability requirements.

TESTING

The reliability acceptance tests that have been contractually required up to now have been Wald's probability sequential ratio tests or fixed length tests as contained in MIL-STD 781 entitled "Reliability Tests: Exponential Distribution". Both types are time terminated tests commonly specified as a certain number of multiples of the equipment MTBF. Using the RAM trade-off method to state reliability requirements, only a minimum MTBF is specified. The contractor may be required to demonstrate a higher reliability in order to meet the availability requirement. Thus, the MTBF is variable above a certain value. To specify a test in terms of multiples of the required MTBF becomes difficult when the MTBF that must be demonstrated to meet all RAM requirements is not known. A fixed length test was selected as the best approach, with the duration expressed as a multiple of the specified MTBF. The specified MTBF shall be determined by the contractor as long as both the minimum reliability and the availability requirements are achieved. So, the reliability test becomes a variable fixed length test.

The mean corrective time (Mct) for each level of maintenance is demonstrated by selecting an appropriate number of maintenance tasks for each level of the demonstration. The consumer's risk is usually selected to be 10%. The maintainability demonstration is similar to the test plans in MIL-STD 471 entitled "Maintainability/ Verification/Demonstration/Evaluation"

This inherent availability is then calculated using the results of the reliability and maintainability demonstrations.

The way the three requirements are structured, it is possible to successfully demonstrate any two requirements, yet fail the third. The minimum reliability and the maximum maintainability can be demonstrated. In this instance, the availability would not be achieved. A high MTBF can be demonstrated so that the availability requirement can be met without achieving even the maximum MTTR required. Or a very low MTTR can be demonstrated so the availability is met with an MTBF below the minimum required. Thus, the specification must make it clear that achievement of the minimum reliability or of the maximum maintainability requirements is not sufficient to satisfy the availability requirement. Furthermore, meeting the availability requirement without meeting the reliability and maintainability requirements is not sufficient. All three RAM requirements must be successfully demonstrated before the equipment can be considered acceptable.

The order of the testing can be important for this methodology. As long as the minimum reliability and the maximum maintainability and the availability requirements are met, the contractor is not required to demonstrate any higher values, although the equipment may be capable of a better reliability or maintainability. Since the number of maintenance tasks to be performed during the maintainability demonstration is a fixed requirement, the true maintainability indices should be demonstrated as far as the test is concerned. The maintainability test should be performed before the reliability test. Then the reliability test need be run only until the higher of the two requirements, the minimum reliability, or the reliability necessary to meet the availability, is demonstrated. Performing the testing in this order allows the shorter, less expensive test to be performed first and allows the minimum amount of testing on the longer, more expensive reliability demonstration.

The objective of any good procurement is to get the best equipment for the lowest cost. The question is, what is the cost that should be minimized? The equipment life cycle cost is the one all-encompassing cost. But it is the most difficult to determine and many factors of it are uncontrollable. Therefore, the unit production price was selected as the cost to be optimized. When the unit production price is controlled, the life cycle cost is controlled to a large degree, and the equipment design and performance is controlled. The lower the unit production price is set, the lower the performance of the equipment and the more uncontrolled the RAM characteristics will be. As the unit production price is raised, additional higher grade components may be used, and the RAM characteristics will be improved. More performance and functions may be added to the equipment to affect the gain in RAM characteristics. These trade-offs are left to the contractors. The unit production price must be established before the development phase so that the equipment can be designed to meet the price.

PROBLEMS

The use of flexible requirements creates special problems in selecting and monitoring contractors. When using a specification which allows the contractor to make such far reaching trade-offs as components selection, quality control, and configuration management, evaluation of competitive development proposals becomes more difficult. It is no longer simply a matter of insuring that all bidders meet the minimum requirements and of selecting a contractor on the basis of the lowest price. Most of the specifications "requirements" are actually goals which can be traded-off. And the price we are most interested in meeting is the not-to-exceed unit production price, which will affect the production contract. The bidder with the highest price on the development contract may be performing the most design effort and production engineering to achieve the highest performance and best RAM characteristics at the not-to-exceed unit production price, thus achieving the lowest life cycle cost. The bidders must attempt to demonstrate to the proposal evaluators that they are aware of the objectives of this method of procurement. They must show that they will make all the necessary trade-off studies in detail to attain as many of the specification goals as feasible. Finally, they must meet all the firm requirements such as reliability, maintainability, and availability and must show that they can afford to produce the equipment they have designed for the unit production price.

Monitoring the contract becomes increasingly difficult when firm requirements do not exist for measuring the contractor's performance. Certain reporting is required during the contract to aid the monitoring. Each contractor must submit a reliability and a maintainability program plan, reliability and maintainability predictions and updates, and test procedures and reports. The program plans can not contain firm commitments because the contractor is allowed to make many trade-offs. He may reduce incoming inspection or screening or decide to use commercial grade components in certain applications. Allowing the contractors to use whatever reliability level of parts he feels are adquate makes it difficult to perform and evaluate a reliability prediction. There is failure rate data available on military standard parts and established reliability parts, but there is very limited data on commercial grade components used in a military environment. The procuring agency often can not afford to wait until the demonstrations to see if the RAM requirements will be achieved. Waiting until the demonstration can cause costly redesign for the contractor and late delivery to the government. Much more emphasis must be placed on the design review process to maintain the delicate balance between RAM and performance characteristics versus cost.

FAULT TREE ANALYSIS -- CONCEPTS AND TECHNIQUES

J. B. Fussell

Aerojet Nuclear Company
Idaho Falls, Idaho, USA 83401
WORK PERFORMED UNDER THE AUSPICES OF THE
U. S. ATOMIC ENERGY COMMISSION

I. INTRODUCTION

Fault tree analysis is a technique of reliability analysis and is generally applicable to complex dynamic systems. Fault tree analysis provides an all inclusive, versatile, mathematical tool for analyzing complex systems. Its application can include a complete plant as well as any of the systems and subsystems. Fault tree analysis provides an objective basis for analyzing system design, performing trade-off studies, analyzing common mode failures, demonstrating compliance with safety requirements, and justifying system changes or additions.

The logic of the approach makes it a visibility tool for both engineering and management. Conventional reliability analysis techniques are inductive in nature and are primarily concerned with assuring that hardware will reliably accomplish its assigned functions. The fault tree method is concerned with assuring that all critical aspects of a system are identified and controlled. The fault tree itself is a graphical representation of Boolean logic associated with the development of a particular system failure, called the TOP event, to basic failures, called primary events. For example, the TOP event could be the failure of the reactor scram system to operate during an excursion with the primary events being failures of the individual scram system components.

In 1961 the concept of fault tree analysis was originated by Bell Telephone Laboratories as a technique with which to perform a safety evaluation of the Minuteman Launch Control System[1]. At the 1965 Safety Symposium, sponsored by the University of Washington and the Boeing Company, several papers were presented that expounded the virtues of fault tree analysis[2]. The presentation of these papers marked the beginning of a widespread interest in the possibility of using fault tree analysis as a reliability tool in the nuclear reactor industry. In the early 1970's great strides were made in the solution of fault trees to obtain complete reliability information about relatively complex systems [3-7]. The collection and evaluation of failure data is still of the utmost importance [8-11].

Fault tree analysis is of major value in:

(1) Directing the analyst to ferret out failures deductively

(2) Pointing out the aspects of the system important in respect to the failure of interest

(3) Providing a graphical aid giving visibility to those in system management who are removed from the system design changes

(4) Providing options for qualitative or quantitative system reliability analysis

(5) Allowing the analyst to concentrate on one particular system failure at a time

(6) Providing the analyst with genuine insight into system behavior.

Fault tree models do have disadvantages. Probably the most outstanding is the cost of development in first-time application to a system. As in the development of engineering drawings, the cost is somewhat offset by future application of the models in accident prevention, maintenance scheduling, and system modifications. Fault tree analysis is a sophisticated form of reliability analysis and is consequently relatively expensive. The additional expense is justified by detail of the qualitative or quantitative analysis resulting from fault tree analysis. Another aspect of fault tree analysis that limits its application at this time is the relatively few persons skilled in its techniques. Even skilled personnel might develop a fault tree for a given system in different ways.

Although certain single failures that can result in several component failures simultaneously, called common mode failures, can be pointed out by a detailed fault tree analysis, the analyst must be alert to include other common mode failures properly in the fault tree. At any rate, the analyst should be aware that fault tree analysis does not inherently ferret out all common mode failures.

II. FAULT TREE TERMINOLOGY

A system <u>component</u> is a basic system constituent for which failures are considered primary failures during fault tree construction. Consequently the components of a given system can change depending on the TOP event being studied or the detail the analyst wishes to include in the fault tree analysis. Some components have several operating states, none of which are necessarily failed states. For example, relay contacts can be open or closed. The description of these states is called the <u>component configuration</u>.

Fault tree construction is the logical development of the TOP event. As the construction proceeds, each fault event is also developed until primary failures are reached. A <u>fault event</u> is a failure situation resulting from the logical interaction of primary failures. The development of any fault event results in a <u>branch</u> of the fault tree. The event being developed is called the <u>base event</u> of the branch. The branch is complete only when all events in the branch are developed to the level of primary failures. Every event in a branch is in the <u>domain</u> of the base event. In addition, if the base event is an input to an AND gate, every event in the branch is in the domain of every input to that AND gate.

A fault tree <u>gate</u> is composed of two parts: (a) the Boolean logic symbol that relates the inputs of the gate to its output event and (b) the output event description. A gate is equivalent to another gate if, and only if, the logic symbol, the output event description, and another parameter, the "effective boundary conditions" associated with the output event, are identical. These effective boundary conditions modify an event and are imposed by the analyst or are generated by previously occurring fault events. A complete treatment of these effective boundary conditions is beyond the scope of this presentation, but is given in Reference 12. The event description includes two parts: (a) the incident identification and (b) the entity identification. The <u>incident identification</u> defines, as briefly as possible, the fault without indicating any hardware involved. The <u>entity identification</u> specifies the component or subsystem involved. These two parts are both required to describe the fault events.

The graphical symbols used in a fault tree fall, basically, into two categories: logic symbols and event symbols. Logic symbols are shown in Figure 1 and event symbols are shown in Figure 2 [1,8,13].

The logic symbols, or logic gates, are used to interconnect the events that contribute to the specified main event, or TOP event. The logic gates that are most frequently used to develop fault trees are the basic AND and OR Boolean expressions. The AND gate provides an output event only if all input events occur simultaneously. The OR gate provides an output event if one or more of the input events are present.

The more frequently used event symbols are the rectangle, circle, and diamond. The rectangle represents a fault event resulting from the combination of more basic faults acting through logic gates. The circle designates a basic system component failure or fault input that is independent of all other events designated by circles and diamonds. The diamond symbol describes fault inputs that are considered basic in a given fault tree. However, the event described is not basic in the sense that laboratory data are applicable. Rather, the fault tree is simply not developed further,

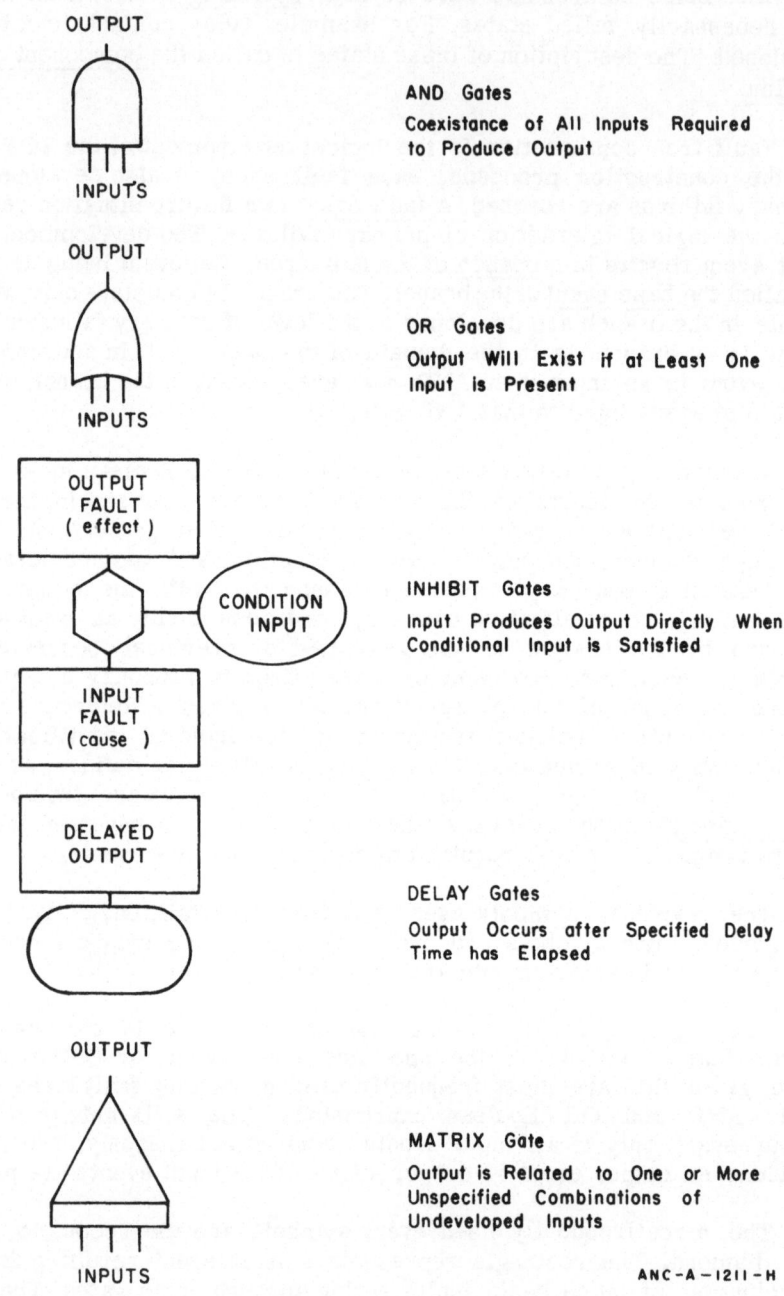

FIG. 1 FAULT TREE LOGIC SYMBOLS.

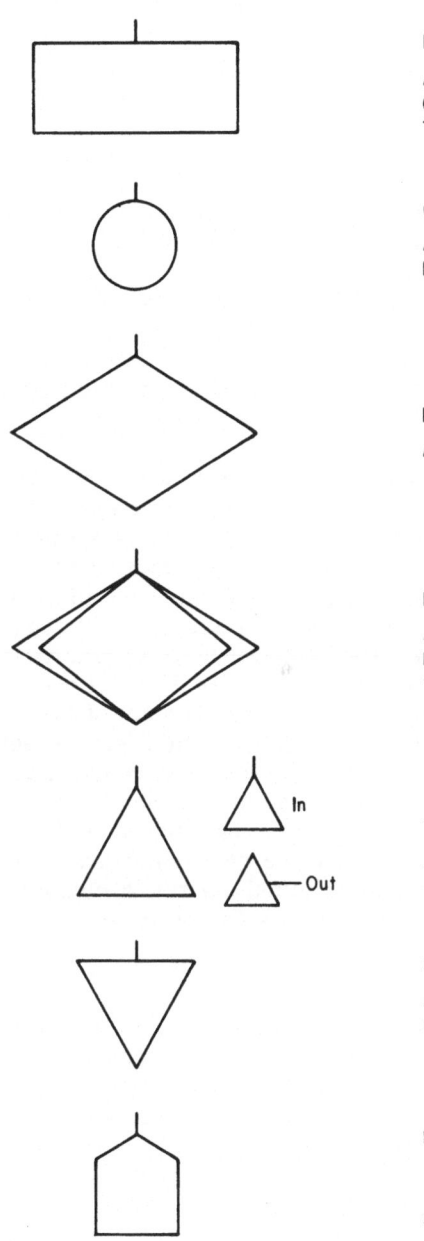

RECTANGLE

A Fault Event Usually Resulting from the Combination of More Basic Faults Acting Through Logic Gates

CIRCLE

A Basic Component Fault — An Independent Event

DIAMOND

A Fault Event Not Developed to its Cause

DOUBLE DIAMOND

A Significant Undeveloped Fault Event That Requires Further Development to Complete the Fault Tree

TRIANGLE

A Connecting or Transfer Symbol

UPSIDE DOWN TRIANGLE

A Similarity Transfer — The Input is Similar But Not Identical to the Like Identified Input

HOUSE

An Event That is Normally Expected to Occur Also, Useful As a "Trigger Event" for Logic Structure Change Within the Fault Tree

ANC-A-1212-H

FIG. 2 FAULT TREE EVENT SYMBOLS.

either because the event is of insufficient consequence or the necessary information is unavailable. In order to obtain a solution for a fault tree, both circles and diamonds must represent events for which reliability information is input to the fault tree. Events that appear as circles or diamonds are treated as primary events.

The triangles shown in Figure 2 are not strictly event symbols although they have traditionally been classified as such. The triangle indicates a transfer from one part of the fault tree to another. A line from the side of the triangle (transfer out triangle) denotes an event transfer out from the associated logic gate. A line from the apex of the triangle denotes an event transfer into the associated logic gate from the transfer out triangle with the same identification number.

The other logic gates and events symbols are shown in Figures 1 and 2 and are explained in those figures.

A minimal cut set is a smallest set of primary events, inhibit conditions, or undeveloped fault events or any combination of these which must all occur in order for the TOP event to occur. The primary events represent the resolution of the fault tree. The minimal cut sets represent the modes by which the TOP event can occur. For example, the minimal cut set A_1A_2 means that both the primary events A_1 and A_2 must occur in order for the TOP event to occur. The occurrence of A_1 and A_2 is a mode by which the TOP event occurs. If either A_1 or A_2 does not occur, then the TOP event does not occur by this mode. The set of events A_1A_2C, where C is another primary event, is not a minimal cut set because C is redundant and is not necessary for the occurrence of the TOP event; C can either occur or not occur, and as long as A_1 and A_2 both occur then the TOP event will occur. A minimal cut set is then a collection of component failures all of which are necessary and sufficient to cause the system failure by that minimal cut set. A complete set of minimal cut sets is all the failure modes for the given system failure.

The minimal cut sets are significant because they depict which failures must be repaired in order for the TOP failure to be removed from the failed state. The minimal cut sets point out the weakest links in the system. The primary events in the one-event minimal cut sets usually are the most important. A single failure analysis is an investigation, or fault tree drawn, in order to obtain only the one-event minimal cut sets (single failures) for the TOP event. For a single failure analysis, the fault tree ends whenever an AND gate is reached that does not have deeper common causes (which effectively transform an AND gate to an OR gate).

III. GENERAL PROCEDURE OF FAULT TREE ANALYSIS

The versatility available to the analyst with regard to the degree of detail reflected in a fault tree analysis is a main feature of the fault tree technique. In the fault tree itself the failures considered as primary failures can be failures of the smallest mechanical linkage in a microswitch or failures of a power generating station. The resolution of the analysis is determined by the needs of the analyst. Having determined the resolution, the analyst has options with regard to the evaluation of the fault tree. Indeed, the fault tree itself can be the final objective. In addition to the system visibility and understanding obtained by studying the fault tree itself, further qualitative analysis of the fault tree can produce all of the system modes of failure. Finally quantitative evaluation is possible; that is, probabilistic failure information can be obtained about the TOP event and minimal cut sets from probabilistic failure information about the components.

The following four steps generally can be present in a fault tree analysis.

(1) System definition

(2) Fault tree construction

(3) Qualitative evaluation

(4) Quantitative evaluation

Each of these steps is discussed. The discussion of qualitative evaluation includes a new approach for obtaining minimal cut sets and also techniques for treatment of mutually exclusive fault events appearing in the same fault tree.

1. System Definition

System definition is often the most difficult task associated with fault tree analysis. Of primary importance is a functional layout diagram of the system of interest showing all functional interconnections and identifying each system component [a]. An example might be a detailed electrical schematic. Physical system bounds are then established focusing the attention of the analyst on the precise area of interest. A common error is failure to establish realistic system bounds and thereby a diverging analysis is initiated.

---

[a] For some systems that are not hardware oriented such a diagram may not exist and, indeed, the fault tree itself can be the only feasible diagrammatic system representation. For the discussion presented the emphasis is directed toward hardware-oriented systems. However, the implications and terminology extend to all fault tree analyses.

Sufficient information must be available for each of the system components to allow the analyst to determine the necessary modes of failure of the components. This information can be available from experience of the analyst or from the technical specifications of the components.

A further step in the system description is to establish the system boundary conditions. These boundary conditions should not be confused with the physical bounds of the system. System boundary conditions define the situation for which the fault tree is to be drawn. A most important system boundary conditions is the TOP event. For any given system, a multitude of possibilities for TOP events exists. The selection of the "correct" TOP event is sometimes a difficult task. It is, however, simply defined as the major system failure of interest. The system initial configuration is described by additional system boundary conditions. This configuration must represent the system in the unfailed state. Consequently these system boundary conditions depend on the TOP event. Initial conditions are then system boundary conditions that define the component configurations for which the TOP event is applicable. All components that have more than one operating state generate an initial condition. System boundary conditions also include any fault event declared to exist or to be not-allowed for the duration of the fault tree construction. These events are called <u>existing system boundary conditions</u> or <u>not-allowed system boundary conditions.</u> An existing system boundary condition is treated as certain to occur and a not-allowed system boundary condition is treated as an event with no possibility of occurring. Neither existing nor not-allowed system boundary conditions appear as events in the final system fault tree. Finally, in certain cases, partial development of the TOP event, called the tree top, is also required as a system boundary condition. If the tree top system boundary condition is required, it is not considered as part of the fault tree construction process because it is obtained by inductive means.

2. Fault Tree Construction

Published information dealing with generalized fault tree construction is quite limited. Haasl[1] has described some general concepts, and Fussell[12] has presented a construction methodology for electrical systems that is verified as being deductive and formal by fault trees being constructed by a computer from electrical schematics. Some of the ramifications of this latter automated fault tree construction are discussed in the conclusions.

An example is given here to demonstrate some of the fundamental aspects of fault tree construction. A sample system schematic is shown in Figure 3. The system physical bounds include this entire system. The system boundary conditions are:

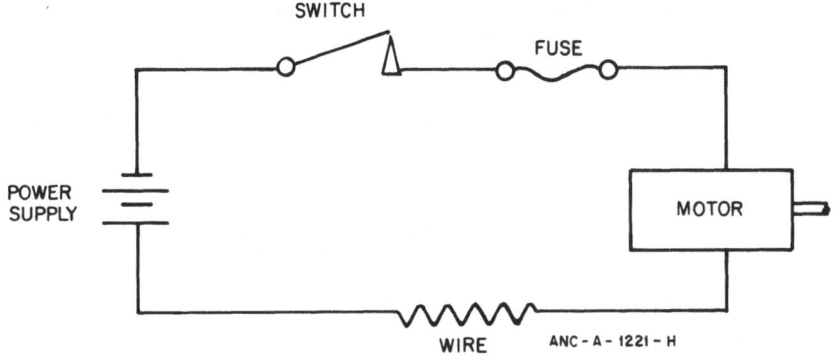

FIG. 3 SAMPLE SYSTEM 1.

TOP Event ≡ Motor overheats

Initial Condition ≡ Switch closed

Not-Allowed Events ≡ Failures due to effects external to system

Existing Events ≡ Switch closed

Tree Top ≡ Shown in Figure 4.

Figure 4 reflects the inductive reasoning that the motor overheats if an electrical overload is supplied to the motor or a primary failure within the motor causes the overheating, for example, if bearings lose their lubrication or a wiring failure occurs within the motor.

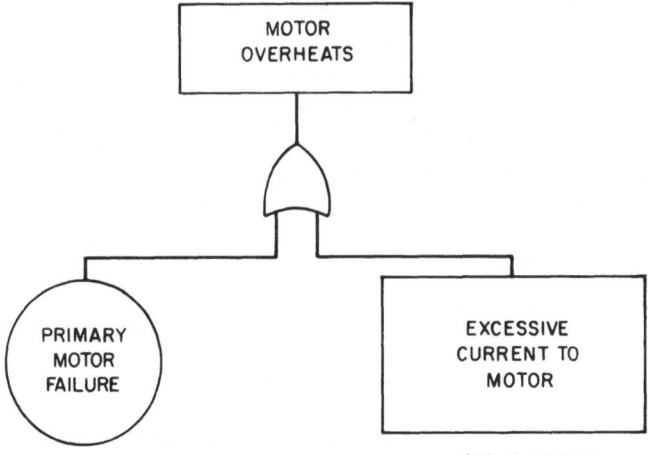

FIG. 4 FIRST TREE TOP SYSTEM BOUNDARY CONDITION FOR SAMPLE SYSTEM 1.

From a knowledge of the components, the fault tree shown in Figure 5 is constructed. The event "excessive current to motor" occurs if excessive current is present in the circuit and the fuse fails to open. The event "excessive current in circuit" occurs if the wire fails shorted or the power supply surges. The fault tree is now complete to the level of primary failures.

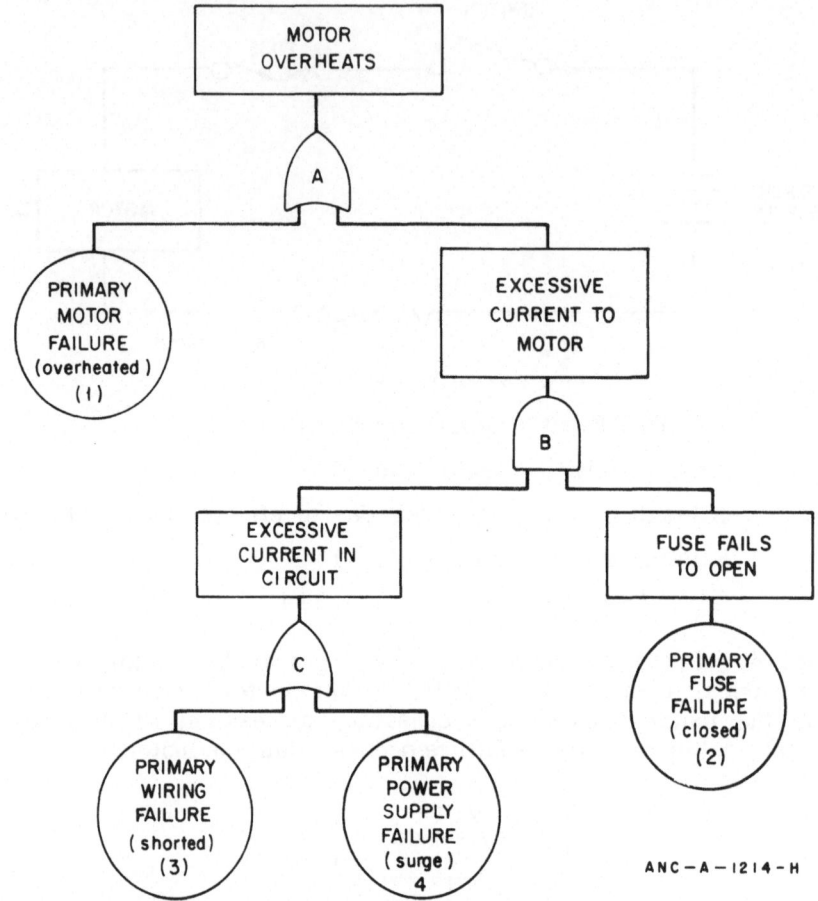

FIG. 5 FIRST FAULT TREE FOR SAMPLE SYSTEM 1.

For the same sample system but with different system boundary conditions a second example is given that illustrates the treatment of secondary failure, that is, failures possibly caused by failure feedback between components. For this example, the system boundary conditions are:

TOP Event ≡ Motor does not operate

Initial Condition ≡ Switch closed

Not-Allowed Events ≡ Failures due to effects external to system (operator failures not included)

Existing Events ≡ None

Tree Top ≡ Shown in Figure 6.

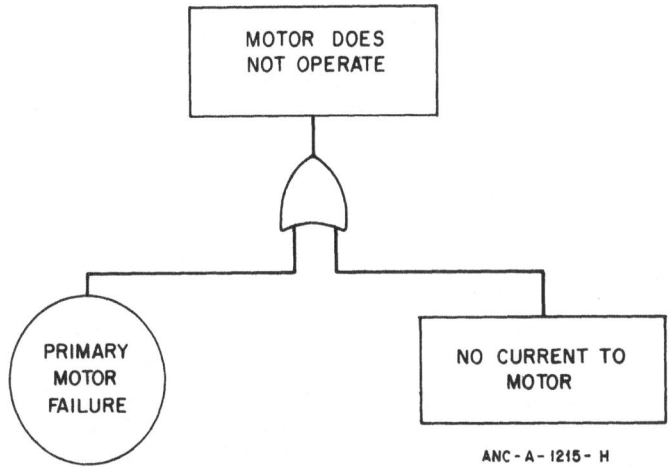

FIG. 6 SECOND TREE TOP SYSTEM BOUNDARY CONDITION FOR SAMPLE SYSTEM 1.

The completed fault tree is shown in Figure 7. Here the diamond symbol is used to indicate the event "switch opened" is not developed to its causes. The switch being opened is a failure external to the system bounds and, in this analysis, insufficient information is available for development of the event.

The event "fuse fails open" occurs if a primary or secondary fuse failure occurs. Secondary fuse failure can occur if an overload in the circuit occurs because an overload can cause the fuse to open. The fuse does not open, however, every time an overload is present in the circuit because all conditions of an overload do not result in sufficient overcurrent to open the fuse. The inhibit condition is then used as a weighing factor applied to all the fault events in the domain of the inhibit condition. Since the inhibit condition is treated as an AND logic gate in a probabilistic analysis, to be discussed later, it is precisely a probabilistic weighting factor. The inhibit condition has many variations of usage in fault tree analysis but in all cases it represents a probabilistic weighing factor.

3. Qualitative Evaluation

A minimal cut set is a collection of primary failures all of which are necessary and sufficient to cause the system failure by that minimal cut set. A complete set of minimal cut sets are all the failure modes for a given system and TOP event. For the fault tree in Figure 5 the minimal cut sets are, by inspection, the sets of primary events:

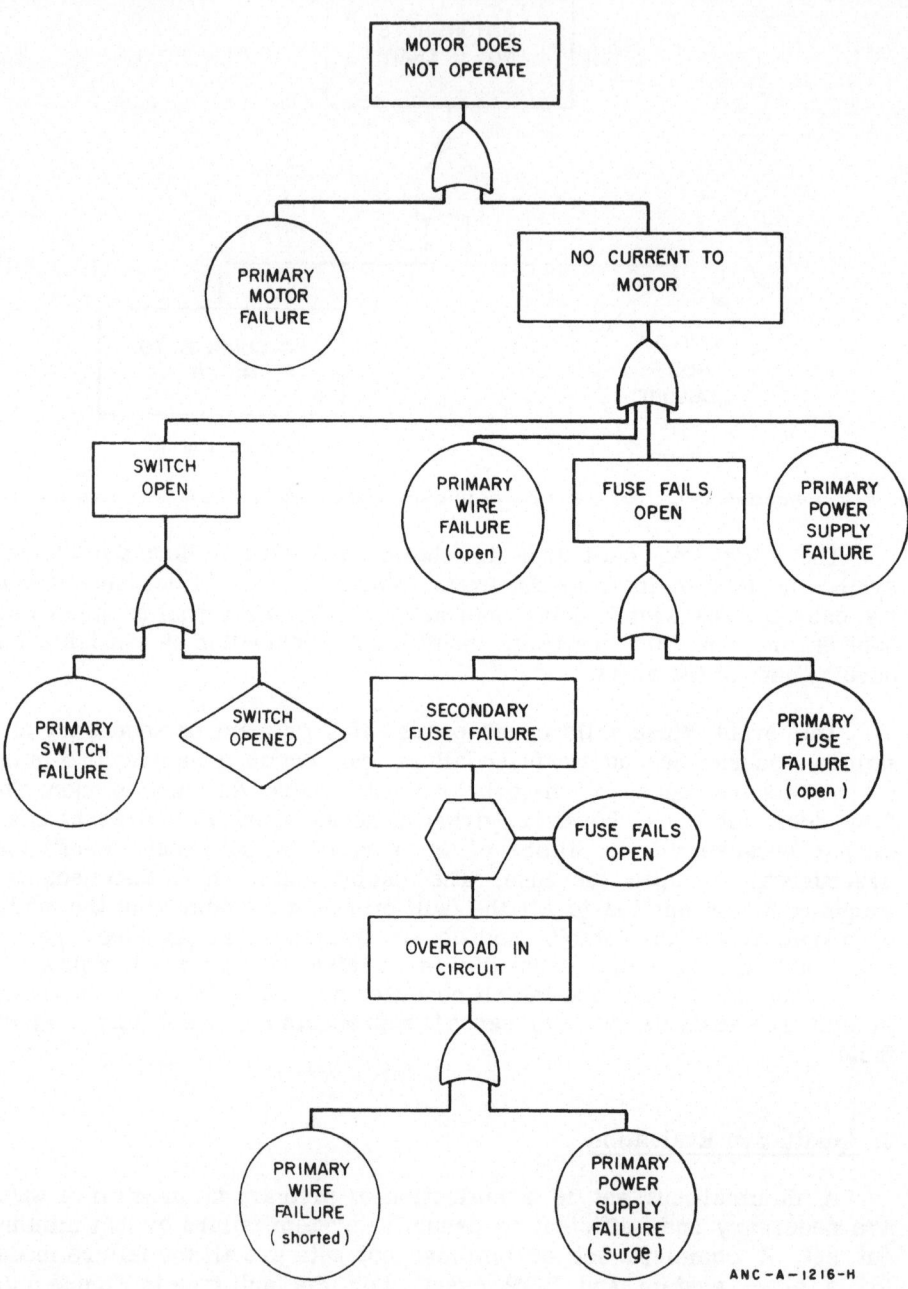

(1) Motor Failure (overheated)

(2) ⎡ Fuse Failure (closed)
    ⎣ Wiring Failure (shorted)

(3) ⎡ Fuse Failure (closed)
    ⎣ Power Supply Failure (surge).

Although these minimal cut sets were determined by examination of the fault tree, with the increased complexity of realistic fault trees, the need for a more logical approach is implied. One such computerized approach has been suggested by Vesely and Narum [14]. The Boolean equation implied by the fault tree is constructed by a computer. The primary events are then "turned on" one at a time. Each time, a check is made to determine whether the equation is "true". Next, all possible combinations of two primary events are turned on and again the equation is checked each time to determine whether it is true. Each time the equation is true, the collection of primary events that were turned on is shown to be a cut set. After these cut sets are determined, all cut sets that are supersets of other cut sets are discarded so as to winnow the minimal cut sets. Also, Vesely and Narum [14] have suggested a Monte Carlo approach whereby appropriate weighing of the primary events is used to accelerate the process of determining the minimal cut sets. However, doubt that all the minimal cut sets have been found is always present when the Monte Carlo approach is used. In practice, both of the preceding methods generally require excessive computer time to obtain cut sets containing more than three primary events.

3.1 <u>New Approach for Obtaining Minimal Cut Sets</u>. A new approach to locating the minimal cut sets is introduced. This method is unique in that, in order to locate the cut sets, it starts at the TOP event and proceeds to the primary events without the use of simulation, Boolean manipulation, or the Monte Carlo approach. Rather, the fault tree is resolved into the minimal cut sets. The execution time is, thereby, not an exponential function as it is with other methods, but rather is approximately a linear function of the average length of the cut sets. A key point of this methodology is that an AND gate alone always increases the size of a cut set while an OR gate alone always increases the number of cut sets. To obtain the minimal cut sets, this method requires that the <u>Boolean indicated cut sets (BICS)</u> be first obtained. The BICS are defined such that, if all the primary events are different, the BICS will be precisely the minimal cut sets. This definition of the BICS does not mean that the method is limited to fault trees with primary events appearing only once in the fault tree.

To obtain the BICS, first, each gate in the fault tree is randomly named with a value ω and each primary event with a value ϕ. The following definitions apply to this approach:

$\rho_{\omega,i} \equiv i^{th}$ input to gate ω

$\lambda_\omega \equiv$ number of inputs to gate ω

$x \equiv$ the xth BICS

$y \equiv$ the yth entry in a BICS

$\Delta_{x,y} \equiv$ variable representing the yth entry in the xth BICS

xmax $\equiv$ largest value of x yet used

ymax $\equiv$ largest value of y yet used in the xth BICS.

The values ω, ϕ, $\rho_{\omega,i}$, λ_ω and the gate type (AND or OR) are assumed as input, where values of $\rho_{\omega,i}$ are discernible values of ω or ϕ. $\Delta_{1,1}$ is first set equal to the ω value representing the gate immediately under the TOP event. From this point on, the goal is to eliminate all ω values from the $\Delta_{x,y}$ matrix. When this elimination is complete, only ϕ values remain and the BICS are determined. To accomplish this elimination, an ω value is located in the $\Delta_{x,y}$ matrix, the values of x, y, and ω are noted, and

$$\Delta_{x,y} = \rho_{\omega,1}. \qquad (1)$$

For ω being an AND gate:

$$\Delta_{x,ymax+1} = \rho_{\omega,\pi} \qquad \pi = 2,3,\ldots,\lambda_\omega, \qquad (2)$$

where ymax is incremented when π is incremented.

For ω being an OR gate:

$$\begin{bmatrix} \Delta_{xmax+1,n} = \Delta_{x,n} & n = 1,2,\ldots,ymax & n \neq y \\ = \rho_{\omega,\pi} & & n = y \end{bmatrix} \qquad (3)$$

$$\pi = 2,3,\ldots,\lambda_\omega$$

where xmax is incremented when π is incremented.

Processes 1 and 2 or 3 are repeated until all the entries in the $\Delta_{x,y}$ matrix become values of ϕ. The BICS are then determined. A simple search procedure is now used to determine the minimal cut sets.

The number of BICS, that is, the number of rows to be required by the $\Delta_{x,y}$ matrix, for a fault tree can generally be determined in a reasonable time by hand. The number of BICS is an upper bound to the number of minimal cut sets. If $x_{i,j}$ is a parameter associated with the $j^{\underline{th}}$ input to the $i^{\underline{th}}$ gate, where $x_{i,j} = 1$ for all primary events, then

$$X_i = x_{i,1} \cdot x_{i,2} \cdot x_{i,3} \cdots x_{i,jmax} \quad \text{if i is an AND gate} \quad (4)$$

$$= x_{i,1} + x_{i,2} + x_{i,3} \cdots x_{i,jmax} \quad \text{if i is an OR gate} \quad (5)$$

$$x_{k,\ell} = X_i$$

where k is the gate into which Gate i is the $\ell^{\underline{th}}$ input. If i is the logic gate number directly under the TOP event then $X_i = X_{TOP}$ is the number of BICS for the fault tree. The value x_k is determined only when all its input parameters are determined; hence, gates that only have primary events ($x_{i,j} = 1$ for all j) as input are the beginning points.

The computation is quite simple, as can be seen from examining the fault tree in Figure 5. From Equation (5), $X_C = (1 + 1) = 2$ and then from Equation (4)

$$X_B = (x_{B,1})(x_{B,2})$$

$$= (X_C)(x_{B,2})$$

$$= (2)(1)$$

$$= 2$$

and, finally, since A is an OR gate

$$X_A = X_{TOP} = (x_{A,1}) + (x_{A,2})$$

$$= (x_{A,1}) + X_B$$

$$= 3.$$

Therefore, the $\Delta_{x,y}$ matrix contains three rows. The maximum number of primary events in any BICS for a fault tree can also generally be determined in a reasonable time by hand. This maximum is an upper bound to the maximum number of primary events in any minimal cut set for that fault tree. The determination is similar to that for the number of BICS. If $y_{i,j}$ is a parameter associated with the $j^{\underline{th}}$ input to the $i^{\underline{th}}$ gate, where $y_{i,j} = 1$ for all primary events, then

$$Y_i = y_{i,1} + y_{i,2} + y_{i,3} \cdots y_{i,\text{jmax}} \quad \text{if i is an AND gate} \quad (6)$$

$$= \max(y_{i,1}, y_{i,2}, y_{i,3} \cdots y_{i,\text{jmax}}) \quad \text{if i is an OR gate} \quad (7)$$

$$y_{k,\ell} = Y_i$$

where k is the gate into which Gate i is the ℓth input and the function max($y_1, y_2 \ldots$) means the maximum argument is chosen. If i is the logic gate number directly under the TOP event then $Y_i = Y_{\text{TOP}}$ is the maximum number of primary events in any BICS for the given fault tree. Y_i is determined only when all its input parameters are determined; hence, the analyst must begin with gates that only have primary events ($y_{i,j} = 1$ for all j) as input.

For an example, the fault tree in Figure 5 is again considered. From Equation (7), $Y_C = \max(1,1) = 1$ and from Equation (6),

$$Y_B = Y_C + y_{B,2}$$

$$= 2.$$

And, finally,

$$Y_A = Y_{\text{TOP}} = \max(1,2)$$

$$= 2.$$

Therefore, the largest BICS contains two primary events. The $\Delta_{x,y}$ matrix for the fault tree shown in Figure 5 is a 2 x 3 matrix. This method can easily be extended to determine the maximum number of one-, two-, three-, ... event BICS, hence an upper bound on the one-, two-, three-, ... event minimal cut sets, respectively, is determined.

An example best illustrates the method of determining minimal cut sets. The fault tree in Figure 5 is used in this example. Each gate is, in that figure, labeled with a letter and each primary event with a number. The input to the methodology is then

ω	Gate Type	λ_w	$\rho_{\omega,i}$	
A	OR	2	1	B
B	AND	2	C	2
C	OR	2	4	3

The solution is begun by setting a $\Delta_{x,y}$ matrix, as:

$\Delta_{x,y}$

	y →	
x	A	
↓		

Since A is an OR gate, Equations (1) and (3) are used to give

$\Delta_{x,y}$

	y →	
x	1	
↓	B	

To eliminate B, Equations (1) and (2) are used to obtain

$\Delta_{x,y}$

	y →	
x	1	
↓	C	2

Finally, since C is an OR gate, Equations (1) and (3) are used again to obtain

$\Delta_{x,y}$

	y →	
x	1	
↓	4	2
	3	2

From the preceding matrix, the minimal cut sets are:

Minimal Cut Set	Primary Events
1	1
2	4, 2
3	3, 2

The results agree precisely with the results obtained previously by inspection. Since all the primary events in the fault tree are different, the BICS in the preceding $\Delta_{x,y}$ matrix are the minimal cut sets. If some of the BICS contain duplicate events, this duplication is eliminated by discarding redundant events. Also, if some of the BICS are supersets of other BICS, all supersets are discarded. The minimal cut sets remain.

The advantage of the method presented here lies in the speed with which it can determine large cut sets. As a typical example, for a fault tree with 2000 BICS, the smallest of which contain 20 primary events and the largest of which contain 25 primary events, the time required by the UNIVAC 1108 computer to locate all the BICS is less than 16 seconds.

3.2 Treatment of Mutually Exclusive Fault Events. All present methods for obtaining minimal cut sets, including the one described here, must be modified somewhat to handle mutually exclusive fault events that appear in the domain of the same AND logic gate. If this modification is not implemented, erroneous minimal cuts sets can result. The manner in which erroneous minimal cut sets may be obtained is best illustrated by an example. The system schematic in Figure 8 is considered for this example. The purpose of the system is to provide light from the bulb. When the switch is closed, the relay contacts close and the contacts of the circuit breaker, defined here as a normally closed relay, open. If the relay contacts transfer open, the light will go out and the operator will immediately open the switch which in turn causes the circuit breaker contacts to close and restore the light. The system boundary conditions include:

TOP Event -- No light

Initial Conditions -- Switch closed
Relay contacts closed
Circuit breaker contacts open

Not-Allowed Events -- Operator failures
Wiring failures
Secondary failures.

Operator failures, wiring failures, and secondary failures are neglected to simplify the resulting fault tree. This fault tree is shown in Figure 9.

Table I gives the collection of primary events that are declared to be minimal cut sets by conventional methods of determining minimal cut sets for the system shown in Figure 8. As can be reasoned from Figure 8, Sets (6), (8), (10), and (12) will not cause the TOP event. Only Set (12), being erroneous, could have been detected from the minimal cut sets themselves.

The reason for these erroneous minimal cut sets is that the fault events "power removed from Circuit Path C", hereafter called $\overline{X}$, and the fault event "power not removed from Circuit Path C", hereafter called $\overline{Y}$, are mutually exclusive fault events. Consequently, collections of component failures that reflect certain combinations of the primary events

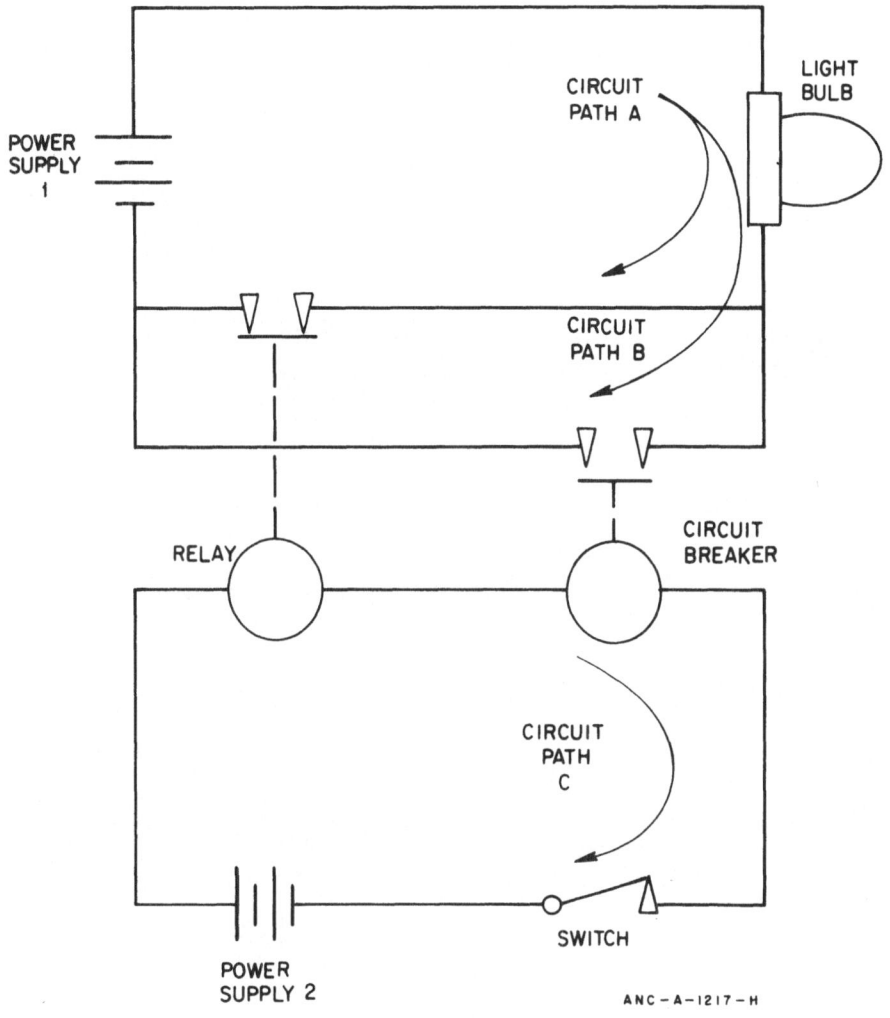

FIG. 8 SAMPLE SYSTEM 2.

used to develop these events will not cause TOP failure. Since these events, $\overline{X}$ and $\overline{Y}$, are both in the domain of an AND logic gate they are, indeed, combined by existing methodologies of determining the minimal cut sets. This explanation using $\overline{X}$ and $\overline{Y}$ is extended to include any mutually exclusive events X and Y.

Mutually exclusive events X and Y, that is, $X \cap Y = \phi$, both in the domain of an AND logic gate are considered. If x_i and y_i represent the ith set in the complete collection of minimal cut sets for X and Y, respectively, where a minimal cut set for a fault event is defined analogously to those of the TOP event, then $x_i \cap y_j = \phi$ for all i and j because each x_i is simply a re-expression of X and each y_j is simply a re-expression

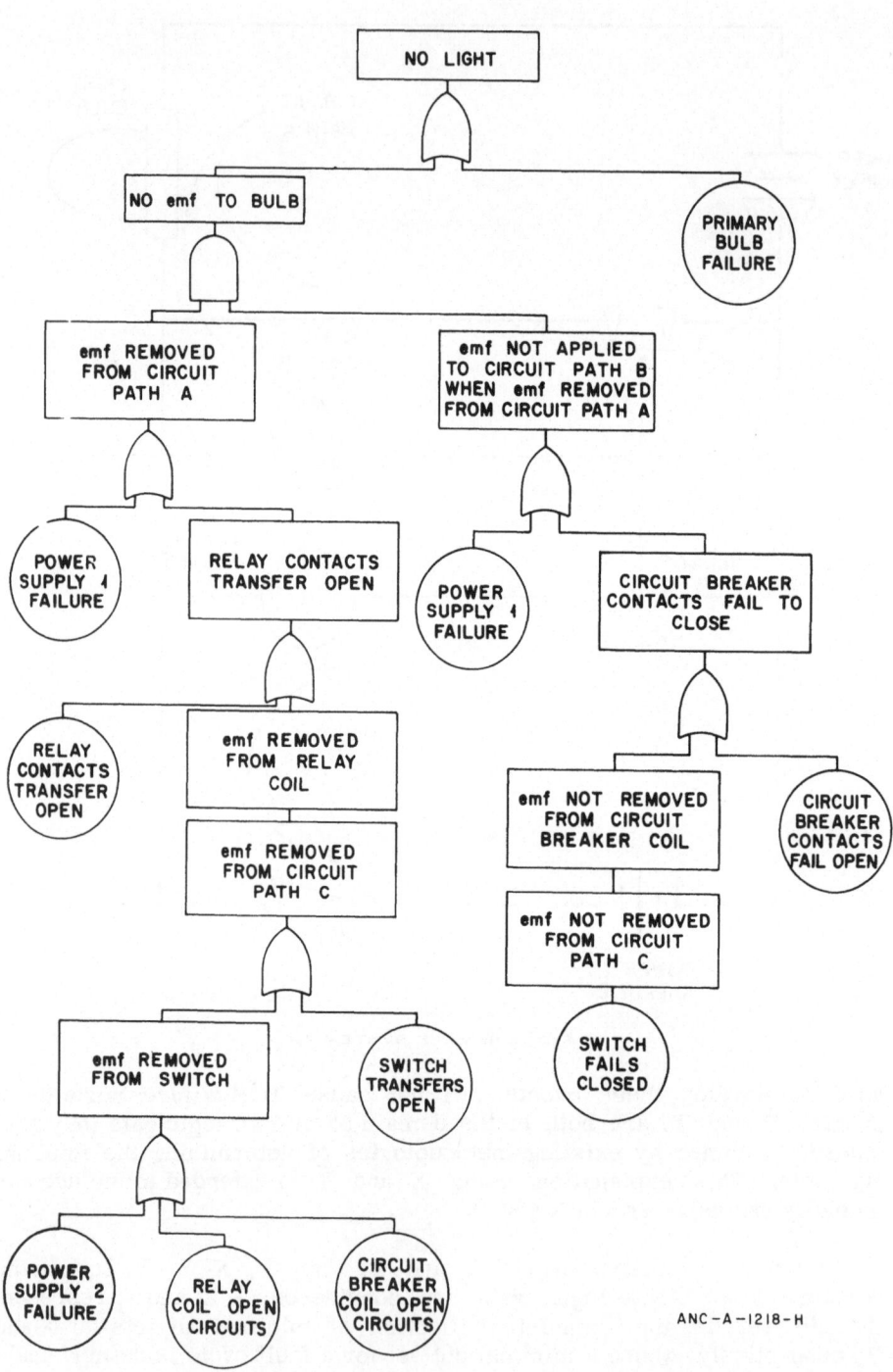

FIG. 9 FAULT TREE FOR SAMPLE SYSTEM 2.

TABLE I

MINIMAL CUT SETS FOR SAMPLE SYSTEM
AS DETERMINED BY CONVENTIONAL MEANS

(1) Primary bulb failure

(2) Primary Power Supply 1 failure

(3) ⎡ Relay contacts transfer open
    ⎣ Circuit breaker contacts fail open

(4) ⎡ Relay contacts transfer open
    ⎣ Switch fails closed

(5) ⎡ Power Supply 2 failure
    ⎣ Circuit breaker contacts fail open

(6) ⎡ Power Supply 2 failure
    ⎣ Switch fails closed

(7) ⎡ Relay coil open circuits
    ⎣ Circuit breaker contacts fail

(8) ⎡ Relay coil open circuits
    ⎣ Switch fails closed

(9) ⎡ Circuit breaker coil opens circuit
    ⎣ Circuit breaker contacts fail open

(10) ⎡ Circuit breaker coil opens circuit
     ⎣ Switch fails closed

(11) ⎡ Switch transfers open
     ⎣ Circuit breaker contacts fail open

(12) ⎡ Switch transfers open
     ⎣ Switch fails closed

of Y. Next, if Z_k represents the kth set of primary events that results from lumping together all the events in x_i with all the events in y_j for each i with each j over all i and j, then i x j such sets will result. If M_ℓ is the ℓth collection of primary events suspected to be a minimal cut set, that is, the output from conventional methods of determining minimal cut sets, then all $M_\ell = \emptyset$ if $M_\ell \supset Z_k$ for all ℓ and k. $M_\ell \supset Z_k$ means that Z_k is a subset of M_ℓ. Hence, any collection of primary events that contains any Z_k is not a minimal cut set and should be discarded.

As an illustration Table I is again considered. This table forms a list of M_ℓ sets, M_1 thru M_{12}, for the fault tree shown in Figure 9. The x_i and y_i sets for events $\overline{X}$ and $\overline{Y}$ are:

x_1 = Power Supply 2 failure

x_2 = Relay coil open circuits

x_3 = Circuit breaker coil open circuits

x_4 = Switch transfers open

y_1 = Switch fails closed.

In general, the x_i's and y_j's are collections of primary events, that is, minimal cut sets for the events X and Y.

Then,

$Z_1 = \begin{bmatrix} \text{Power Supply 2 failure} \\ \text{Switch fails closed} \end{bmatrix}$

$Z_2 = \begin{bmatrix} \text{Relay coil open circuits} \\ \text{Switch fails closed} \end{bmatrix}$

$Z_3 = \begin{bmatrix} \text{Circuit breaker coil open circuits} \\ \text{Switch fails closed} \end{bmatrix}$

$Z_4 = \begin{bmatrix} \text{Switch transfers open} \\ \text{Switch fails closed.} \end{bmatrix}$

From Table I and the Z_i's, Sets M_6, M_8, M_{10}, and M_{12} are discarded because $M_6 \supset Z_1$, $M_8 \supset Z_2$, $M_{10} \supset Z_3$, and $M_{12} \supset Z_4$. In general, the M_ℓ sets will be larger than the Z_i sets.

Implementation of this technique for handling mutually exclusive events is considered by the author to be the best approach for the methodology to determine minimal cut sets described in Reference 14. For the methodology of determining minimal cut sets described herein a more straight forward approach is possible. Before determining the minimal cut sets, mutually exclusive events are flagged. These events are then never combined so erroneous minimal cut sets are not obtained. However, if these erroneous additional minimal cut sets are considered the error is generally conservative; that is, a higher system failed probability results in a quantitative analysis.

4. Quantitative Evaluation

Since the introduction of fault tree analysis, the area receiving the most research and development effort has been the evaluation of fault trees[5,7,15]. The evaluation of a fault tree is obtaining reliability information about the TOP event and perhaps the minimal cut sets from the data supplied for the failure of the basic components. Basically three methods for solutions to fault trees have been presented to date: (a) the direct simulation approach[15], (b) the Monte Carlo methods[7], and (c) direct analytical solutions[6].

The direct simulation approach basically uses Boolean logic hardware similar to that used in digital computers in a one-to-one correspondence with the fault tree Boolean logic to form an analog circuit. Immediately this method was seen to be prohibitively expensive. An effort was then made to obtain information from the fault tree by a hybrid method wherein parts of the solution were obtained using the analog technique and parts from a digital calculation in an effort to obtain a costwise competitive technique of solution. Because of the expense involved, this method has received relatively little attention.

Monte Carlo methods are perhaps the most simple in principle but in practice become outstandingly complex. Since Monte Carlo is not practical without the use of a digital computer, it is discussed in that framework. The most easily understood Monte Carlo technique is called "direct simulation". The term "simulation" is used in conjunction with Monte Carlo methods frequently because Monte Carlo is, indeed, a form of mathematical simulation. This simulation should not, however, be confused with the direct analog simulation as discussed previously. Probability data are provided as input and the simulation program represents the fault tree on a computer to provide quantitiative results. In this manner, thousands or millions of trial years of performance can be simulated. A typical simulation program involves the following steps:

(1) Assignment of failure data to input fault events within the tree, and if desired, repair data

(2) Representation of the fault tree on a computer to provide quantitative results for the overall system performance, subsystem performance, and the basic input event performance

(3) Listing of the failure that leads to the undesired event and identification of minimal cut sets contributing event results

(4) Computation and ranking of basic input failure and availability performance results.

In accomplishing these steps, the computer program simulates the fault tree and, using the input data, randomly selects the various parameter data from assigned statistical distribution parameters, and then tests whether or not the TOP event occurred within the specified time period. Each test is a trial, and a sufficient number of trials is run until the desired quantitative resolution is obtained. Each time the final event occurs, the contributing effects of input events and the logical gates causing the specified final event are stored and listed as computer output. The resultant output provides a detailed perspective of the system under simulated operating conditions and provides a quantitative basis to support objective decisions.

The third method of solution is direct analytical solution. To illustrate how this method might be applied to a simple fault tree for static conditions, the fault tree shown in Figure 10 is considered that contains independent, primary events A, B, C, and D with constant probabilities of failure 0.1, 0.2, 0.3, and 0.4, respectively. This assumption of constant failure probabilities distinguishes this example from realistic fault tree evaluation. However, the fault tree, as shown in Figure 10, is not in convenient form because Events X1 and X2 are not independent because they both are functions of Primary Event B [a]. By Boolean manipulation the fault tree shown in Figure 11 is equivalent to the one shown in Figure 10. This equivalence can be verified by noting that the minimal cut sets for both fault trees are identical. The fault tree shown in Figure 11 is in convenient form for calculating the probability of the TOP event.

Two basic laws of probability used in a fault tree evaluation are:
$$P(A1 \cup A2) = P(A1) + P(A2) - P(A1 \cap A2)$$
$$P(A1 \cap A2) = P(A1) \, P(A2|A1).$$

The first law simply states that the probability of a union $A1 \cup A2$ is the sum of the probabilities of the individual events minus the probability of their intersection. In terms of the fault tree, the probability of a two-event OR gate is the sum of probabilities of the two events attached to the gate minus the probability of the two events both occurring. The second law states that the probability of an intersection of events $P(A1 \cap A2)$ is equal to the probability of one, $P(A1)$, times the probability of the other, given the occurrence of the first, $P(A1|A2)$. In terms of the fault tree in Figure 11, the probability of a two-event AND gate is the product of the probabilities of the two attached events, because primary events of a fault tree are independent.

---

[a] This fault tree could be solved directly using elements of conditional probability theory.

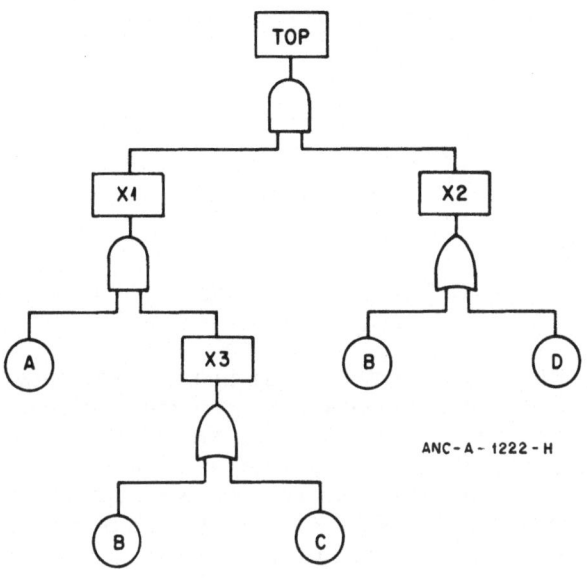

FIG. 10 SAMPLE FAULT TREE FOR PROBABILITY EVALUATION.

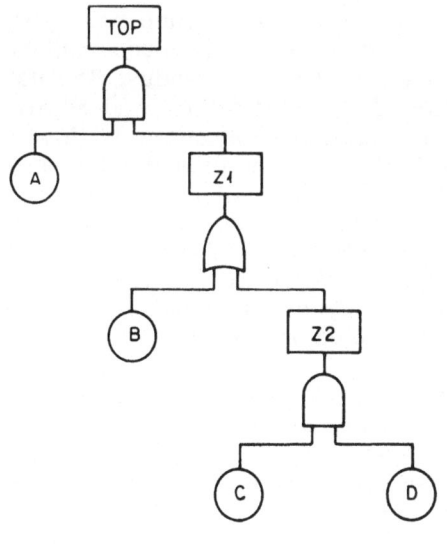

FIG. 11 BOOLEAN EQUIVALENT OF SAMPLE FAULT TREE SHOWN IN FIGURE 10.

Since all events are independent in the fault tree shown in Figure 11, unlike the events of the tree shown in Figure 10, the event probabilities are as follows:

$$P(Z2)=P(C)P(D)$$
$$P(Z1)=P(B)+P(Z2)-P(B)P(Z2)$$
$$P(TOP)=P(Z1)P(A).$$

Upon substitution,

$$P(TOP)=P(A)P(B)+P(A)P(C)P(D)-P(A)P(B)P(C)P(D)$$
$$P(TOP)=0.0236.$$

The probability of the system being in the failed state constant with respect to time is 0.0236 for the given primary event failure probabilities. This fault tree has two minimal cut sets, AB and ACD. Primary Event A appears in both minimal cut sets and hence is most crucial to the system. If the probability of Event A can be reduced to one-half of its original value, that is, from 0.1 to 0.05, the system failure probability is reduced to 0.0118, or one-half its original value.

In spite of the seeming simplicity of this example, until recently, a practical method for solving complex fault trees analytically was not known for trees containing primary failures demonstrating failure probabilities as complex function of time and repair possibilities. With the advent of Kinetic Tree Theory [6] in 1970, such analytical solutions requiring relatively small amounts of computer time were possible for complex trees. The solution of the fault tree itself is accomplished through a blend of probability theory and differential calculus. The use of AND, OR, and INHIBIT gates is allowed. General failure and repair distributions are handled. Complete probabilistic information is first obtained for each primary failure of the fault tree, then for each minimal cut set, and finally for the TOP failure itself. The information is obtained as a function of time and, hence, with regard to reliability, complete kinetic behavior is obtained. The expressions developed are in simple form, and application to yield numerical results is both efficient and straightforward, with an average computer time on the order of one minute required for a 500 primary failure fault tree (on the IBM 360/75 computer) [6].

As an elementary example of a fault tree solution with failure and repair probabilities as functions of time, the case is considered of two identical, independent system units, A and B, operating such that the simultaneous failure of both is required to cause system failures as shown in the following fault tree.

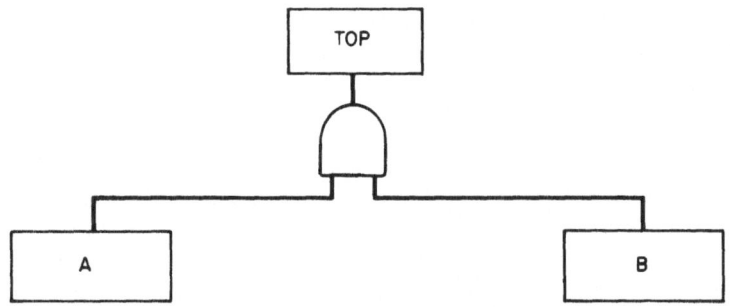

For Events A and B, F(t) represents the time-to-failure distribution function. A repair facility is used such that the time-to-repair distribution function is represented by G(t). These functions are:

$$F(t) = 1 - e^{-\lambda t}$$

$$G(t) = 1 - e^{-\mu t}.$$

The quantity λ is termed the failure rate for a primary failure and μ is termed the repair rate. Both are assumed constant for this example. If q(t) is the probability of the primary failure existing at time t, then from Reference 16:

$$q(t) = \frac{\lambda}{\lambda + \mu} - \frac{\lambda}{\lambda + \mu} e^{-(\lambda + \mu)t}.$$

Now Q(t) is defined as the probability that the TOP event exists at time t. Since the TOP failure exists at time t if and only if all the primary failures exist at time t,

$$Q(t) = \prod_{j=1}^{2} q_j(t)$$

$$= [q(t)]^2$$

$$= \frac{\lambda^2 - 2\lambda^2 e^{-(\lambda+\mu)t} + \lambda^2 e^{-2(\lambda+\mu)t}}{(\lambda + \mu)^2}.$$

The availability of the system, A(t), then is given by

$$A(t) = 1 - Q(t)$$

$$= \frac{\mu^2 + 2\lambda\mu}{(\lambda + \mu)^2} - \frac{\lambda^2 e^{-2(\lambda+\mu)t}}{(\lambda + \mu)^2} + \frac{2\lambda^2 e^{-(\lambda+\mu)t}}{(\lambda + \mu)^2}.$$

Of interest is the fact that these results are precisely the results obtained in Reference 16 for a parallel redundant system configuration using the theory of Markov processes.

IV. CONCLUSIONS

Fault tree analysis is a versatile reliability tool that has rapidly won favor with those involved in reliability and safety calculations. Fault tree analysis is a relatively new subject, continuously developing and expanding. Concepts and techniques of fault tree analysis have been developed over the past decade and now predictions from this type analysis are important considerations in the design of many systems such as aircraft, ships and their electronic systems, missiles, and nuclear reactor systems.

Routine, hardware-oriented fault tree construction can be automated; however, considerable effort is needed in this area to get the methodology into production status. When this status is achieved, the entire analysis of hardware systems will be automated except for the system definition step. This automated, quantitative prediction should be thought of as a distinct type of approach that could never replace conventional fault tree analysis. A value of the fault tree technique is that the analyst is forced to understand the system. Many system weaknesses are corrected while the fault tree is being constructed. A value of the technique is thus the construction process as well as the tree itself and resulting probability numbers. This automated analysis is a hardware-oriented approach that does not include environmental and human effects that can cause failures and, therefore, is apart from a true indepth fault tree analysis.

Automated analysis is not undesirable; to the contrary, when verified on adequately complex systems, automated analysis could well become a routine analysis. It could also provide an excellent start for a more indepth fault tree analysis that includes environmental effects, common mode failure, and human errors. The automated analysis is extremely fast and frees the analyst from the routine hardware-oriented fault tree construction, as well as eliminates logic errors and errors of oversight in this part of the analysis. Automated analysis then affords the analyst a powerful tool to allow his prime efforts to be devoted to unearthing more subtle aspects of the modes of failure of the system.

Fault tree analysis can be the most simple or most sophisticated analytical reliability tool depending on the needs of the analyst. At the same time, it is perhaps the least generally understood. Although the concepts of fault tree analysis are outstandingly simple, constructing suitable fault trees for complex systems requires thorough understanding of the principle of fault tree construction and evaluation as well as an understanding of the system being analyzed.

V. REFERENCES

1. D. F. Haasl, "Advanced Concepts in Fault Tree Analysis", System Safety Symposium, June 8-9, 1965, Seattle: The Boeing Company.

2. "System Safety Symposium", Proceedings of symposium sponsored by the University of Washington and the Boeing Company, Seattle, Washington, June 8-9, 1965.

3. S. N. Semanderes, "ELRAFT A Computer Program for the Efficient Logic Reduction Analysis of Fault Trees", IEEE Transaction on Nuclear Science, Seattle, Washington, November 1970, p 79.

4. P. Nagel, "Importance Sampling in System Simulation", IEEE Transactions on Nuclear Science, Seattle, Washington, November 1970, p 101.

5. W. E. Vesely, Analysis of Fault Trees by Kinetic Tree Theory, IN-1330, (October 1969).

6. W. E. Vesely, "A Time-Dependent Methodology for Fault Tree Evaluation", Nuclear Engineering and Design, 13(2), August 1970.

7. P. A. Crosetti, Computer Program for Fault Tree Analysis, DUN-5508, (April 1969).

8. P. A. Crosetti, "Fault Tree Analysis with Probability Evaluation", IEEE Transactions on Nuclear Science, Seattle, Washington, November 1970, p 132.

9. G. E. Greger, D. A. Snyder, P. D. Gross, "Description and Uses of a Critical Systems Data File for Nuclear Plants", The American Society of Mechanical Engineers, United Engineering Center, 345 East 47th Street, New York, New York 10017, January 9-12, 1972.

10. M. M. Yarosh, "Accident Analysis", Nuclear Safety, 3(4), 1962.

11. L. Leonardini, "The Third Reliability Meeting at Riso", Nuclear Safety, 11 (4), July-August 1970.

12. J. B. Fussell, Synthetic Tree Model-A Formal Methodology for Fault Tree Construction, ANCR-1098 (March 1973).

13. A. B. Mearns, "Fault Tree Analysis: The Study of Unlikely Events in Complex Systems", System Safety Symposium, June 8-9, 1965, Seattle: The Boeing Company.

14. W. E. Vesely and R. E. Narum, PREP and KITT: Computer Codes for the Automatic Evaluation of a Fault Tree, IN-1349 (August 1970).

15. J. M. Michels, "Computer Evaluation of the Safety Fault Tree Model", System Safety Symposium, June 8-9, 1965, Seattle: The Boeing Company.

16. G. H. Sandler, System Reliability Engineering, Prentice-Hall, Inc., Englewood Cliffs, N. J., 1963, pp 112-132.

OPTIMAL AVAILABILITY OF A COMPLEX SYSTEM

S. L. Gandhi and E. J. Henley

University of Houston
Cullen College of Engineering
Houston, Texas

ABSTRACT

A method for the evaluation of system availability of complex non-series parallel systems is proposed. The system availability is then optimized with respect to the number of redundant units in each module and the preventive maintenance schedule, using a modified mixed integer gradient-method. Solutions to example problems have been obtained.

INTRODUCTION

In an earlier paper [1] the authors developed a method and a computer algorithm to calculate the reliability of a complex system, given the reliability of each unit and the system configuration in the form of a reliability graph. This method can be used to obtain another important design parameter namely the sensitivity of the overall reliability to individual system units. The sensitivity function can also be used for maximizing system reliability using a modified mixed integer gradient-method [2,3] and for obtaining optimum redundancies in a reliability network.

In this paper we present an application of the above method to the problem of finding optimum redundancies and preventive maintenance schedules and maximizing system availability or system profit of a complex system having linear or non-linear constraints.

MODULE REPRESENTATION OF RELIABILITY GRAPHS

A reliability graph consists of nodes and modules with directed arrows between them, as shown in Figure 1. A module is defined to be a single unit or simply connected, identical parallel units some of which are in standby. More than one module between two arbitrary nodes is permitted. We distinguish between three standby redundancy modules. A unit whose failure probability in standby is zero is called a cold standby. One whose failure probability is the same in standby as in service is called a hot standby. Cases that lie in between these two extremes are called warm standby. It is assumed that all units in a module are independent and identical. Depending on the type of redundancy, we have different expressions for module reliability.

SYSTEM RELIABILITY

Consider a system of N modules with an input and an output node. The system reliability definition used here is the probability of successful operation of all the molules in at least one minimal path. A path is a group of branches which form a connection between input and output when traversed in a stated direction. A minimal path is one which contains a minimum number of non-identical modules to form a path.

Let P_i, $i = 1, 2, \ldots M$ denote the minimal paths M in number. The system reliability R can be expressed by

$$R \equiv \text{Probability \{at least one path is successful\}}$$
$$= P_r \{ \bigcup_{i=1}^{M} P_i \} \qquad (1)$$

where $\cup$ denotes the union.

By use of the expansion rule for the probability of the union of M events [4], we have the following expression

$$R = \sum_{i=1}^{M} P_r P_i - \sum_{i=1}^{M} \sum_{j>i}^{M} P_r(P_i \cap P_j)$$
$$+ \sum_{i=1}^{M} \sum_{j>i}^{M} \sum_{k>j}^{M} P_r(P_i \cap P_j \cap P_k) + \ldots + (-1)^{M-1} P_r \{ \bigcap_{i=1}^{M} P_i \} \qquad (2)$$

where ∩ denotes the intersection.

The total reliability R is thus given in terms of module reliabilities by

$$R = \sum_{i=1}^{M} \prod_{\ell \in P_i} R_\ell - \sum_{i=1}^{M} \sum_{j>i}^{M} \prod_{\ell \in P_i \cup P_j} R_\ell$$

$$+ \sum_{i=1}^{M} \sum_{j>i}^{M} \sum_{k>j}^{M} \prod_{\ell \in P_i \cup P_j \cup P_k} R_\ell + \ldots + (-1)^{M-1} \prod_{\ell \in (\cup_{i=1}^{M} P_i)} R_\ell \qquad (3)$$

where R_ℓ is the module reliability and the members of the ith path, the union of the ith and j^{th} paths, etc. are denoted by $\ell \in P_i$, $\ell \in P_i \cup P_j$, etc.

SENSITIVITY CALCULATION

The system reliability expression given by Equation (3) is a bilinear function of each modules reliability. The sensitivity of the system reliability R to the module reliability R_i can therefore be obtained by the simple formula

$$S_{R_i} = \frac{\partial R}{\partial R_i} = R/_{R_i=1} - R/_{R_i=0} \qquad i=1,2,\ldots N \qquad (4)$$

where N is the total number of modules. S_{R_i} is called a module sensitivity.

The sensitivity of the system reliability R to the number of units in a module can then be obtained as:

$$S_{n_i} = \frac{\partial R}{\partial n_i} = \frac{\partial R}{\partial R_i} \frac{\partial R_i}{\partial n_i} = S_{R_i} * \frac{\partial R_i}{\partial n_i} \qquad (5)$$

where n_i is the number of units in the ith module. S_{n_i} is called a unit sensitivity.

Sensitivity is a measure of the system reliability improvement and provides important information for maximizing the reliability of complex systems.

SYSTEM AVAILABILITY AND PREVENTIVE MAINTENANCE

System availability can be improved by adding redundancy to the system and/or by performing preventive maintenance on the system according to some prescribed schedule [5,6].

In the absence of scheduled preventive maintenance, the system availability A is given by

$$A = \frac{MTBF}{MTBF + MDT} = \frac{\int_0^\infty R(t)dt}{\int_0^\infty R(t)dt + MDT} \qquad (6)$$

where MTBF is the mean time to system failure and MDT is system mean downtime or mean repair time.

Now suppose preventive maintenance is performed on the system every 'T' hours of continuing operation. If the system fails before T hours have elapsed, emergency maintenance is performed at that time. Preventive maintenance is then rescheduled. We assume that the system is as good as new after any type of maintenance, scheduled or emergency, and that the system either operates at full capacity or is down for maintenance. Let,

$f(t)$ = system failure time probability density function

T_E = mean time to perform emergency maintenance on the system

T_S = mean time to perform scheduled maintenance on the system.

Thus, mean time between system maintenance is given by:

$$MTBM = T * R(T) + \int_0^T tf(t)dt = \int_0^T R(t)dt \qquad (7)$$

Also, the mean down time for the system is:

$$MDT = T_E * (1-R(T)) + T_S * R(T) = T_E - (T_E - T_S) * R(T) \qquad (8)$$

System availability A is then defined as

$$A = \frac{MTBM}{MTBM + MDT} = \frac{\int_0^T R(t)dt}{\int_0^T R(t)dt + [T_E - (T_E - T_S) * R(T)]} \quad (9)$$

where, $R(t) = f_1(R_i(t), i=1,2,...N)$ as given by Equation (3) and $R_i(t) = f_2(\hat{n}_i, t)$.

System availability is thus a non-linear function of the module redundancies, n_i's, and the maintenance interval T. In the present analysis it is assumed that T_E and T_S are independent of n_i and T.

SYSTEM PROFIT

In many situations, system profit is closely related to system availability. To examine this relationship, we define the following terms

T^+ = total lifetime of the system in hours

T_u = total time during which the system is operating at full capacity

$$T_u = A * T^+ \quad (10)$$

T_D = total time during which the system is down

$$T_D = (1 - A) * T^+ \quad (11)$$

Also, $T^+ = T_u + T_D \quad (12)$

C_{f_i} = fixed capital cost of a unit in the ith module

n_i = number of units in the ith module

C_{TF} = total fixed capital investment

$$C_{TF} = \sum_{i=1}^{N} n_i C_{F_i} \quad (13)$$

C_M = system maintenance costs per hour of system downtime. This can be taken as some percentage, x, of fixed capital investment

$$C_M = \left(\sum_{i=1}^{N} n_i C_{F_i} \right) * x/100 \tag{14}$$

C_{TM} = total maintenance costs for a system lifetime of T^+ hours

$$C_{TM} = C_M * T_D \tag{15}$$

C_I = net income per hour; the difference between the selling price of the product and all expenses other than the maintenance costs and capital costs.

C_{TI} = total net income during system lifetime of T^+ hours

$$C_{TI} = C_I * T_u \tag{16}$$

We can now write a cash balance for the system lifetime period of T^+ hours.

Profit = Net income - capital cost - maintenance cost, i.e.,

$$P = C_{TI} - C_{TF} - C_{TM}$$

$$P = C_I * T_u - \sum_{i=1}^{N} n_i C_{F_i} - C_M * T_D$$

$$P = C_I * A * T^+ - \sum_{i=1}^{N} n_i C_{F_i} - \left(\sum_{i=1}^{N} n_i C_{F_i} \right) * x/100 * (1 - A) T^+ \tag{17}$$

This equation does not take into account cost of capital, time value of money or any other of the widely used economic methods.

OPTIMAL DESIGN

Two types of optimization problems are considered.

Availability Maximization

The problem can be stated as follows: Maximize the system availability A

$$A = \frac{\int_o^T R(t,n_1,n_2,\ldots n_N)dt}{\int_o^T R(t,n_1,n_2,\ldots n_N)dt + T_E(1-R(T,n_1,n_2\ldots n_N))} \quad (18)$$

$$+ T_s * R(T,n_1,\ldots n_N)$$

Subject to the cost constraint

$$C_{TF} + C_{TM} \leq C_{AO}, \text{ i.e.,}$$

$$(\sum_{i=1}^{N} n_i C_{F_i}) (1 + \frac{x}{100} * (1 - A) * T^+) \leq C_{AO} \quad (19)$$

and

$$n_i \geq r_i$$

n_i : integer $\qquad i=1,2,\ldots N$ \qquad (20)

$$T \geq 0$$

where r_i is the minimum number of units in the ith module that must operate for the module to function successfully. Other constraints, if present, can also be included.

This is a non-linear mixed integer programming problem.

Profit Maximization

This problem can be stated as follows: Maximize the system profit over its lifetime of T^+ hours

$$P = C_I * A * T^+ - (\sum_{i=1}^{N} n_i C_{F_i}) (1 + (1 - A) * \frac{x}{100} * T^+) \quad (21)$$

Subject to the cost constraint

$$C_{TF} \leq C_{PO}$$

that is

$$\sum_{i=1}^{N} n_i C_{F_i} \leq C_{PO} \tag{22}$$

and

$$n_i \geq r_i$$

$$n_i : \text{integer} \qquad i = 1, 2, \ldots N \tag{23}$$

$$T \geq 0$$

Other constraints, if present, can also be included.

This is also a non-linear mixed integer programming problem.

METHOD OF SOLUTION

A method proposed by Reiter and Rice [3] based on the modified gradients of the objective function was used. The method is applicable to the optimization problems formulated above, since we can readily generate the sensitivity function and hence the gradients of the objective function.

The search algorithm has three phases:

Phase I: Setting up an initial or starting point $\underline{X}_I$ using uniformly distributed random numbers.

Phase II: Searching for a feasible solution $\overline{X}^+$ of the problem, starting with X_I. A search by the method of weighted perpendiculars is performed. If the feasible solution is not obtained in a given number of iterations, the search is terminated, and we go back to phase I of the search algorithm.

Phase III: Locating a point $\underline{X}_p$ which maximizes the objective function in the feasible region around $\overline{X}^+$. This involves calculating the modified gradients of the objective function which define the path on which we search for improved values of the objective function.

X_I, X^+ and X_P denote vectors of variables, real and/or integer.

The method provides local optima. Repeated use of the algorithm, with improved lower and upper limits on the system variables, n_i and T, gives better results, and hopefully the global optimum.

ILLUSTRATIVE EXAMPLE

An example is now considered, the solution to which is obtained both for availability maximization and profit-maximization. Consider the module reliability graph of the system shown in Figure 1. Assume all modules have 1 out of n_i hot standby redundancy and that units in each module have exponential failure pdf. Table 1 gives the capital cost and failure rate data for a single unit of each module.

Other data are:

C_I = 10.00 $/hr.

X = 0.02%

T^+ = 80,000 hrs.

T_E = 400 hrs.

T_S = 100 hrs.

C_{AO} = $350,000

C_{PO} = $200,000

DISCUSSION OF RESULTS

The results for the system availability maximization and the system profit-maximization are tabulated in Tables II and III respectively. Also the various steps involved in the optimization procedure are shown for the Result-1 of each problem in Tables IV and V respectively. In all, ten initial points were used for each problem. The results indicate that different initial points in the case of system availability maximization give different local optima as far as n_i's are

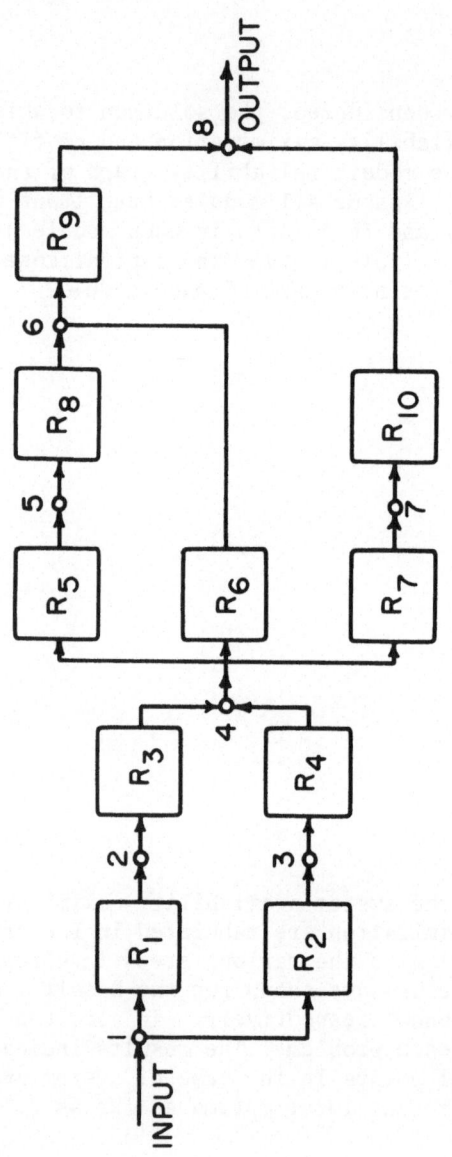

Figure 1 Reliability Graph

concerned. This is due to the fact that the system availability is a monotonically increasing function with respect to the n_i's. The computer time for a single result is about 1 minute on an IBM 360 and 15 seconds on UNIVAC 1108.

Although this method does not insure a global optimum, it does find various near-optimum solutions. From a practical consideration, the number of near optimum solutions provide a wider choice for design purposes.

TABLE I DATA

MODULE (i)	BRANCH	CAPITAL COST/UNIT (C_{F_i}) $/UNIT	FAILURE RATE (λ_i) HOUR^{-1}
1	1 - 2	5000.00	$3.0 * 10^{-4}$
2	1 - 3	15000.00	$1.0 * 10^{-4}$
3	2 - 4	10000.00	$2.2 * 10^{-4}$
4	3 - 4	15000.00	$1.2 * 10^{-4}$
5	4 - 5	5000.00	$3.2 * 10^{-4}$
6	4 - 6	15000.00	$1.1 * 10^{-4}$
7	4 - 7	5000.00	$3.0 * 10^{-4}$
8	5 - 6	15000.00	$1.0 * 10^{-4}$
9	6 - 8	5000.00	$3.0 * 10^{-4}$
10	7 - 8	10000.00	$2.2 * 10^{-4}$

TABLE II SYSTEM AVAILABILITY MAXIMIZATION

MODULE NUMBER	OPTIMUM NUMBER OF UNITS IN EACH MODULE				
	RESULT 1	RESULT 2	RESULT 3	RESULT 4	RESULT 5
1	5	1		2	1
2	2	3		2	2
3	2	1		1	1
4	2	3	FEASIBLE POINT CAN NOT BE LOCATED	3	3
5	1	2		2	1
6	2	2		2	2
7	2	1		2	4
8	1	1		1	1
9	6	8		8	5
10	1	1		1	2
Fixed Capital Cost ($)	205,000	215,000		210,000	205,000
Total Maintenance Cost ($)	140,822	134,648		139,485	142,286
Optimum Maintenance Interval (hours)	3,190	3,575		3,490	3,380
Optimized System Availability	0.957066	0.960858		0.958487	0.956620
System Profit ($)	419,831	419,038		417,304	418,010

TABLE II CON'T   SYSTEM AVAILABILITY MAXIMIZATION

MODULE NUMBER	OPTIMUM NUMBER OF UNITS IN EACH MODULE				
	RESULT 6	RESULT 7	RESULT 8	RESULT 9	RESULT 10
1	2	1	1	3	1
2	2	2	3	2	3
3	1	1	1	1	1
4	3	3	3	2	3
5	3	1	1	1	1
6	1	2	2	2	2
7	4	3	2	4	3
8	1	1	1	1	1
9	4	8	7	7	6
10	2	1	1	1	1
Fixed Capital Cost ($)	200,000	205,000	210,000	200,000	210,000
Total Maintenance Cost ($)	146,922	141,368	135,422	145,765	137,502
Optimum Maintenance Interval (hours)	3,030	3,510	3,490	3,260	3,430
Optimized System Availability	0.954087	0.956900	0.959696	0.954448	0.959077
System Profit ($)	416,347	419,153	422,334	417,794	419,759

TABLE III SYSTEM PROFIT MAXIMIZATION

MODULE NUMBER	OPTIMUM NUMBER OF UNITS IN EACH MODULE				
	RESULT 1	RESULT 2	RESULT 3	RESULT 4	RESULT 5
1	3	3	1	3	3
2	1	1	2	1	1
3	2	2	1	2	2
4	1	1	2	1	1
5	1	1	1	1	1
6	1	1	2	1	1
7	2	2	1	2	2
8	1	1	1	1	1
9	3	3	5	3	3
10	1	1	1	1	1
Fixed Capital Cost ($)	135,000	135,000	165,000	135,000	135,000
Total Maintenance Cost ($)	155,112	155,138	143,377	155,116	155,130
Optimum Maintenance Interval (hours)	2,145	2,170	2,840	2,150	2,170
Optimized System Profit ($)	452,439	452,404	448,176	452,433	452,415
Availability	.928189	.928177	945691	928187	928181

TABLE III CON'T SYSTEM PROFIT MAXIMIZATION

MODULE NUMBER	OPTIMUM NUMBER OF UNITS IN EACH MODULE					
	RESULT 6	RESULT 7	RESULT 8	RESULT 9	RESULT 10
1	3	1	1	3	3
2	1	2	2	1	1
3	2	1	1	2	2
4	1	2	2	1	1
5	1	1	2	1	2
6	1	1	1	1	1
7	2	3	2	2	1
8	1	1	1	1	1
9	4	3	4	4	4
10	1	2	1	1	1
Fixed Capital Cost ($)	140,000	160,000	155,000	140,000	140,000
Total Maintenance Cost ($)	153,017	148,688	150,276	153,045	152,779
Optimum Maintenance Interval (hours)	2,240	2,560	2,580	2,290	2,190
Optimized System Profit ($)	452,334	444,847	446,248	452,297	452,657
Availability	.931689	.941919	.939405	.931676	.931795

TABLE IV

STEP	MOD 1	MOD 2	MOD 3	MOD 4	MOD 5	MOD 6	MOD 7	MOD 8	MOD 9	MOD 10
Initial Point (1)	5	2	6	6	1	3	2	3	10	4
Feasible Point (2)	5	1	1	2	1	1	1	1	2	1
(3)	5	1	1	2	1	1	1	1	3	1
(4)	5	1	1	2	1	2	1	1	3	1
(5)	5	2	1	2	1	2	1	1	3	1
(6)	5	2	1	2	1	2	1	1	4	1
(7)	5	2	2	2	1	2	1	1	4	1
(8)	5	2	2	2	1	2	1	1	5	1
(9)	5	2	2	2	1	2	1	1	6	1
(10)	5	2	2	2	1	2	1	1	6	1
(11)	5	2	2	2	1	2	2	1	6	1
(12)	5	2	2	2	1	2	2	1	6	1
(13)	5	2	2	2	1	2	2	1	6	1
(14)	5	2	2	2	1	2	2	1	6	1
(15)	5	2	2	2	1	2	2	1	6	1
Optimum Point (16)	5	2	2	2	1	2	2	1	6	1

TABLE IV CON'T

STEP	MAINTENANCE INTERVAL (hours)	COST CONSTRAINT ($)	SYSTEM AVAILABILITY ($)
Initial Point (1)	2131	+338,723	--
Feasible Point (2)	2484	- 11,563	0.911412
(3)	2481	- 24,423	0.922165
(4)	2478	- 14,032	0.931263
(5)	2478	- 6,748	0.939910
(6)	2480	- 13,996	0.945832
(7)	2490	- 4,856	0.948966
(8)	2503	- 5,588	0.952112
(9)	2539	- 2,652	0.953954
(10)	2639	- 4,518	0.954537
(11)	2690	- 271	0.955875
(12)	2790	- 1,734	0.956321
(13)	2890	- 2,824	0.956654
(14)	2990	- 3,574	0.956882
(15)	3090	- 4,016	0.957017
Optimum Point (16)	3190	- 4,178	0.957066

180

TABLE V

STEP	MOD 1	MOD 2	MOD 3	MOD 4	MOD 5	MOD 6	MOD 7	MOD 8	MOD 9	MOD 10
Initial Point (1)	5	2	6	6	1	5	3	5	10	6
Feasible Point (2)	3	1	2	1	1	1	1	1	8	2
(3)	3	1	2	1	1	1	1	1	8	1
(4)	3	1	1	1	1	1	1	1	8	1
(5)	3	1	1	1	1	1	1	1	7	1
(6)	3	1	1	1	1	1	1	1	6	1
(7)	3	1	1	1	1	1	1	1	5	1
(8)	3	1	1	1	1	1	1	1	4	1
(9)	3	1	1	1	1	1	1	1	3	1
(10)	2	1	1	1	1	1	1	1	3	1
(11)	2	1	2	1	1	1	1	1	3	1
(12)	3	1	2	1	1	1	2	1	3	1
(13)	3	1	2	1	1	1	2	1	3	1
(14)	3	1	2	1	1	1	2	1	3	1
Optimum Point (15)	3	1	2	1	1	1	2	1	3	1

TABLE V CON'T

STEP	MAINTENANCE INTERVAL (hours)	COST CONSTRAINT ($)	SYSTEM PROFIT
Initial Point (1)	1131	+285,000	--
Feasible Point (2)	1131	- 35,000	330,717
(3)	1211	- 45,000	363,167
(4)	1298	- 55,000	366,674
(5)	1380	- 60,000	386,472
(6)	1448	- 65,000	403,457
(7)	1507	- 70,000	418,106
(8)	1558	- 75,000	429,769
(9)	1606	- 80,000	435,455
(10)	1640	- 85,000	439,440
(11)	1771	- 75,000	445,012
(12)	1870	- 70,000	448,389
(13)	1945	- 65,000	451,396
(14)	2045	- 65,000	452,222
Optimum Point (15)	2145	- 65,000	452,439

REFERENCES

1. Inoue, K., and Gandhi, S. L., (1973) Chapter 12 in, "Graph Theory in Modern Engineering", by Henley, E. J. and Williams, R. A., Academic Press, New York.

2. Gandhi, S. L., Inoue, K. and Henley, E. J., Proceedings of the IFIPS Conference, Eindhovan, Holland, Oct., 1972, Norm Holland Publishing Co., (In Press).

3. Reiter, S. and Rice, D. B. (1966), Mgt. Sci. $\underline{12}$, No. 11, 829.

4. Feller, W. (1967), "An Introduction to Probability Theory and its Applications"; Vol. I., John Wiley & Sons, New York.

5. Rau, J. G. (1970), "Optimization and Probability in Systems Engineering", Van Nostrand Reinhold Company, New York.

6. Bazøvsky, I., (1961), "Reliability Theory and Practice", Prentice Hall, Inc., New Jersey.

7. Benning, C. J., (1967), IEEE Trans. on Reliability, R-16, No. 3, 136.

8. Polovko, A. M., (1968), Fundamentals of Reliability Theory, Academic Press, New York.

9. Shooman, M. L., (1968), "Probabilistic Reliability on Engineering Approach", McGraw-Hill, New York.

A REVIEW OF SYSTEM RELIABILITY ASSESSMENT

A. E. Green

Systems Reliability Service, ASRD, UKAEA,
Culcheth, Warrington, Lancashire, England

INTRODUCTION

The problem of reliability dates back to antiquity and there tend to be well worn phrases which indicate this. The old German saying "Der Krug so lange zum Wasser geht bis er zerbrocken wird", which roughly means that the earthenware jug used to transport water from well to house continues in service until it inevitably meets with an accident; the jug then, obviously, needs to be replaced. Another well worn phrase in French is "Ça ne marche pas", which could mean it does not work.

The ancient Egyptians in building the pyramids obviously intended to have a reliable structure. But who would have thought that today somebody would have the audacity to say that the failure-rate of the Great Pyramids is about 1 in 5,000 years. Such a statement was made from a structural point of view. However, a reliability analyst may define reliability as:

> "That characteristic of an item expressed by the probability that it will perform its required function in the desired manner under all the relevant conditions and on the occasions or during the time intervals when it is required so to perform."

Hence, it could be argued that the Great Pyramids have not been successful in meeting the "required function" as funerary edifices since they were plundered of their treasures a long time ago.

The comment made by Pope in 1959[1] may also apply to this statement of failure-rate, viz. "This is obviously a field of experimentation where the engineer must have had the assistance of the statistician ... In all cases the testing of complex components should be planned statistically and the results analysed by a competent statistician". There again, Poisson's derivation might have remained a mathematical curiosity but for L. Von Bortkiewicz[2] in 1898 showing the statistical meaning and importance of Poisson's derivation, which was in connection with the number of Prussian cavalrymen killed each year by the kick of a horse.

This causes one to wonder on the extent to which engineering judgement together with limited test experience may be used to give reasonable estimates of current reliability. One may have supposed that the paper published by the Rev. T. Bayes in 1763 entitled "Essay towards solving a problem in the doctrine of chance", may have helped by being able to take an unknown probability P of the event happening a number of times and making a statement as to what is the chance of P lying between two probability values. Unfortunately, this seems to have led to the complete split between statisticians into 'Bayesians' and 'Classicists' which does not always help the engineer.

Developments during the Second World War caused inroads to these problems of reliability. In the development of the VI missile, started in 1942, the joint work of Robert Lusser and Erich Pieruschka[3] combining engineering judgement and statistics, produced the well-known Lusser product law of reliabilities,[4] i.e. R system = $R_1 R_2 R_3$ - - - . The application of this thinking brought the realisation that a large number of fairly 'strong links' can be more unreliable than a single 'weak link' if reliance is being placed on them all. This approach gave rise to a great improvement in the reliability of the VI missile.[5]

Subsequently, a keen interest in reliability emerged as military equipment became more complex. The lack of reliability and maintainability has, in many cases, caused large amounts of money and effort to be spent on maintenance rather than operations. This has been seen in wheeled transport where Wesley Stout[6, 7] has written about the American reliability of tanks during World War II - "Our tanks were better because the Germans never learned to think in terms of reliability as we use the word, i.e. maximum performance and minimum care and replacement. Just as the European pursues science for science sake, so is he prone to design and make machines for machinery's sake. A captured Panzer Commander grumbled that his tanks appeared to have been built by watchmakers". It appears that the French expression "Ça ne marche pas" applied, but also, from the work of Lusser it suggests that there could have been a communication problem. On the other hand, from a reliability point of view, one would not have thought of decrying watchmakers.

It appears that from time to time the Army has carried out surveys on its fleets of wheeled vehicles. Numerical evaluations of the initial costs, the maintenance costs and the degree of availability achieved, have enabled them to plan the most economical replacement periods. Industrial commercial users of vehicles have commenced to undertake similar reliability calculations. The whole of this experience emphasises the importance of reliability from the cost and maintenance point of view.

In the 1950's there were parallel developments going on in the space industry as well as the nuclear and other high risk industries. Special reliability assessments became necessary due to the consequences involved. The use of numerical reliability rather than subjective statements such as "it is highly reliable, made of good quality materials", became the order of the day and people started to be required to say what they meant. This had its impact on the training of management personnel where one USA administrator suggested[8] that "We are now facing a period where a separation of the men from the boys will occur. The glib tongue and the slick brochure is inadequate in the atmosphere now pervading government circles". However, product assurance had spread its wings in the USA and it was difficult to backtrack. In the nuclear industry, backtracking was not possible; accidents had taken place and people interested in maintaining development of nuclear reactors found themselves suggesting that the overall chance of any untoward occurrence was of the order of 1 in 1,000,000. Systems for safeguarding such situations started to have their chance of failing to perform the required functions on demand, stated as, perhaps, once in every 10,000 occasions.

Such a critical approach also became apparent in the electrical supply industry in determining costs of supply configurations and the availability of supplies to consumers. In the chemical industry, technology had produced larger plant, in some cases with greater potential energies available, with escalating costs both from capital and running points of view. This led to people concerning themselves with money-making systems and to take a criterion of how much money could be earned by the system. This, in turn, caused the reliability of more complex systems to be assessed. One can consider a money-making system in which is involved financial system reliability at one hierarchical level and a technological system reliability at a lower hierarchical level. Trade-offs between cost and technical performance must be studied.

An important outcome of this type of consideration was that a cheap technological system with a low technological reliability could lead to the same financial system reliability as a costly technological system with a very high technological reliability. In recent years, the development of safety and ecological problems have imposed, in many cases, direct reliability requirements, and techniques of cost benefit analysis have proved difficult when giving a monetary value to human life. All this has presented to the reliability analyst a problem of carrying out reliability assessments which, due to the rate of the development of technology and the increased consequences of failure, both from the economic and safety points of view, have caused greater emphasis to be placed upon methods of prediction in reliability assessment.

CRITERIA FOR ASSESSMENT

The introduction of quantified reliability techniques of system assessment has caused particular industries to carefully examine criteria for judging adequacy. This is a logical consequence as would be expected from the definition of reliability given in the Introduction to this paper, the essential process of reliability assessment being to give a measure of the relationship between some required performance and the achieved performance.

Examples may be found as in the aircraft industry where papers a decade ago were dealing with the reliability requirements for automatic landing of aircraft.(9) The British Civil Aviation Authority have quantified the reliability for automatic landing systems by saying that the probability of a fatal landing should not be worse than 10^{-7}. An interesting example of the breakdown of this risk is given by Warren (10) Other considerations, such as the pilot in the loop and out of the loop, in providing a safe landing capability are considered in Ref. 11.

In the nuclear field a proposed requirement has been expressed as in references (12,13) Along one axis is shown increasing consequence and the other axis is related to the probability of occurrence. The consequence is measured by the release of a radioactive fission product expressed in curies of I131 and the other axis is measured in reactor years or the events occurring in a complete reactor programme. It is of interest to note that the starting point was taken as a programme of 30 reactors operating for 30 years, which represents 900 reactor years. This criterion logically permits a system reliability assessment to be undertaken and for it to be demonstrated that all consequences and their occurrence produce points which lie below the criterion line.(14) Earlier attempts were made to derive similar types of criteria, as described in Ref. 15, and recent thoughts on reliable, economic and safe nuclear plant are given in Ref. 18. Similarly, in the chemical industry criteria have been emerging and have arisen out of the application of quantified techniques to the assessment of systems.(16)

In parallel with such developments, the space industry has produced a large amount of literature which paved the way for requirements to be met by reliability assessment and for the utterance of "That's one small step for a man, one giant leap for mankind" by Neil Armstrong as he stepped from Apollo's lunar module to the surface of the moon on the 20th July, 1969.(17)

The general literature shows that the techniques of reliability assessment and the use of quantified reliability methods have required to be put on to a logical footing by statements of numerical criteria, as previously mentioned. A basic technology has emerged which is a method of closed loop thinking to show whether or not the system will have adequate reliability, as opposed to the open loop method of thinking where one just makes things better and better.

Generally, the techniques of reliability assessment have developed along similar lines with each industry having its own environmental and particular problems, depending upon the extent of field data existing and the ease by which testing data may be obtained. However, it may be noted that the reliability assessment of the complex systems have been heavily dependent upon the prediction techniques using system synthesis, due to the need to carry out a reliability evaluation in the early stages of the design because reliability is at such a premium.

METHOD OF ASSESSMENT

The methods of assessment developed over the past two decades have been generally system orientated and initially concentrated on deterministic solutions.(19,20) Mechanistically, a basic process could be followed by analysing failures, defining the generic problems and then developing managerial and technical solutions. This approach could be carried out by field experience where it is immediately available, or some sample testing could be instituted and the results used to give estimates of the reliability characteristics. This approach is akin to the development of the wheel and axle which has been used over long periods of time and evolved by means of trial and error. In the large complex systems the information may only be derived at some sub-system level, perhaps, or even component part level and the question as to how the whole could be put together and a synthesis made of the system reliability became an outstanding issue.

This method of synthesis poses a problem in the particular disciplines of the overall design and engineering which requires a definition of the system. This can be given in engineering terms such as drawings, specifications, operational modes, maintenance procedures, and the system may be defined within its own boundaries but, in the subsequent analysis, mathematical modelling can produce an abstractness which is devoid of the practical connotations. Furthermore, the entrenchment of the determinists during the last decade has been shaken by the probabilistic nature of the world in which we live and the whole range of engineers find themselves with a common demand for a probabilistic approach. Safety and economic demands of modern society have further required us to promote methods of efficient conservation in the use of our resources as pursued by the assurance technologists.(21) Hence, the deterministic and probabilistic approaches have become combined in carrying out a reliability assessment and a reliability model which involves a multi-disciplinary approach is prepared with a view to estimating the correlation between the performance achievement function Q and the performance requirement function H. Various methods have been developed to expand the reliability model to a system depth which permits a solution based on the availability of the appropriate data. Broadly speaking, the reliability is measured by some overall function f(Q,H) which represents the probability of the achieved performance meeting the required performance, i.e. the reliability.

It is considered that the capability of a system requires to be assessed and defined and this represents a singular solution of the overall reliability function and is a measure of the deterministic capability of achievement which is invariant and forms a reference for the modelling of the variability of the system. Hence, the matrix representation of the transfer characteristic h for the overall system achievement leads to a deterministic evaluation of the elements ϕ_{ij} in the equation

$$[\phi_{ij} | t, z] = \frac{O_h}{I_q}$$

Techniques of analysis have been appropriately developed by finding means of interconnecting partial solutions of large systems.(22) Methods have been chosen so that the overall system may be "torn up" at any section and if any changes are made in one section of the system then they may be easily programmed in the process of solution. This type of work has been well developed from a topological point of view and is a generic problem of dealing with, for example, nodes, meshes, branches and planes.(23) As an example, a simplified diagram of tensor analysis developed by Kron   as   applied   to large electrical and electronic systems of the Lagrangian type. Here, Kron's approach was to tear up the model rather than the equations in order not to lose the interconnecting constraints between the torn up sections of the system.(24) An example of the modelling techniques for electrical machines and networks is given in Ref. 25. Matrices and graph theory with, for example, Boolean functions, have been developed which adequately deal with certain types of systems involving transport, signal and information flow, switching, automation and logic.

This approach by the system reliability analyst enables the capability of the system to be assessed and, in general terms, involves modelling in co-ordinates in generalised spaces. Obviously, the question of variability then involves multi-dimensional statistics and probability in determining whether the system will have the capability to function when it is required to do so.

RELIABILITY ANALYSIS

Basically, two methods of reliability analysis have been finding favour. One may be called a performance spectrum approach and the other one is a fault tree approach. The performance spectrum method is an attempt to make a probabilistic statement on the variability and complete failure of the equipment or system in the following general form:(26,27)

$$\text{Probability of failing} = (1 - p) \int_Q^\infty f(x)dx + p$$

where: $f(x)$ = the density function for the variability of performance of the equipment in time

Q = the required performance level which should not be exceeded

p = the probability of complete failure of the equipment

which may be illustrated in simplified form where some design estimate of performance exists about which there is variability which extends through the whole spectrum of performance. The variations in performance are more likely to occur near to the design estimate or in the catastrophic failure region. Superimposed upon this pattern is a requirement shown as Q but itself may also be distributed in space and time.

Typically with this technique, the following process of analysis is necessary:

1. A logical process of establishing the precise required performance of a system under all the relevant conditions.

2. A logical process for analysing a system in order to determine its inherent capability.

3. A variation analysis of all software aspects of the system in order to arrive at a combined or overall variability function. This involves methods for combining distributions of variation.

4. A variation analysis of all hardware aspects of the system in order to arrive at a combined or overall variability function for the system hardware. Again, this involves methods for combining distributions.

5. A logical process of fault mode analysis to determine the system availability in the context of freedom from catastrophic modes of failure in the hardware of the system.

6. A process of combining the results of 3, 4 and 5 together and ultimately arriving at a particular reliability value or a reliability value averaged over the appropriate time and space domains.

The fault tree techniques were originally introduced by the Bell Telephone Laboratories in the Minute-man launch safety evaluation which proved powerful in evaluating critical safety hazards. In the fault tree approach all the various factors leading to a particular outcome are considered, which is the essential difference between the analytical approach of arriving at a spectrum of performance. This fault tree approach leads to the requirements for systematic examination of systems in a logical fashion and the optimum methods of doing this are still in development.(28) However, the early fault tree diagram technique, which was a manual process, has developed into a computerised process.(29) Not all reliability analysts favour fault tree diagrams because of their complexity, although there are advantages over conventional analytical methods.

Both methods of approach have found favour in the analysis of large system configurations and various computer programs have been evolved. Useful stepping stones in computer reliability analysis of complex systems have been the American ARMM Program(30) and the NOTED Program(32) which has a strategy for calculating the probability, as a function of time, of the successful operation of a system and caters for up to 200 different probability laws.

Such techniques are, however, sensitive to the types of distributions used and their combination. It has been noted that due to the versatility in fitting time to failure distributions, the Weibull distribution has, in recent years, assumed a position of importance.(31,91) However, its usage does give rise to

difficulty due to the estimation of its parameters and the
calculations involved are not always simple. It may be noted
from the density function of the Weibull distribution for the two
parameter case, viz.

$$f(x) = (\gamma/\theta)x^{\gamma-1} \exp(-x^{\gamma}/\theta)$$

that if $\gamma = 1$, then the density function becomes the well-known
one-parameter exponential distribution. This distribution has
probably been of the greatest value in applying the various
techniques that are available and, at the same time, maintaining
a certain simplicity but requiring careful examination of the
results.

These methods need techniques for the analysis of failure
modes and effects on equipment and systems. The field of safety,
both in the aerospace and nuclear industries, have contributed
greatly to such analysis and developed techniques for hazards and
safety analysis. In analysing such failure modes at the equipment
level, it has been found that simple designations of failure such
as safe or dangerous are not sufficient and various coding systems
have been derived. The effect of each component part fault is
analysed in sufficient detail to be given a fault category.
Typically, a four character code has been used where the first
character describes the main effect of the fault, the second
character indicates which equipment function is affected by the
fault, the third character shows whether or not a warning is
given when the fault has occurred and the fourth character
describes which equipment function or facility reveals the fault.
These are discussed in more detail in Ref. 33. From this
analysis of the equipment failure rates, not just the overall
failure rate is required but failure rates for the various
modes.(34,35) Although this type of failure mode analysis has
tended to grow up from safety work, it is also being utilised in
availability studies. The analysis is laborious and has been
the subject of computerisation.(36)

From a system point of view, it has become the general
practice to do a sensitivity analysis for reasons of limitation of
data and in order to minimise the amount of work that is required.
In many cases of system analysis, accurate data is not always
required and it has been useful to analyse masses of data from
different sources and to use this data in a generic fashion in
order to obtain rough estimates for system reliability. A
typical example of rough data taken from various sources and for
different environments is given in reference (37). This gives
failure rates for component parts, equipments and systems. This
type of rough approach can also form the basis of envelope or
boundary approaches which have become particularly useful,

especially in safety analysis. Quite often, the problem is to
demonstrate that some boundary value will not be exceeded and the
true or exact value is not always of particular interest.(38)
Another use of this data is when carrying out a sensitivity
analysis on a system in order to find which data are critical.
It is not always necessary to have accurate data for all items of
a system in order to complete a satisfactory analysis.

It should be noted that computer based techniques have become
necessary in many cases due to the increased complexity and size
of the systems as they exist today. Nevertheless, the systems
which have higher reliability often can make demands such that
long computer times can be involved if inappropriate techniques
are applied.(39)

In the past few years, attention has been given to particular
stochastic processes of the Markov type.(40) In the theory
associated with Markov chains a simple generalisation consists of
permitting the outcome of any trial to depend on the outcome of
the directly preceding trial and not on any other. This can be
a limitation when considerations of time dependency enter into the
problem. The application of Markov chains in reliability
assessment has been applied in detail to processes that are
discrete or continuous in time and space.(57,58,59,60) Where the
hazard function is not constant the Markov approach can lead to
difficulties of analytical solution. Monte Carlo techniques
have been used when the problem is intractable by analytical
techniques or, perhaps, computation of the probabilistic equations
in the reliability analysis of the system.(61)

A particularly difficult area for the reliability analyst is
during the commissioning and early operation of new complex
systems. Here it may be necessary to take into account initial
failures and errors which are corrected in order to improve the
system performance. This process modifies reliability
characteristics such that the analysis may have to deal with such
characteristics as failure rate decreasing with time. When the
failure rate reaches the equilibrium or constant value, then early
failure or debugging may be considered as completed.(41)

TESTING

Testing methods have been useful means by which direct
reliability assessments may be undertaken or have enabled estimates
to be made. Sequential sampling methods of the AGREE type have
been useful in this connection in the earlier stages of
production. Other types of testing, such as accelerated life
tests, have been well reported. Broadly, these sample tests fall
into two categories; first, those samples which already exist
and from which some inferences may be drawn; secondly, those

samples which are deliberately created in order to check some pre-
determined influence. It is of interest to note the type of test
undertaken in which a consensus of opinion is obtained as to the
failure rate of equipment by using large numbers of people to
participate.(37) This enables information to be derived where
techniques are used to employ uncertainty distributions.

In the rapidly developing technology, tests may be considered
to have been undertaken for the first time in history. For
example, when close-up pictures of the planet Mars were produced
by Mariner 6 and 7 and man landed on the moon by Apollo 11, and
similar first time events. It has been reported that Mariner 6
and 7 had disconcerting events during their flights and it will be
recalled that unexpected events took place with Apollo 11 and 12
and this was also seen with Skylab. In such aerospace
applications it is suggested that the fundamental items which
contributed to some of these problems are a lack of proper mission
simulation in ground tests and poorly integrated test planning and
administration. This information also points the way to the
direction in which the reliability analyst must undertake
reliability assessment and a greater understanding of human
influence factors is often needed. This applies to many industries.
The demands for the equipment in such projects requires to have
its reliability evaluated with high confidence levels and various
techniques have been investigated, for example, design margin
techniques for measuring performance degradation during storage.(42)
A further interesting example is the system development and testing
done in connection with the automatic landing system in the Boeing
jet transport prototype.(44)

Non-destructive testing has been found to play an important
role in various programmes by permitting critical areas to be
monitored and can apply to original testing or maintenance
programmes.(92) In the nuclear industry, as an example, non-
destructive tests have been carried out on fuel elements prior to
charging in a nuclear reactor. The data obtained have been used
as a sample of normal production fuel elements and correlations
made of post-irradiation measurements to interpret quantitatively
non-destructive test results in the light of reactor
performance.(43)

RELIABILITY DATA

In the past few years centralised reliability data banks
have evolved(45,46) such as FARADA,(52) RADC Reliability Analysis
Centre, French Reliability Centre, Swedish Data Bank(47) and the
Systems Reliability Service Data Bank (SYREL).(48) In order to
meet the needs of the reliability analyst and other interested
parties, various types of 'failure rate' data have been
accumulated arising from testing and field experience. In the

case of the SYREL Data Bank, the basic data flow is as
follows. Basically, there are two stores, one for event data and
the other for reliability data. The event data is collected from
the plant and coded in accordance with an agreed convention and
then stored in computer files. There is a requirement for a high
degree of flexibility in the presentation of the subsequent output
from this store which can be just a routine feedback, for the
guidance of plant management, or the derived generic reliability
data, which is one of the several inputs to the Reliability Data
Store.

The trend for this type of data bank has been to concentrate
on electronic data(90) and to extend into other fields such as
mechanical equipment. Currently, indications are that efforts
are being made to amass more data on human error rates and
associated probabilities. Obsolescent data can be a common
problem of data banks and it is of current concern as to how far
it is necessary to have accurate data and to what extent use can
be made of data which cannot be considered as very new.

The indications are that there never appears to be sufficient
data to enable the work of the reliability analyst to be undertaken
easily. It may be seen from the general literature(49) that there
is a gap which has developed between the theory and techniques of
developing elaborate mathematical models and practice, where one
does not have enough data to apply the most simple formula.
Various techniques have been investigated for goodness-of-fit tests
for data analysis. Computational techniques have been appropriately
developed(50) and the arguments of confidence and the use of such
techniques using chi-square distribution have been well documented.
The derived information can be of importance in the synthesis
process and is considered in Ref. 51.

The conflict between the Bayesians and the Classicists could
be mentioned here.(53) For instance, the important question of
"Do we make use of our engineering judgement together with limited
test experience as reasonable estimates of current reliability
status and potential problem areas" is considered in Ref. 54. It
emerges that for a truly Bayesian reliability measurement
programme then, obviously, it requires the formulation of a
statistical model with the quantification of priors and, of course,
a belief in the results.

FEEDBACK OF DATA

Whilst the designer may feed forward his ideas and requirements for manufacturers to make hardware, etc., and for the system
to be commissioned and made operational, it is essential for feed
back of data to be given. This requires appropriate data
collection to be initiated from the design stage. The recording

of operational performance is an essential part of reliability
analysis and assessment. As an example, the information given
on the first day space malfunctions(55) indicates over 50%
total malfunction of an electrical type. This information is
valuable and allows the examination of whether the launch
environments are a possible cause, for example, vibration/acoustic
environment. In other industries such as the nuclear and chemical
industries and aerospace generally, more attention is being paid
to the need for this type of operational data to have the
appropriate data logging. This is seen in such items as the
Pareto distribution curves and information for future designs in
Ref. 56.

This feedback of data allows the designer and other
interested parties to investigate the correlation between
prediction and practice. It is important in any technology to
have an in-built 'learning technique' of this type.

In the past, a number of systems, or elements of systems,
have been analysed at the design stage or production stage and
the data on the reliability performance subsequently collected
during a reasonable sample period of operation. One can visualize
the results of taking the ratio r, of the value of the particularly
observed failure rate, to the corresponding predicted failure
rate for each item system element studied. This particular
example was for about 50 different system elements and shows the
trends usually indicated. That is that r is found to be randomly
variable and the variation is approximately lognormal. From this
particular plot, which is on log-normal probability paper, it can
be seen that the median of r is 0.76 and that the chance of the
ratio r being within a factor of 2 of the median value is 70%
and that the chance of the ratio being within a factor of 4 is
96%. It may be deduced that if the techniques are applied with
due care and attention, then the numerical values yielded can
reasonably be expected to be within a factor of 2 (either way) of
the actual figures.

APPLICATION OF RELIABILITY ASSESSMENT

The reliability assessment techniques which have been
discussed have had diverse applications. They have been used
intensively in nuclear applications, particularly for systems which
safeguard high risk situations. This has included risk
analysis(62,63) and the associated systems.(64,65) These
techniques have been applied to similar systems in other industries
such as electrical power systems.(66,67,68,69) It emerges that
the assessment of conventional protective systems for plant is in
the range of 10^{-2} to 10^{-4} probability of failure on demand. In
the case of fairly sophisticated automatic sprinkler systems for
fire, the probability of failure on demand is about 10^{-2}, mainly

because of human influence, but automatic fire detection systems with self monitoring techniques and redundancy show some improvement. In these systems the mathematical modelling techniques appear to be adequate but the reliability data available is just barely sufficient. The techniques have been applied to chemical plant systems(70) and the benefits of reliability assessment have been used typically for improvements not only in safety but also in maintenance downtime of plant.(71)

In the application of assessment techniques to computers, the hardware problems have been dealt with but the software problems still present problems in undertaking quantified reliability techniques. However, useful contributions have been made for online computers in power stations.(72) Developments of reliability analysis are required for assessing the software of such computers and research is being undertaken.

Limited application of these assessment techniques has been made in the machine tool industry, for example, on conventional high pressure diecasting machines.(73) In this type of application the basic data is not readily available and it is necessary to utilise generic data from data bank sources and apply environmental and stressing factors and, in parallel, to set up the appropriate data collection. Useful availability and design studies have been undertaken and information can be collected fairly quickly due to high failure rates. The importance of the breakdown maintenance and the human factor, such as the operator, emerges from such studies. Medical equipment is another area of study and here the safety aspects are most important because the direct consequence of failure may be the loss of life.

Generally, these systems permit existing mathematical modelling techniques and sufficient data to be brought together in order to allow the techniques of reliability assessment to be applied in a satisfactory manner. The reliability analyst must carefully study the extent of redundancy and diversity of the system and must ensure that dependency which exists must be modelled to take into account such aspects as common mode failures. Studies of these have shown that for many systems, for example, control and instrumentation systems, even where a system may be redundant a probability of failure on demand of 10^{-3} may be limited by common mode failure and the assessment must carefully take into account diverse means of overcoming such dependency or, if lower probabilities of failure are to be demonstrated, the reliability analyst may find himself dealing with situations involving tails of distributions where there is little or no data. Ref. (74) gives a guide to such possible ranges.

The main difficulties in assessing the foregoing types of

systems, as they become larger and more complex, are in the organisation of the engineering data with all its practical details into a form which may be handled by a computer and amassing the appropriate reliability data. Research is being undertaken in the areas of this logical handling of engineering information and a greater emphasis is needed on the use of centralised data banks which already have the capability of handling the appropriate data if it is fed back from the field or other sources.

The techniques of assessment have already been applied to devices of a mechanical nature but as they have increased in size and become more dependent upon structural evaluation, as in the case of pressure vessels and aerospace hardware, the literature shows the need for the further development of the appropriate techniques of reliability assessment. As the reliability assessment techniques are multi-disciplinerary, so this equally involves bringing together the appropriate disciplines as in the case of estimating by the distribution of time-to-damage method which has been applied, for example, to the reliability estimation of the cracking of pipes.(75) This brings together earlier techniques such as safety margins based on stresses and strengths(76) requiring the understanding of material properties, a knowledge of a damage prediction equation together with uncertainties which may even involve subjective judgements so that a probabilistic statement may be made on pipe failure. In the case of pressure vessels, data has been collected and analysed.(77, 78) Many of the methods have been found to be difficult due to the lack of data but the application of extreme values theory which has been developed and based on the statistics of extremes as developed by Gumbel(79), particularly in combination with Freudenthal, has been finding favour. This has given us a useful approach to structural reliability analysis based on such concepts as the time to the first failure.(80,81) A recent example of this type of reliability assessment considers the probabilistic approaches to design and considers the probability and consequences of a river flooding a nuclear power plant.(82) This type of reliability assessment is dealing with rare events which extends the reliability analyst to the ultimate.

Reliability assessment requires, to some extent, human influence factors to be modelled into a system analysis. This influence may be due to a designer, an operator or a maintenance man. Although there is extensive literature on human beings, the modelling of many human tasks is not well known.(84) The application of reliability assessment has been applied in some instances but the appropriate techniques of evaluation are still subject to much development. A combination of techniques has been used and tests have been undertaken to obtain the appropriate data on operators in the laboratory and in actual environments.(83,26)

Methods have been suggested for comprehensive approaches involving estimates of time and reliability of human beings in various roles.(85,86,87,88,89)

Where there is a need to demonstrate high reliability by means of assessment, quite often this situation is also allied to high risk situations in the event of failure. It has been the developing practice to ensure in such situations that the assessment is independent, that is, the design or operating group is separate from that of the assessing group. This is an attempt to minimise the risk of common errors in demonstrating the adequacy of the system.(38,16)

CONCLUSIONS

The general review undertaken shows that over the past two decades there has been a trend to extend from deterministic to probabilistic reliability assessment. The general modelling techniques are, in many applications, adequate for their purpose, provided the necessary co-ordination of the feedback of data via centralised data banks is undertaken. The development of modern technologies has also brought about larger and more complex systems which require a synthesis of the reliability, particularly where the data is not available at the complete system level. Careful consideration must be paid to common mode failures of systems and human influence factors must be taken into account and modelled accordingly. The independent reliability assessment has become complimentary to the design and operating process where high reliability is required to be predicted.

1. Pope, J. A. (Editor), "Metal Fatigue", Chapman and Hall, London, 1959.

2. Keynes, J. M., "A Treatise on Reliability", MacMillan and Co. Ltd., 1921.

3. Pieruschka, E., "Principals of Reliability", Prentice-Hall, 1963.

4. Bazovsky, I., "Reliability Theory and Practice", Prentice-Hall, 1961.

5. Green, A. E., and Bourne, A. J., "Reliability Considerations for Automatic Protective Systems", Nuclear Engineering, August, 1965.

6. Stout, Wesley, Winals, "Tanks are Mighty Fine Things", Chrysler Corp., 1946.

7. Kivenson, Gilbert, "Durability and Reliability in Engineering Design", Topics in Engineering Design, Pitman Publishing.

8. Tall, M. M., "Product Assurance Training", 10th National Symposium on Reliability and Quality Control", January, 1964, Page 260.

9. Tye, W., "Degree of Reliability Required - Especially in Relation to Future Aircraft", The Royal Aeronautical Society and the Inst. of Elect. Engineers Joint Conference on the Importance of Electricity in the Control of Aircraft, 1962.

10. Warren, D. V., "Safety Assessment of Systems for Landing Airplanes in Bad Visibility", Proceedings of the CREST Specialist Meeting on Applicability of Quantitative Analysis of Complex Systems and Nuclear Plants in its Relation to Safety, Munich, 26 - 28 May, 1971, Issue November 1971.

11. Adkins, L. A. Jnr., and Thatro, M. C., "Reliability Requirements for Safe All Weather Landing", 1968 Annals of Assurance Sciences, 7th Reliability and Maintainability Conference, San Francisco, Calif. July 14 - 17, 1968.

12. Farmer, F. R., "Reactor Safety - The Carrot or the Stick" National Academy of Engineering, Symposium on Public Safety - A Growing Factor in Modern Design", Washington, 1st May, 1969.

13. Farmer, F. R., "Experience in the Reduction of Risk", Proceedings of the Symposium on Major Loss Prevention in the Process Industries, Inst. of Chemical Engineers, 1971.

14. Beattie, J. R., Bell, G. D., and Edwards, J. E., "Methods for the Evaluation of Risk", UKAEA Report AHSB(S)R.159.

15. Siddall, Ernest, "Statistical Analysis of Reactor Safety Standards", Nucleonics 17(2): 64-69, February 1959.

16. Stewart, R. M., and Hensley, G., "High Integrity Protective Systems on Hazardous Chemical Plants", Proceedings of the CREST Specialist Meeting on Applicability of Quantitative Reliability Analysis of Complex Systems and Nuclear Plants in its Relation to Safety, Munich, 26-28 May, 1971.

17. Losee, John, Preface "Assurance Technology Relates to Today's World", Annals of Reliability and Maintainability, 1971, Vol. 10.

18. Ramey, James T., "Quality Assurance: Translating Doctrine into Action", 35th Annual American Power Conference, Chicago, May 8th, 1973.

19. Chestnut, H., "System Engineering Tools", John Wiley, New York, 1965.

20. Deutch, R., "System Analysis Techniques", Prentice-Hall, Englewood Cliffs, New Jersey, 1969.

21. Levy, Lionel, Preface "Assurance Technology Spinoffs", Annals of Reliability and Maintainability, 1970, Vol. 9.

22. "Decomposition as a Tool for Solving Large Scale Problems" NATO Advanced Study Institute, July 17-26, 1972, Pembroke College, Cambridge, England, Proceedings published by North Holland Publishing Co.

23. Balasubrannanian, N. V., Lynn, J. W., and Sen Gupta, D. P., "Differential Forms on Electromagnetic Networks", Butterworths, 1970.

24. Kron, Gabriel, "Diakoptics - The Piecewise Solution of Large-Scale Systems", Macdonald and Co. Ltd., London, 1963.

25. Lynn, J. W., "Tensors in Electrical Engineering", Edward Arnold Ltd., 1963.

26. Green, A. E., "Safety Assessment of Automatic and Manual Protective Systems for Reactors", Instrument Practice, 24, 109.

27. Green, A. E. and Bourne, A. J., "Reliability Technology", John Wiley, 1972.

28. Crosetti, Paul A., and Bruce, Richard A., "Commercial Application of Fault Tree Analysis", Annals of Reliability and Maintainability, 1970, Vo, 9.

29. Kanda, Kazuo, "Computerisation of System/Product Fault Trees", 1969 Annals of Assurance Sciences, p.325.

30. McKnight, et al, "Automatic Reliability Mathematical Model", North American Aviation Inc., Report No. NA66-838.

31. Cohen, A. Clifford, "Maximum Likelihood Estimation in the Weibull Distribution Based on Complete and on Censored Samples", Technometrics, November 1965, Vol. 7, No. 4.

32. Woodcock, E. R., "The Calculation of the Reliability of Systems - The Program NOTED", UKAEA Report AHSB(S)R.153.

33. Green, A. E. and Bourne, A. J., "Safety Assessment with Reference to Automatic Protective Systems for Nuclear Reactors", UKAEA Report AHSB(S)R.117.

34. Eames, A. R., "Reliability Assessment of Protective Systems" Nuclear Engineering, March 1966.

35. "Reliability Stress and Failure Rate Data for Electronic Equipment", Washington Department of Defence, 1965, Report No. MIL-HDBK-217A.

36. REDAP 25 - Computer Program Prepared by Racal Research Ltd., Tewkesbury, England.

37. Green, A. E., "Reliability Prediction", Conference on Safety and Failure of Components, University of Sussex, 3-5 September, 1969.

38. Green, A. E., "Safety Assessment of Systems", IAEA Symposium on Principles and Standards of Reactor Safety, Julich, Germany, 5-9 February, 1973.

39. Bourne, A. J., "Considerations in the Use of System Reliability Models", UKAEA Report No. SRS/GR/4.

40. Feller, W., "An Introduction to Probability Theory and its Applications, John Wiley, New York, 1957.

41. Barlow, et al, "System Debugging Model", Operations Research and Reliability, edited by Daniel Grouchko, published by Gordon and Breach.

42. Hecht, H., "Interpretation of Test Data by Design Margin Techniques", Annals of Reliability and Maintainability, 1970 Vol. 9.

43. Jaech, J. L., "A Program to Estimate Measurement Error in Nondestructive Evaluation of Reactor Fuel Element Quality", Technometrics, Vol. 6, No. 3, August 1964.

44. Clifford, D. R., "All-Weather Automatic Landing System Development and Testing", Annals of Reliability and Maintainability, 1970, Vol. 9.

45. Tiger, Bernard, "Reliability Techniques and their Role in Program Optimisation", 1968, Annals of Assurance Sciences, 7th Reliability and Maintainability Conference, San Francisco, Calif., July 14-17, 1968.

46. Ireson, W. G., "Reliability Handbook", McGraw Hill, 1966.

47. Birger, Olsson, "Swedish Failure Rate Data Bank", Electronic Components, pp. 555-556.

48. Ablitt, J. F., "An Introduction to the 'SYREL' Reliability Data Bank", UKAEA Report SRS/GR/14.

49. "Preface - The Theory and Practice of Reliability", Operations Research and Reliability, edited by Daniel Grouchko, published by Gordon and Breach, Science Publishers, Paris.

50. Shooman, Martin L., Johnsen, James and Straub, Robert, "Generation of Failure Data Fitting a Specified Hazard Function", Annals of Reliability and Maintainability, 1970, Vol. 9.

51. Myhre, J. M. and Saunders, Sam C., "Comparison of Two Methods of Obtaining Approximate Confidence Intervals for System Reliability", Technometrics, Vol. 10, No. 1, Feb. 1968.

52. "Failure Rate Data (FARADA) Program", Naval Fleet Missile Systems Analysis and Evaluation Group, Corona, Calif., USA, June 1967.

53. McKiggan, I. F., "Bayesian Logic in Quality Assurance", The Quality Engineer, May 1973, Vol. 37, No. 5.

54. Olsson, John E., "Implementation of a Bayesian Reliability Measurement Program", Annals of Assurance Sciences 1968, 7th Reliability and Maintainability Conference.

55. Losee, John, "Assurance Technology Relates to Today's World, pp. 93, etc., Annals of Reliability and Maintainability, 1971 Vol. 10.

56. Marsh, W., and Ferguson, H. R. M. S., "The Development of a System for Collecting and Analysing Service Failures", Nuclear Engineering and Design, Vol. 13 (1970), No. 2, North Holland Publishing Co.

57. Sandler, G. H., "System Reliability Engineering", Prentice-Hall, Space Technology Series 1963.

58. Roberts, Norman H., "Mathematical Methods in Reliability Engineering", McGraw-Hill Book Co.

59. Lee, Christopher T. H., "Mean Times of Interest in Markovian Systems", IEEE Trans. on Reliability, Vol. R-20, No. 1, Feb. 1971.

60. Buzacott, John A., "Markov Approach to Finding Failure Times of Repairable Systems", IEEE Trans. on Reliability, Vol. R-19, No. 4, November 1970.

61. Garrick, B. J., "Unified Systems Safety Analysis", Nuclear Engineering and Design, Vol. 13 (1970) No. 2, North Holland Publishing Co.

62. Beattie, J. R. and Bell, G. D., "A Possible Standard of Risk for Large Accidental Releases", Symposium on Principles and Standards of Reactor Safety, IAEA, Julich, Germany, 5-9 February, 1973.

63. Sinclair, C., Marstrand, P., and Newick, P., "Innovation and Human Risk", published by the Science Policy Research Unit, University of Sussex, 1972.

64. Green, A. E., "Assessment of Sensing Channels for High Integrity Protective Systems", Instrument Practice, Vol. 22, No. 10, October 1968, pp 862-866, 869.

65. Bourne, A. J., "A Criterion for the Reliability Assessment of Protective Systems", Control, October 1967.

66. Billinton, Roy, "Power System Reliability Evaluation", Gordon and Breach Science Publishers.

67. De Sieno, C. F., and Stine, L. L., "A Probability Method for Determining the Reliability of Electric Power Systems", IEEE Trans. on Reliability, March 1965, Vol. R-14, No. 1.

68. Bourne, A. J., Snaith, E. R. and Fletcher, D., "Reliability Analysis of Electrical Supply Systems for Nuclear Power Stations", 1968, Paper presented at the CREST Meeting of Specialists on the Reliability of Electrical Supply Systems and Related Electro-Mechanical Components for Nuclear Reactor Safety - Ispra, Italy, 27-28 June.

69. Sayers, B., "An Approach to the Design of Fire Detection Systems for Quantitative Reliability", Fire Research Station Symposium, London, March 1972.

70. Hensley, G., "Plant and Process Reliability", Instrument Practice, Nov. 1971.

71. Cason, Roger L., "Estimate the Downtime your Improvements will Save", Hydrocarbon Processing, Jan. 1972.

72. Jervis, M. W., "On-line Computers in Power Stations", Proc. IEE, IEE Reviews, Vol. 119, No. 8R, August 1972.

73. Bourne, A. J., "General Results of an Investigation into the Reliability of High Pressure Die-casting Machines" presented as a paper to the Diecasting Society UK in 1971/1972 and also published in the Die-Casting Engineer, USA, September/October 1972, Vol. 6, No. 15, pages 32-40.

74. Eames, A. R., "Principles of Reliability for Nuclear Reactor Control and Instrumentation Systems", UKAEA Report SRD R1, 1971.

75. Wilson, S. A., "Pipe Reliability Estimating by the Distribution of Time-to-Damage Method", Trans. of the American Nuclear Society, 15th Annual Meeting, June 15-19, 1969, Seattle, Vol. 12, No. 1.

76. Lusser, Robert, "Reliability Through Safety Margins", Research and Development Division Army Rocket and Guided Missile Agency, Redstone Arsenal, Alabama, October 1958.

77. Phillips, C. A. G., and Warwick, R. G., "A Survey of Defects in Pressure Vessels Built to High Standards of Construction and its Relevance to Nuclear Primary Circuit Envelopes", UKAEA Report AHSB(S)R.162, 1969.

78. Kellermann, O. and Seipel, H. G., "Analysis of the Improvement on Safety Obtained by a Containment and by Other Safety Devices for Water-cooled Reactors", IAEA Symposium on the Containment and Siting of Nuclear Power Reactors, Vienna, April 1967.

79. Gumbel, E. J., "Statistics of Extremes", Columbia University Press, New York, 1958.

80. Freudenthal, Alfred M., "Reliability Analysis Based on Time to the First Failure", Reprinted from Aircraft Fatigue, Pergamon Press, 1972.

81. Freudenthal, Alfred M., (Editor) "International Conference on Structural Safety and Reliability", Smithsonian Institute Washington DC, April 1969, Pergamon Press, 1972.

82. Wall, Ian B., "Probabilistic Assessment of Flooding Hazard for Nuclear Power Plants", Presented at American Nuclear Society 19th Annual Meeting, Chicago, Illinois, June 14, 1973.

83. Ablitt, J. R., "A Quantitative Approach to the Evaluation of the Safety Function of Operators on Nuclear Reactors", UKAEA Report AHSB(S)R.160.

84. Roush, S. Larry, "Human Factors, A Constant-Change State", Annals of Reliability and Maintainability, 1967, Vol. 6, All Systems Go?, 6th Reliability and Maintainability Conf. Florida, July 1967.

85. Swain, A. D., "A Method for Performing a Human-Factors Reliability Analysis", Sandia Corporation, SCR-685, August 1963.

86. Payne, Altman and Smith, "An Index of Electronic Equipment Operability", Report No. AIR-C-43-1/62-RP (1-5).

87. Adkins, L. A., et al., "Development of All Weather Landing System Reliability Analysis and Criteria for Category III Airborne Systems Phase III", Report FAA-RD-67-20 and Report AD-655-240, Lockheed Georgia Co., Marietta, Georgia.

88. Askren, William B. and Regulinski, T. L., "Quantifying Human Performance Reliability", AFHRL-TR-71-22, Psychology in the Air Force Symposium, Air Force Academy, Colorado, 20-22 April, 1971.

89. Swain, A. D., "Overview and Status of Human Factors Reliability Analysis", Proc. of the 8th Reliability and Maintainability Conf., American Inst. of Aeronautics and Astronautics, New York, July 1969, pp 251-254.

90. Charles, T. G., "Elektronik Componenter, Tillforlitlighet", IRU AB, Lidingo, Sweden, 1972.

91. Dubey, Satya D., "Asymptotic Properties of Several Estimators of Weibull Parameters", Technometrics, Vol. 7, No. 3, August 1965.

92. Weldon, Warren J., "Nondestructive Testing - A Condition Monitored Maintenance Tool", Annals of Reliability and Maintainability, 1970, Vol. 9.

SECURITY ASSESSMENT OF POWER SYSTEMS

H. H. Happ

General Electric Company
Schenectady, New York, U. S. A.

ABSTRACT

This paper describes security aspects of power systems from an on-line operating viewpoint. The stages in system security assessment, from monitoring and displaying to automatically controlling the system for security, will be presented. The four states of a power system, the Normal State, the Alert State, the Restorative State, and the Emergency State, will be described, as well as examples of operating modes that correspond to these states. A hierarchy of action centers will be presented by means of which the security of the systems are controlled and piecewise methods can be executed for determining the security state of power systems.

INTRODUCTION

Power System Security is an integral part of what is commonly called System Automation[1]. Two closely allied functions to System Security are "Automatic Generation Control" (AGC), and the second "Supervisory Control and Data Acquisition" (SCADA). AGC is often called load frequency control; its purpose is to control the real generation of the system so as to maintain frequency, and total net interchange power (summation of tie powers) at the desired levels; AGC also encompasses economic dispatch, generation reserve calculations and other necessary functions associated with interfacing with the system dispatcher.

Supervisory Control and Data Acquisition (SCADA) refers to the collecting of information throughout the system, displaying that information at the control center, and the ability to operate devices at remote locations from the control center. The SCADA function, although not considered a System Security function, is clearly the most important ingredient, or step 1 toward Power System Security. Figure 1 below portrays data telemetering and display, the SCADA level of automation. A form of AGC is assumed to be present, maintaining frequency and net interchange at the desired levels. At this level in the automation technology, the operator does not have the capability of predicting how his system will perform for any other load or system condition except the present one. The predicting function, which represents the next level or step in technology, is shown in Figure 2, where a predictor has been added, as well as a contingency evaluator. In a power system, the system predictor takes the form of a computer program which models the system; the modelling in the steady state is called a load flow program if done digitally. The operator by the use of the load flow has the capability of studying future load and system conditions, either an hour ahead, a day ahead, etc. Contingency evaluation tests whether or not a generator outage and or a line outage, or outages, can be tolerated. The function, as shown in Figure 2, are performed at certain intervals, usually every few minutes, on the real system as well as on a future system as represented by a load flow.

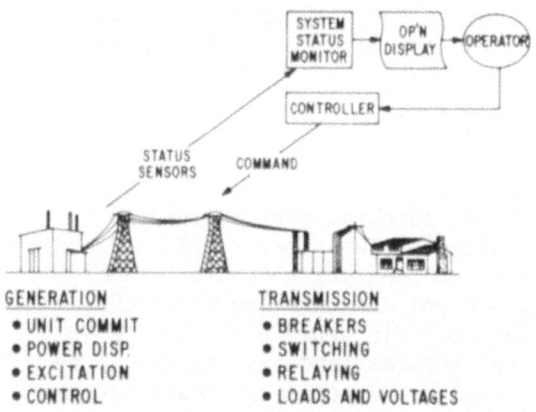

Figure 1 - First step in system security - data telemetering and display

ASPECTS OF SYSTEMS SECURITY

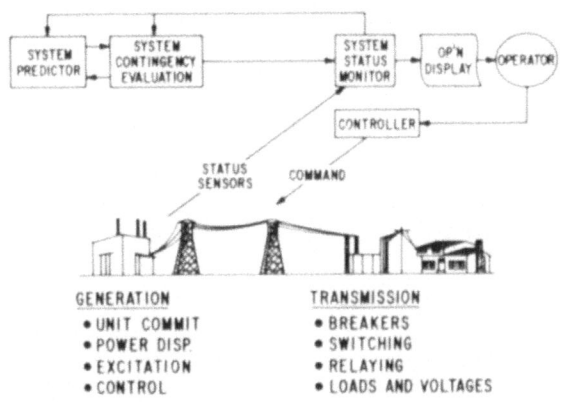

Figure 2 - Second step in system security - system predictor

Figure 3 shows one additional block entitled "System Corrective Strategy." This is an all encompassing word and denote functions, algorithms, etc. for aiding the dispatcher in overcoming a present or an expected problem in the present system; or an expected problem in a future system.

In Figure 4 a single line has been added between the system status monitor function and the controller. This line denotes the circumventing of the operator function and thus automatic control. We have as yet not arrived at the stage of automatic control depicted by Figure 4.

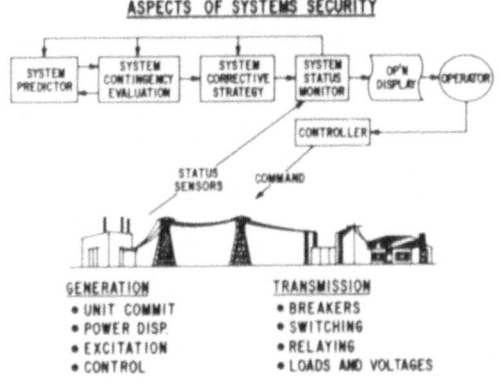

Figure 3 - Third step in system security - system corrective strategy

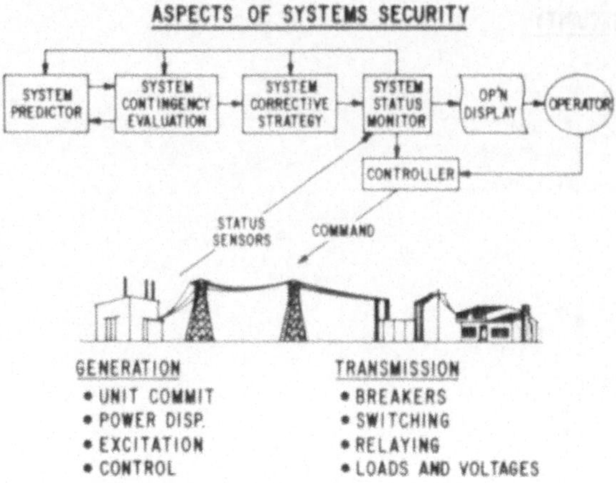

ASPECTS OF SYSTEMS SECURITY

Figure 4 - Fourth step in system security - automatic control

SECURITY STATES OF A SYSTEM

With the above security functions outlined, we will next describe the various states by which security can be categorized[2,3], as shown in Figure 5.

The Normal State is the state where all loads are met, and at the proper frequency, voltage and security. No impending emergencies are detected by the contingency evaluator when operating in the Normal State by definition. The objective in this state is to operate the system so as to remain in this

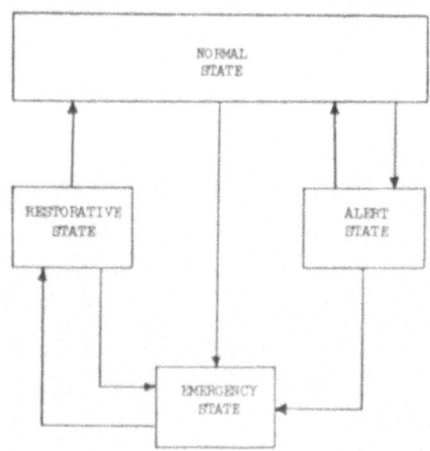

Figure 5 - Four power system security states

state, in terms of quality and security of operations; and to allocate generation such as to minimize total costs or total fuel expended within security and other constraints present, such as total emission.

The Alert State is one where an impending emergency has been detected, either by the contingency evaluation function or by some other manner, but in which all loads are met and no apparatus is overloaded. The operating objective would of course be one to return to the Normal State as rapidly as possible by a reallocation of real and reactive generation, of interchange, and other possible steps. The arrows in Figure 5 correspondingly point from the Alert State to the Normal State as well as from the Normal State to the Alert State.

The Emergency State is one where a deterioration in the service is taking place; where some customer loads are not met, and or where the quality of the service in terms of frequency and voltage is deteriorating. The operating objective here is to stop the emergency from spreading, and to obtain a generation load balance by tripping necessary loads or ties. The arrows in Figure 5 show that an Emergency State can result from any of the other states.

The Restorative State is the state of the system following an emergency, but where the emergency has been stabilized and previous to the establishment of the Normal State. This state denotes that the deterioration of the system condition has been halted, although the operating conditions are far from normal, in that not all customer demands are met and the quality of service may not be normal. The operating objective is to restore service as rapidly as possible, and to return to the Normal State. The arrows from the Restorative State indicate that the system can possibly slip back into an Emergency State.

The four steps outlined previously toward improved system operations can be combined with the four states in a flowchart manner as shown in Figure 6. As indicated in the figure, no control action is required for security in the Normal State. In the Alert State, as diagnosed directly from the system or through contingency evaluation, a corrective strategy function has to be executed; it is assumed, as shown in Figure 6, that it is carried out through the operator without automatic control. If an Emergency State is detected, automatic control is initiated through corrective strategy logic to alleviate the emergency. Improvement in the data in the first block in Figure 6 stands for state estimation, in case sufficient data is telemetered, limit checking, smoothing, etc. These functions are implied with all telemetered inputs in this text, although not explicitly shown.

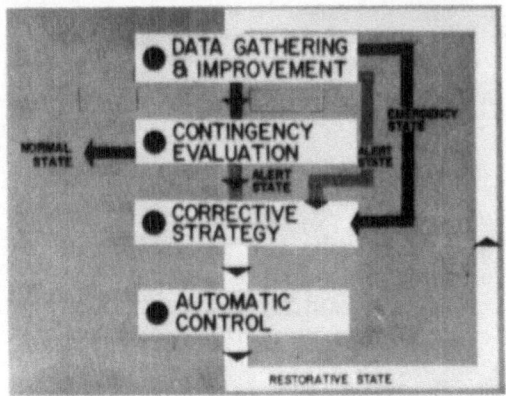

Figure 6 - Flowchart combining steps in system security with security states

PRESENT SYSTEM SECURITY FUNCTIONS

Although the above logic appears feasible, the industry is not yet at the point where logic blocks as complete as those described above exist. Functions more commonly in existence are shown in Figure 7. "AGC" in Figure 7 stands for "Automatic Generation Control" including economic dispatch; as indicated previously, its function is to control the allocation of generation so as to maintain frequency and net interchange, and to minimize the total fuel costs. It is executed about every two seconds. The Contingency Evaluation function is executed every few minutes, as has been previously described. The reserve determination function checks to see that there is sufficient reserve in the system; it can be either deterministic or stochastic. In the latter case the probability of loss of generation or of a line outage is taken into consideration as

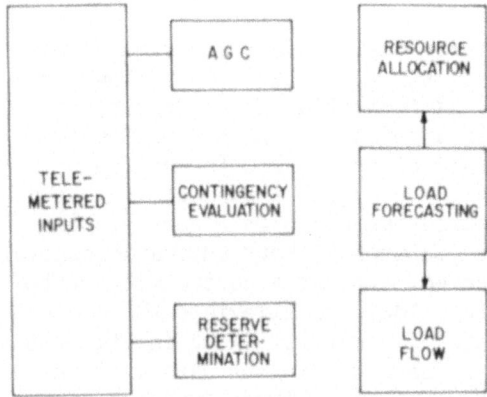

Figure 7 - Present system security functions

well as the probability of a load change beyond that anticipated.
The above three functions are operated on-line in that the system
itself is used in the computational process. The three functions to the right in Figure 7 are operated off-line in that
system conditions are studied which are usually different from
those in the present real system. Resource allocation, the
first of the three, is either a unit commitment program[4] or it
is performed by a load dispatcher manually; the latter function
includes nuclear, and is therefore extending in time from a day
up to two years; maintenance scheduling is included in the
resource allocation function. The load flow, the work horse
network flow model, is listed at the bottom in Figure 7 with
the load forecasting function in the middle. All three functions interact, in that they depend on each other.

One area in system security that so far has not been
specifically considered on line, except for system operators, is
the problem of voltage. Voltage is of course synonimous with
the var resources available or planned within the system[5].

An example of an Emergency State condition where cascading
loss of generator voltage control occurs, is as follows:
a limiting condition on the transmission capability between a
company and its neighbors, immediately following a loss of
generation or similar type disturbance, is often the voltage
and reactive power supply present in the affected area before
local generation is increased, and the flow on the tie lines
is reduced[6]. The combination of the increased power flow over
the tie lines and the reduction in the var supply due to the
lost generation often means increased var flow over the ties
with a resulting lower voltage in the receiving area. The
local generators will attempt to supply reactive power by
operating near ceiling excitation; after half a minute or so
the automatic voltage regulators on the local generators must
return the field excitation to normal ratings; this will result
in yet lower voltage, increased excitation on other machines,
a spreading of the low voltage condition and a cascading loss
of generator voltage control. The capability of real power
import capacity is also lowered as voltage reductions occur
in the load area; or to put it in another way, as KVAR import
increases to the load area, the capacity for real power import
is reduced.

The solution to the Emergency State problem is to plan for
sufficient emergency reactive to support system voltage during
that state, so that the proper amount of emergency real power
can be provided by the neighbors. One method of providing
emergency reactive is to correct local load by KVAR to obtain
near unity power factor. The KVAR available from the generators

then represent emergency KVAR. Voltage control equipment available includes machine excitation, synchronous condensers, transformer tap changing, shunt capacitors, reactors, series capacitors and other devices.

The voltage scheduling procedures presently used largely consist of developing voltage schedules by means of load flows for future load and system conditions including contingency conditions.

Telemetering and on-line computing capabilities are increasing rapidly; therefore, on-line control of both real as well as reactive resources is expected in the near future. Logic blocks are presented below containing an optimal load flow, complex contingency evaluation and corrective strategy for dispatching both real and reactive power during the various security states outlined above, while satisfying both line flow and voltage constraints during contingency conditions. A computer hierarchy for their implementation is also presented.

NEAR TERM FUTURE SYSTEM SECURITY FUNCTIONS

An optimal load flow, as will be illustrated later in this paper, is a procedure by which all watt and var resources committed are scheduled such that the constraints of both hardware and system constraints are satisfied. It is a key element in the economy security logic shown in Figure 8 below. Additional logic blocks employed in Figure 8 are security checks under both operating and contingency conditions, and corrective strategy blocks which will not be described in detail. A description of the logic in Figure 8 follows: system variables are continuously checked from the system, as indicated by the direct line from the system input around the "Optimal Load

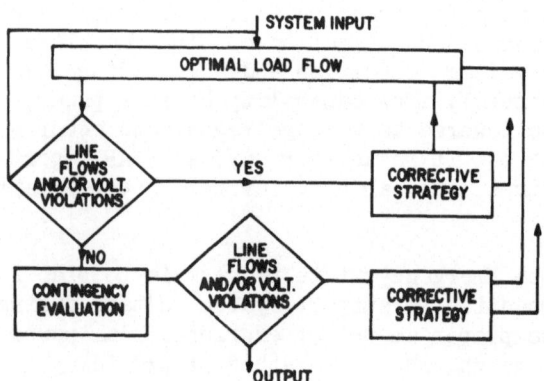

Figure 8 - Economy-security logic

Flow" function. If a line flow, voltage violation, etc. is detected (possible emergency status), "Corrective Strategy" and subsequent control is initiated with or without the aid of the optimal load flow.

The contingency evaluation function is executed at frequent intervals (every few minutes) from the actual system readings; if no violations are detected, the system is in the Normal State; if a violation is detected, the system is in the Alert State and corrective strategy, with or without the aid of the optimal load flow and control to execute the corrective strategy, is initiated. Every few minutes the optimal load flow is executed for either obtaining a proper watt and/or var allocation or for security purposes, and the very same logic loops in Figure 8 outlined for the real system are traced out, except with the optimal load flow remaining within the loop. The function of an optimal load flow and of contingency evaluation will briefly be discussed below, in conjunction with Figures 9-11 inclusive.

OPTIMAL LOAD FLOW

In an ordinary load flow, the power output of all generators (except one) are specified as well as all generator voltages. The load flow program determines the generator vars, all bus voltages, angles, and all line flows.

An optimal load flow, just like an ordinary load flow, solves the network equations with the exception being that the generator powers are not specified, and instead min-max ranges for each generator are given. The optimal load flow, from the cost curves of the units, determines the economic allocation of generation and thus the power output of the units; note that it performs an economic dispatch along with solving the total network, and it can therefore function in the role of economic dispatch (system dispatch).

The data requirements, at a minimum, are the same as those of the ordinary economic dispatch and consist of the generator and tie powers. It can additionally accommodate the load real and reactive powers which can be formed from the line telemeterings, bus voltages, and the status of all lines.

Comparing the classic B matrix technique approach with the rigorous, we can state that the classic technique yields a dispatch which theoretically is not quite as economic as that produced by the optimal load flow. Although recent studies show them to be very close on the systems studied[7]. An additional major benefit realized by the optimal load flow, is that an ordinary load flow is realized in addition to the dispatch which closely resembles the actual system conditions; a security benefit is thus obtained.

The optimal load flow described so far optimizes the allocation of real generation only. The desired voltage at each generator terminal is specified which yields a unique var output of each unit. If instead of the desired generator terminal voltage, a satisfactory range in terminal voltage of each unit is also given, as shown in Figure 9, the possibility exists to optimize both watts and vars[8].

Instead of minimum cost being the aim of the allocation of power among the units, the system goal can also be one of emission. Minimum Emission Dispatch applied to a total power system minimizes the total stack emissions for the entire power system.

The optimum load flow with additional logic blocks can be used to avoid operating conditions where either line flows or voltage constraints are violated during normal operating conditions or during contingency conditions. Corrective strategy logic can be incorporated to avoid line limits by, for example, shifting generation away from the economic optimum; the optimum load flow achieves the shift by establishing maximum and minimum limits for all the units on-line which are imposed by the control dispatch program. Notice that the imposition of line limits and corrective strategy will move the system away from its optimum operating point, and security is thus traded for economy.

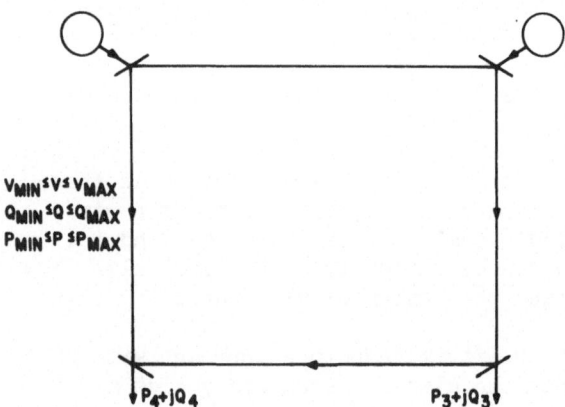

Figure 9 - Optimal load flow - with generator limits shown

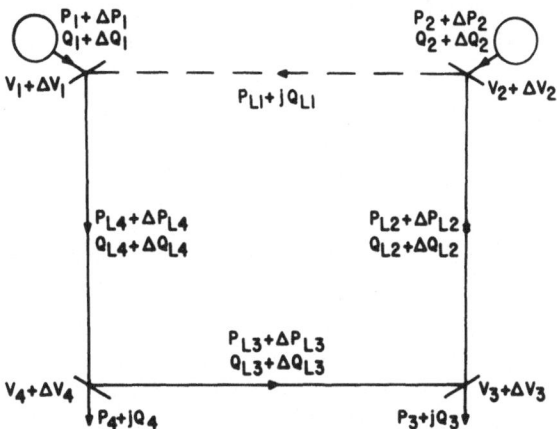

Figure 10 - Contingency evaluation - illustration of line loss

Other corrective strategies can be used also, as for example a change in interchange, and changes in the transmission configuration through switching, etc.

CONTINGENCY EVALUATION

The optimal load flow, when used as an economic dispatch tool represents a useful security tool. But security has to be measured. Contingency Evaluation provides such a measurement. Up to the present time, contingency evaluation has been restricted to real power flow only. In the latter case the change in real power flow in the transmission network is determined for the event a line is lost or when a generator is lost as illustrated in Figures 10 and 11. In the complex case, we determine not only the expected real power change in the lines of the network,

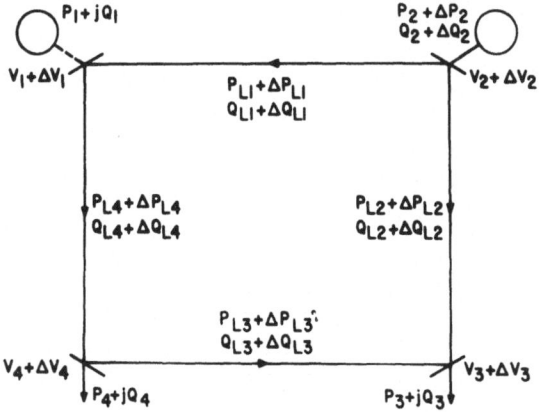

Figure 11 - Contingency evaluation - illustration of generator loss

but the changes in var flows as well. The complex case is more
exact since line limits are given in terms of MVA; futhermore,
the expected change in bus voltages can also be calculated in
the complex case, and compared to the change allowed, as
illustrated in Figures 10 and 11. As in most matters of this
kind, a price has to be paid for the additional accuracy and
refinement obtained, and represented by increased running time
and larger core requirements compared to the more limited real
power flow case. Note that contingency evaluation as discussed
here assumes that the system, when encountering a line or generator outage, will reach a new steady-state, i. e., will be
transiently stable.

MORE GENERAL SECURITY FUNCTIONS

Figure 12 contains general function blocks, including the
operations planning of all energy and var resources and their
control. It should not be inferred that the watt and var control in the AGC block, nor in other blocks, are required to be
simultaneous.

Energy Resource Management listed in Figure 12 is concerned
with the longer term aspects of all energy and var resources
that are available within a system. Today's systems consist of
a mixture of energy resources such as steam units, peaking
units, conventional hydro units, both pondage hydro as well as
run-of-river hydro, pumped hydro and nuclear units.

The problem faced by operating personnel is how to allocate
the energy resources so as to satisfy the future hourly load
plus whatever interchanges are being contracted, to maintain
the reserve requirements specified; to satisfy the many operating

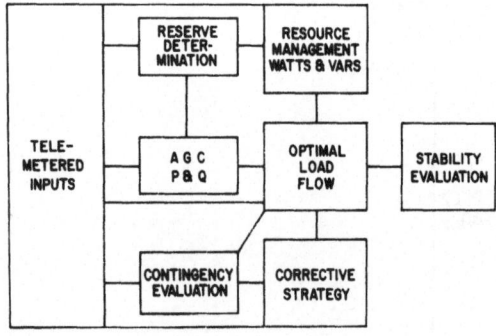

Figure 12 - Security functions

constraints, and finally to minimize the cost or usage of fuel. Notice that several hierarchies in time are involved in the energy resources listed, and they interlink with each other. Nuclear units typically operate on a yearly or longer fuel cycle schedule; pondage hydro operate on a weekly to yearly basis depending upon the size of the reservoir; pumped hydro operate on a daily to weekly refill cycle. Steam units are committed for a good part of the day and gas turbines are used hourly over the peak periods.

Several programs that treat the various time periods are required. A yearly program for the nuclear and large pondage hydro, a weekly program for pondage hydro and pumped hydro, and a daily program for steam and gas units. The longer term programs provide the input to the shorter term programs; that is, the operation of the large pondage hydro is determined on a yearly basis with the amount of energy allocated per week for the year determined and used as input to the weekly program. The weekly program, in more detail, allocates the energies daily, and inputs them to the daily program; the daily program allocates the energies hourly and determines the start and stop times of the steam and gas units taking the run-of-river hydro into consideration.

ACTION CENTERS

A multicomputer configuration consists of hierarchy of computers which are linked to each other by communication channels. Figure 13 is a general computer configuration which is rapidly evolving in the U. S. A., and Figure 14 represents a more detailed view of all action centers through which security control is accomplished.

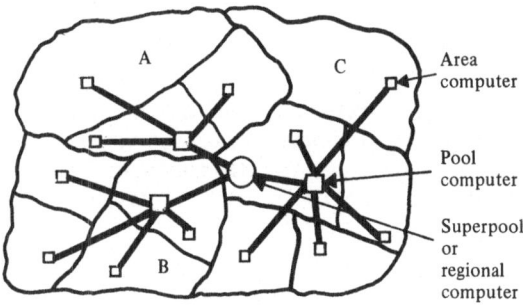

Figure 13 - Computer hierarchy consisting of dispatch - security computers

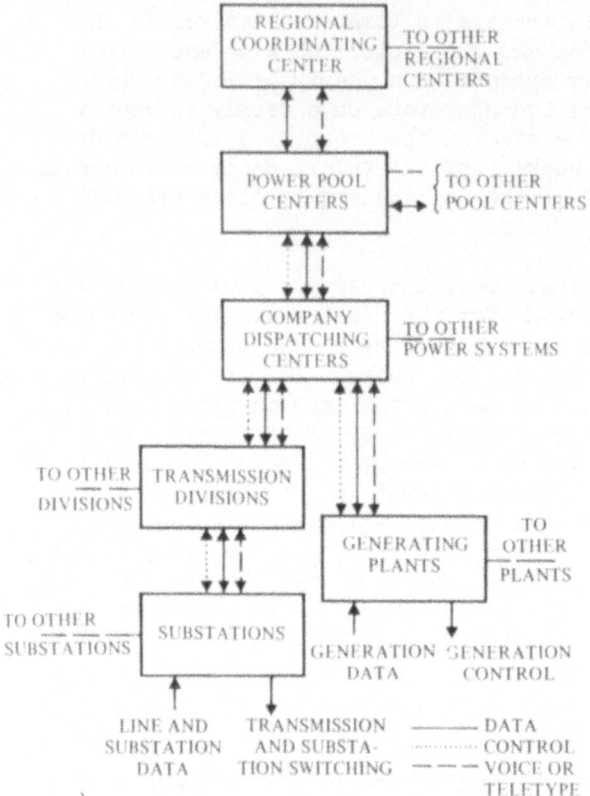

Figure 14 - Hierarchy of action centers

Unique piecewise methods[9-12] can be executed in the above multicomputer configuration for such applications as multiarea AGC[13-15], and power system load flow and stability[16, 17] that make use of the computer capabilities present in pools and superpools to enhance the overall security of power system operations.

CONCLUSIONS

This paper has described what may be called the theory of power system security, present practices, and near term developments.

REFERENCES

1. D. N. Ewart, H. J. Fiedler, H. H. Happ, L. K. Kirchmayer, "Modern Dispatch Techniques of Interconnected Power Systems," IEEE Publication 69M70-PWR, 1970, 263 pages

2. T. E. Dyliacco, "The Adaptive Reliability Control System," Vol. PAS 86, No. 5, 1967, pp. 517-528

3. T. C. Chihlar, J. H. Wear, D. N. Ewart, L. K. Kirchmayer, "Electric Utility System Security," Proc. of the APC, Vol. 31, 1969, pp. 891-908

4. H. H. Happ, R. C. Johnson, W. J. Wright, "Large Scale Hydro-Thermal Unit Commitment - Method and Results," IEEE Trans. on PAS, Vol. 90, No. 3, 1971, pp. 1373-1384

5. H. H. Happ, "Management of Power System Reactive," A talk presented before the "North American Interconnection Committee" (NAPSIC), St. Louis, Mo., May 10, 1973

6. C. Concordia, "System Compensation - An Overview," Transmission, March, 1973, pp. 3-6

7. H. H. Happ, "Optimal Power Dispatch," IEEE Trans. on PAS, presented at 1973 Summer Power Meeting, Vancouver, B. C., July, 1973

8. J. Carpentier, "Conbribution a l'Etude du Dispatching Economique," Bulletin de la Societe Francaise des Electriciens, Ser. 8, Vol. 3, August, 1962

9. G. Kron, "Diakoptics - The Piecewise Solution of Large-Scale Systems," McDonald, London, 1963

10. H. H. Happ, "Diakoptics and Networks," a book, Academic Press in Series Mathematics for Science and Engineering, New York, London, 1971, 312 pages

11. H. H. Happ, Editor, "Gabriel Kron and Systems Theory," Union College Press, 1973

12. A. Brameller, M. N. John, M. R. Scott, "Practical Diakoptics for Electrical Networks," Chapman and Hall, London, 1969

13. H. H. Happ, "Diakoptics and System Operations: Automatic Generation Control in Multi-Areas," Proc. of the IEE Vol. 120, No. 4, 1973, pp. 484-490

14. H. H. Happ, "Power Pools and Superpools," IEEE Spectrum, March, 1973, pp. 54-61

15. H. H. Happ, "Multi-Computer Configurations and Diakoptics: Dispatch of Real Power in Power Pools," IEEE PICA Proceedings 1967, pp. 95-107; also IEEE PAS, Vol. PAS-88, No. 5, 1969, pp. 764-772

16. H. H. Happ, J. M. Undrill, "Multi-Computer Configurations and Diakoptics: Power Flow in Power Pools," IEEE Trans. on PAS, Vol. No. 6, 1969, pp. 789-796

17. H. H. Happ, "Multi-Computer Configurations and Diakoptics: Stability Analysis of Large Power Systems," IEEE PICA Proc., Minneapolis, Minn., 1973, pp. 101-104

A RELIABILITY/MAINTAINABILITY DATA FEEDBACK SYSTEM

JAMES A. HESS

US Army Electronics Command
Fort Monmouth, New Jersey

In a speech at the Integrated Logistics Support Symposium in Washington, D.C. on July 23, 1969, General F. J. Chesarek, then the Commanding General of the U.S. Army Materiel Command, demonstrated the importance of maintaining a viable reliability and maintainability data base when he said:

> "It is a design requirement to build reliability and maintainability into our equipment. Specified reliability and maintainability goals are becoming contractual requirements that must be met along with other functional performance requirements. AMC now insists that the design requirements formally predict what the reliability and maintainability will be, that it be measured during development, and that it be validated during test and operations. We will then have an audited trail of what we intended to do and what we actually did in the way of performance."

The data system is needed during all phases of equipment life cycle. It must provide failure data showing the failure cause and the specific mode of each failure during design. In order to determine if the equipment is achieving its RAM objective or if serious deficiencies exist during production, the RAM data system must be utilized. Decisions whether or not product improvements are needed when the equipment is in the field should also rely on the data system for evidence.

Since the data system at the U.S. Army Electronics Command (ECOM) was expected to be useful to three principal users: the director of maintenance, the director of research, development, and engineering, and the director of product assurance; a working group was set up to assess the needs of the Command and recommend an approach for the data system. The working group was composed of representatives of all laboratories as well as the functional directorates.

REQUIREMENTS FOR THE DATA SYSTEM

Based on the needs identified by the ECOM working group, a data feedback system was devised that satisfies the prime requirements of a RAM data system for a military material developer. It has already been stated that data feedback is needed during all phases of the life cycle. Specifically, during the conceptual phase, the ability to quickly retrieve past problem information on similiar equipments, or the same equipment if this is a redesign, and provide it to the designer or developer is essential. Thus we can learn from our past mistakes and hopefully avoid developing materiel with the same deficiencies. The availability of a valid baseline for the class of equipment in question also permits better insight into the RAM characteristics of the equipment and allows specification of RAM indices that are consistent with the state-of-the-art.

During the design phase one of the most valuable activities of the data system is the anticipation of potential and incipient problems before the "roof falls in."[1] Another effective utilization of the data system is the supply of information to allow implementation of Integrated Logistics Support planning. This requirement is applicable during all life cycle management phases. The use of RAM data such as failure rates and fault location, correction, and check out times provide necessary data for determining spare part, maintenance shop, and maintenance technician allocation. Timely generation of this data facilitates evaluation of the life cycle cost of the equipment.

Once the equipment is fielded, the data system must ensure that the equipment meets its design requirements in the user environment and that changes in operational performance are detected.[1] Based on the analysis of feedback data, a determination of the need for remedial action is possible. In fact, the performance of all assessment functions are enhanced through use of the data system. Preparation of reliability status reports, identification of least reliable items, verification of the effectiveness of corrective actions, and detection of trends are simplified when the data system is utilized.

The system must be comprehensive enough to allow determination of failure rates and maintenance times on systems, subsystems, components, and parts. Application, stress, and problem data on parts and components identified by vendor must be derived.

DESCRIPTION OF THE DATA SYSTEM

In order to establish an effective and economical reliability and maintainability data system at ECOM, it was determined to prepare and maintain a reliability record for each end item developed and/or procured by ECOM. The essential elements of the data system are depicted in Figure 1. A computer algorithm is required to perform predictions and generate the prediction file. An experience file must be maintained to describe the RAM history, and analysis routines must be available to compare the predicted versus actual experience of each end item.

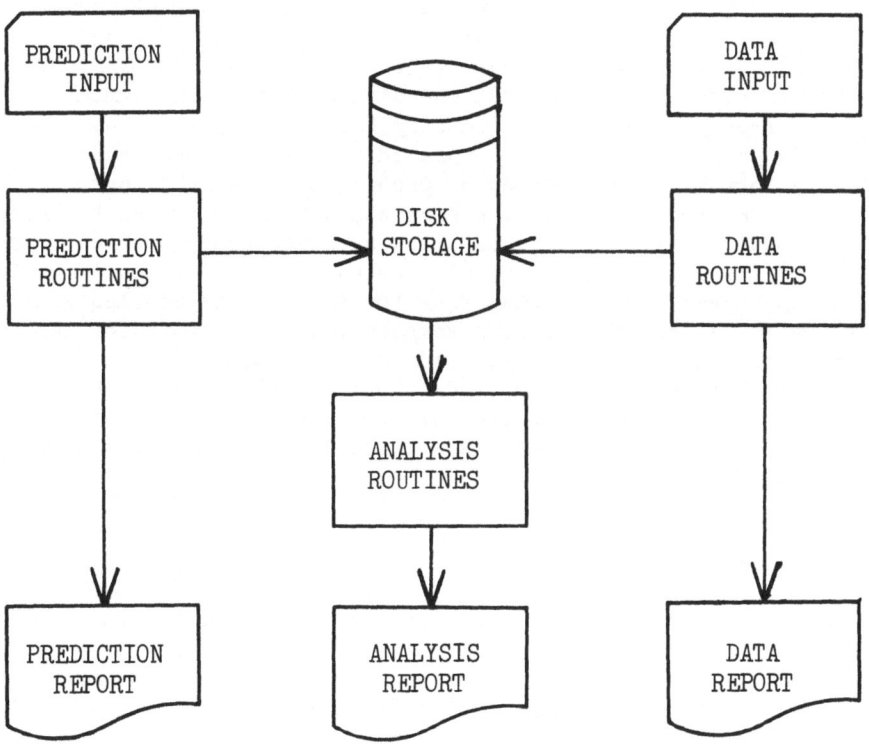

Figure 1. Data System Flowchart

The Prediction

To assess the performance of an end item as well as its components and modules, it is first necessary to determine its composition. To satisfy this need, a complete top-down breakout of the equipment into assemblies, sub-assemblies, modules, and components is necessary. Since verification of projected reliability and maintainability requirements requires that predictions be performed early in the design process, it is obvious that the prediction should be submitted on computer cards or tape and a standardized component data bank of failure rates used to compute the predicted RAM indices.

The approach taken in the development of this data base is to require the submission of a complete reliability prediction as part of each development contract. Every component used in the equipment must be described by providing its identification and electrical and environmental stresses. Components may be combined into modules and modules into equipments. The failure rate for each component is determined from equations in the failure rate library utilizing the electrical and environmental stresses the part sees. Initially, failure rates provided in the RADC Reliability Notebook[4] will be used. As more current failure rates are determined, the data bank will be augmented. Alerts from the Government Industry Data Exchange Program (GIDEP) program will be incorporated so that the Army and its contractors will be informed of problem areas as they are surfaced. Failure rates for the components will be summed to determine module failure rates, assuming a series configuration. The reliability of the equipment will be calculated for the specified mission times from a boolean equation showing the modules necessary for successful mission completion. The inputs for the whole prediction will be prepared by the contractor, run utilizing a government owned program, and submitted for review. This procedure allows quick evaluation of the impact of design changes on reliability and it provides a complete breakdown of the equipment for future assessment.

Data Sources

Reliability and maintainability data is generated during all phases of the equipment life cycle by the developing government agency and during every phase except the operational phase by government contractors. It is imperative that as much of this data as possible be retained by the RAM data system since data on malfunctions and operational difficulties as well as repair and maintenance data form the backbone of any RAM data feedback system.

The data system allows entry of three general types of experience data. Failure and repair data uncovered during tests conducted when the equipment is in the development phase from the largest body of data available.

One of the best sources of RAM data is the contractor's data collection system. MIL-STD-470[2] requires the establishment of a maintainability data collection system that must be integrated with that for reliability, and MIL-STD-785A[3] requires that each contractor must have "a closed loop system for collecting, analyzing and recording all failures that occur during all in-plant tests and those that occur at installation sites or test sites prior to turnover to the procuring agency. The contractor shall describe his proposed system for initiating failure reports, the analysis of failures, and the feedback of corrective action."

Although all contractors may use their existing data collection systems, their output must be compatible with the ECOM RAM data feedback system. ECOM developmental contracts will require that every anomily that occurs after initial satisfactory operation of the equipment be analyzed and reported on either computer cards or tape in a format specified in the contract. The failure reports will show equipment, module, and component identification, time to failure, electrical and environmental stresses, and failure resolution. Each test will be summarized by abstracting the test type, test date, environment, test plan, plus summaries of burn-in characteristics, test length, number of failures and repair actions and mean times. The requirement for submission of data in a computer compatible format will provide a near optimum input procedure since the analyzing, abstracting, and keypunching will only be done once. A thorough review of the data will be made by government RAM engineer assigned to the equipment before the data is entered into the equipment file. His review will assure that the information supplied is complete and correct and that all supporting data are in order.

Data collected from government agencies provides the other source of test data. At the present time, a majority of this data comes from operational and developmental tests conducted by the U.S. Army Test and Evaluation Command (TECOM). The results of TECOM tests are fed back to ECOM in the form of test reports and Equipment Performance Reports. All of the data must be analyzed, abstracted, coded, reviewed, and entered into the system by the responsible RAM engineer.

Another type of test has been initiated at ECOM to assess the effects of age and field usage on an equipment. A pilot program for the Field Exchange and Evaluation Program (FEEP) has just been completed by the R&M Division of the Product Assurance Directorate. The theory of the FEEP program is direct exchange of new equipment for those in use in the field. Evaluation of the field samples allows a determination of the effects of field maintenance and usage on the end item. The collection, analysis, and entry of all RAM data from the program is under control of the FEEP program manager.

Experience data is also collected directly in the user environment. Benefits accruing from the collection of field data over test data include responsiveness to the actual stresses imposed on the equipment, assessment of the effect of administrative and logistical delays on the repair cycle, and awareness of the problems of human error in operating and maintaining the equipment.

Two major problems arise when field data collection is attempted in the Army. Tracking of the repair of any given failure is difficult because of the multi-echelon maintenance utilized. The solution to this problem is relatively straightforward when pre-numbered multi-part tags are supplied to the user. Upon detection of a failure, a tag is completed, one part sent to the data collection agency, and the rest attached to the part that is sent to the next echelon for repair. As each maintenance echelon continues this process, a complete history of the repair is provided for the analyst. A more serious problem, however, is the attitude of the men in the field. Because they don't see the results of their effort and feel that the data system is an attempt by "big brother" to look over their shoulder; the maintenance technician may be apathetic or even hostile to the data collection effort. The solution here is to have engineering representatives of the developing agency in the field to explain the program on a personal basis and to assist in the preparation of the failure reports. Personnel requirements limit this collection effort to a sample data collection program.

The information collected is similar to that collected during the test programs except that data describing administrative and logistics delays must be reported. Since the data will be submitted from several different echelons, a preprocessing of the data is required to assure that each case is complete before it is entered into the system. Otherwise, the processing is identical to test data.

The third type of experience data is generated from deficiency reports prepared by the equipment user. Depending on the seriousness of the deficiency and the location where it is detected, the user may submit an Unsatisfactory Materiel Report (UMR), Equipment Improvement Request (EIR), or Report of Item Deficiency (ROID) upon detection of hazards, substandard quality, non-conformance with standards, or other circumstances which warrant corrective action. Although these data sources do not provide enough data to allow a quantitative reliability and maintainability analysis, they do provide a trend detecting capability that is essential for a viable data feedback system.

These reports will not be entered into the data system in their entirety but will be given an identification number and then abstracted and coded before entry. This will facilitate sorting, searching, and manipulating the file to detect deficiencies and record the corrective actions. Since deficiency reports arrive from many sources, data reduction at one centralized location will ensure consistent treatment for similar reports. Furthermore, a keyword dictionary must be utilized when abstracting the reports.

The Data Structure

A data system based solely on the preparation of "fixed-output" reports will be of limited use and will be quickly outgrown. The structure of the ECOM system is designed to retrieve and present data in any order and it may be tailored to the needs of any user. Multi-list and inverted list structures were considered as possible file systems, but their large overhead for input transactions preclude their use. Similarly, the exclusive use of serial files was discarded because of the excessive cost of preparing varying types of reports or extracting information from the data base for further analysis.

The approach formulated for the data base rests on the definition of a logical record for each equipment as shown in Figure 2. The equipment group is composed of data elements identifying the nomenclature, line number, and name, as well as keyword descriptors of the equipment. An indication of mission definition, failure definition, maintenance philosophy and developing laboratory is also included. The contract group provides identification of the contractor, dollar value, contract number, specification number, development phase, date of award, along with summaries of the specified, predicted, and demonstrated RAM indices. The remaining groups contain the data identified previously under "Data Sources."

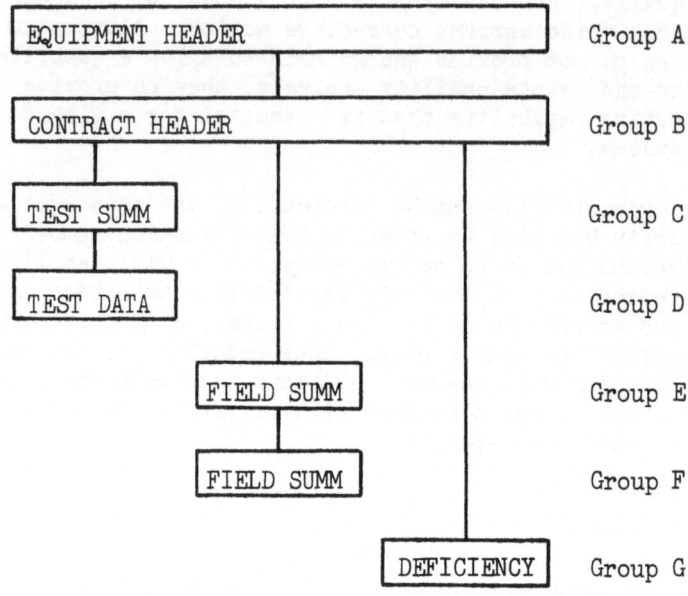

Figure 2. Equipment Logical Record

   The physical structure of the files provides cohesive organization for and includes a referencing mechanism among logical records. A multi-file structure has been chosen with the records of each repeating group in Figure 2 included in separate direct access files. Although the ability to tie the files to functional applications has been retained, the advantages of a data base concept such as non-redundant storage of data, grouping of interrelated data, consistent use and sharing of common data, and the ability to process for more than one application have been realized by chaining the logical records together with pointers. The most frequent access to the logical records will be by nomenclature, line number, or contract number, therefore, an efficient method had to be found to enter new records and delete or search for specified records on these keys. Hash searching was discarded since the memory required for efficient searching would be excessive and because the memory for deleted items can not readily be used. Since the directory must be maintained in ascending order for binary searching, this too was discarded because of the excessive overhead when adding or deleting. The algorithm decided upon is based on the AVL tree. An AVL tree is maintained for each of the keys identified above by establishing a chain of pointers throughout the equipment and contract files. The AVL algorithm assures that the search trees are balanced at all times and that the bound on the longest path, L, is

$$L \leq 1.5 \log_2 (N+1)$$

where N is the number of elements in the tree.[5]

The logical records are chained together with pointers to maintain their integrity. Every element of groups A, B, C, and E, points to its most recently defined element in the next-lower group. A chain is also maintained in Groups B through G linking all filial elements to that element chained to the higher level. All lists are bi-directional.

As was previously stated, this structure allows great flexibility in the retrieval of information and yet does not impose an unrealistic overhead when adding or deleting information. Ordered reports or queries based on nomenclature, line number, and contract number are easy to prepare utilizing the tree structure. The ability to enter the data base at any of the levels and search sequentially enhances the capability to respond to generalized queries and the multifile structure reduces the searching required. Once a "hit" has been made, the pointers provide the capability to transverse the structure both vertically and horizontally to gather data. This approach also allows direct interrogation of the data base with FORTRAN or COBOL analysis programs as discussed in the next section.

DATA ANALYSIS

Although the predicted and demonstrated data are valuable, the real benefit of the data system appears when they can be compared. During the early phases of the equipment life cycle, a number of assumptions must be made to forecast the RAM characteristics. When the design is completed and real system data are collected, the assumptions can be revised and modified to fit the real world. Based on a comparison of predicted versus actual data, the computerized analysis must:

1. Pinpoint critical maintenance items based on early failure and repair data,

2. Derive repair times based on test data and predicted failure rates and maintenance times,

3. Determine system and module failure rates,

4. Find distributions for time-to-repair and time-to-failure, and

5. Assess failures by environment and stress levels to uncover components with abnormal failure characteristics.

The computer programs for the analysis routines are not yet fully defined. Their definition will be accomplished after implementation of the data base.

CONCLUSION

A cost effective methodology to improve the reliability, maintainability and effectiveness of our electronics systems is required. Although there is more involved in this improvement process than the collection and analysis of data, the success of our development programs requires an ability to detect our past mistakes as well as prove our accomplishments. A successful RAM data base is essential to give us the hindsight necessary to move forward.

REFERENCES

1. Hope, J. I, et al A Study of the Army Material Command Reliability Program, a report of the Panel on Accelerated Development of Reliability to the Commanding General, AMC, 28 June 1972.

2. Department of Defense Military Standard, Maintainability Program Requirements, MIL-STD-470, 21 March 1966.

3. Department of Defense Military Standard, Reliability Program for Systems and Equipment Development and Production, MIL-STD-785A, 28 March 1969.

4. Ryerson, C.M., Webster, S.L., and Albright, F.G. Reliability Notebook, Volume II, Technical Report RADC-TR-67-108, Rome Air Development Center, Griffiss Air Force Base, New York.

5. Stone, H.S. Introduction to Computer Organization and Data Structures, McGraw Hill, New York, N.Y., 1972.

DESIGN FOR MAN-MACHINE SYSTEM RELIABILITY IN PROCESS CONTROL

F.P. Lees

Department of Chemical Engineering
Loughborough University of Technology
Leicestershire, England.

INTRODUCTION

Although there have been since 1960 some 20 studies of the process operator, and much other potentially relevant work, these have been carried out mainly by workers in human factors, often university-based, and the results have not been well assimilated by industry. This need is recognized, however, and the Institution of Chemical Engineers has recently commissioned a study of man and computer in process control (14); this work may be referred to for a more extensive treatment of the process operator.

The problem of quantifying man-machine system reliability in process control has been discussed in a previous paper (24). The need to think in terms of overall man-machine system reliability and the varieties of process and control system were emphasized, a taxonomy of errors in process control was presented, areas where existing work has relevance were defined and a reliability data collection scheme which would obtain human error data as well as other reliability data was outlined.

The present paper is concerned with design factors which affect man-machine system reliability in process control. An understanding of these is desirable not only in system design itself, but also in operator modelling and system assessment for reliability.

PROCESSES, CONTROL SYSTEMS AND PROCESS COMPUTERS

It is important to emphasize at the outset the rather wide variety of processes and the effect of this variety on the operator's job. In a chlorine cell-room the operator's function may be largely that of monitoring. In a melting shop the scheduling of a group of electric arc furnaces may be the principal function. On a set of batch reactors it may be the sequential control function which predominates. In other cases there may be a more balanced mix of functions.

The control system provided also greatly affects the operator. Three broad stages of manual, analogue and computer control can be defined, but there are significant distinctions to be made within these stages. Thus, for example, analogue control systems differ widely in the extent to which they offer, in addition to basic feedback control of individual loops, facilities such as sophisticated measurement or automatic alarm scanning or sequential control.

Process computers may be classified as either setpoint control or direct digital control (ddc) systems. In the former the computer alters setpoints of conventional analogue controllers, while in the latter it replaces the controllers, itself receiving measurements, calculating the control algorithm and sending out regulating signals. The control function of some ddc machines is limited to this basic feedback control, though more usually the machine carries out supervisory functions also, such as feedforward or sequential control or optimization, which are of course the raison-d'etre of a setpoint control computer. A machine which carries out only basic ddc thus leaves most of the control to the operator, though it gives him an excellent tool with which to work, while a setpoint control computer with an active supervisory program may reduce the operator's function largely to one of monitoring; the two systems could hardly be more different.

The effect of such factors on the role of the operator has been discussed in greater detail by Edwards and Lees (12,13).

REQUIREMENT FOR RELIABILITY IN PROCESS CONTROL

It should not be assumed that high reliability is a requirement in all processes. Again, it is necessary to emphasize the wide variety of processes. It is worth, therefore, examining what degree of reliability is required.

Reliability is necessary for reasons of safety and of economics. Both factors are important in all processes, but the emphasis differs. On a nuclear reactor considerations of safety are paramount, in a paper mill economic considerations predominate.

Even where economics is the prime consideration there are great differences between processes in the relationship between economics and reliability. On many processes the penalties of a moderate degree of unreliability may not be particularly severe. On other processes, where failures may cause serious damage to equipment or major plant shutdown, high reliability is needed.

In so far as reliability is required in process control systems for economic rather than for safety reasons there is a limit to the degree of reliability which is worth aiming for. Indeed, as this level is exceeded, economies will tend to be made which bring the reliability back to the economic level. For example, if general plant reliability rises it becomes less attractive to install storage capacity so this is eliminated; if the reliability of plant items with installed spares improves such sparing is reduced; and so on. The unacceptably low reliabilities on some large plants in recent years and the resultant push to improve on these should not be allowed to obscure this fact.

Thus, as a consequence of the variety of processes, there are great differences also in the extent to which it is economic to seek to improve the reliability of the control system.

DESIGN FOR RELIABILITY

Some of the main human factors activities in the design of a man-machine system are given in Fig. 1. For clarity only one iterative loop is shown, but the design process involves many such loops.

While all aspects of good practice in man-machine system design contribute to reliability, some are worth particular mention.

System Specification

The starting point for design is the system specification. While the need for proper specifications is in general well understood in process control, this is not so true in respect of reliability. Here the approach is often partial, concentrating on particular features, such as reliability of the process computer or of the protective system.

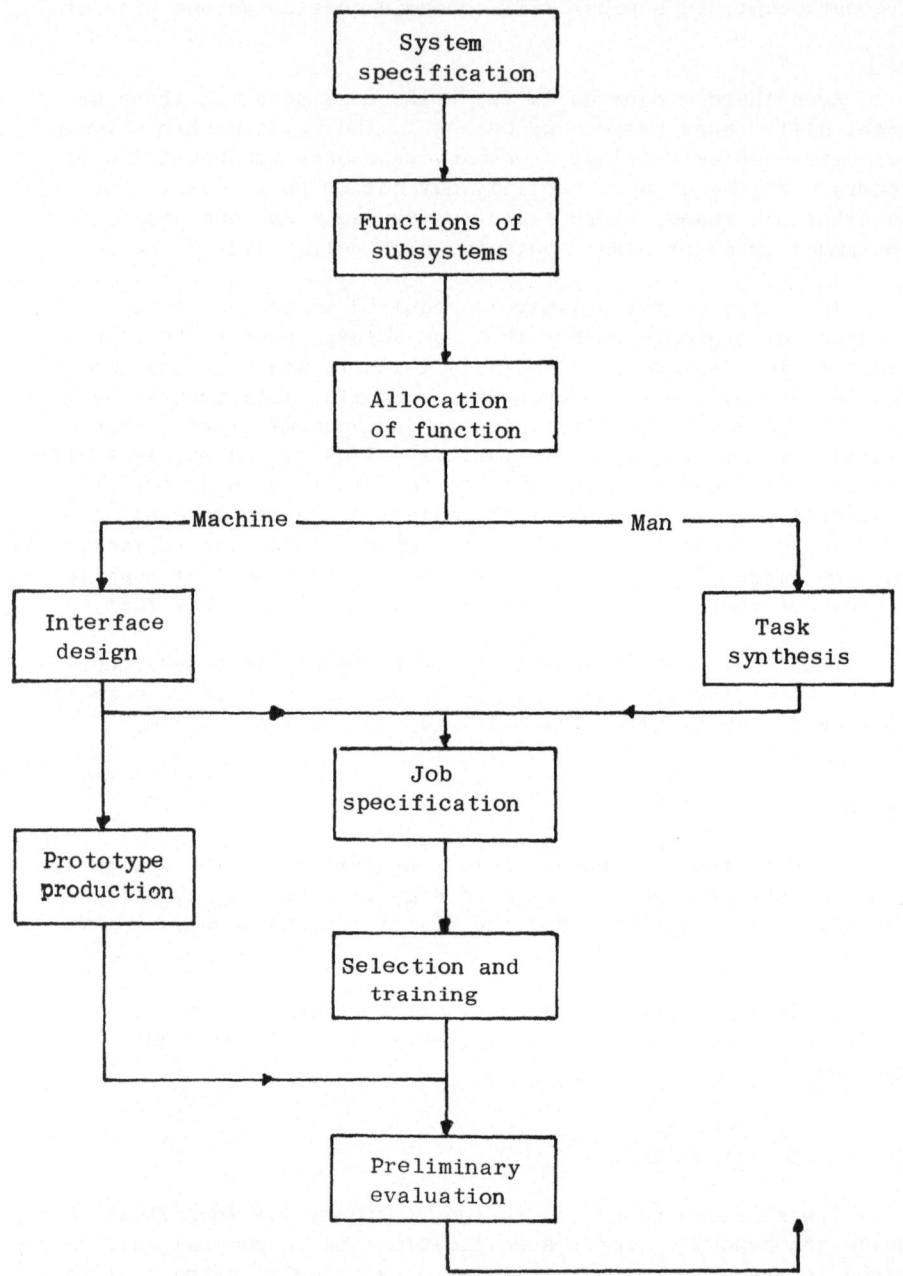

Fig. 1 Human factors activities in system design

Allocation of Function

In deciding the allocation of function it should be borne in mind that some of the most sophisticated control systems in existence, such as the SAGE system for the air defence of the U.S., described by Sackman (36), are not totally automatic systems, but operator-controlled, computer-supported systems. Man is placed at the centre of the system.

Allocation of function is frequently considered in terms of a listing of the functions at which man and machine excel, such as that given by Fitts (16). Such a list is valuable in providing initial orientation to the problem, but needs to be used with care. As Fitts later pointed out, what matters is not what functions man is good at, but what it is best, from the system point of view, that he should perform.

Ability to perform a task is not the only criterion. Thus de Jong and Köster (21) have emphasized the importance of motivation in the allocation of function in process control tasks.

A Fitts-type list serves to underline some of the characteristics of man which make him such a valuable component in respect of system reliability. These include the facts that he can handle information coded in many different ways, that he has unique problem-solving and heuristic capabilities and that he degrades gracefully.

Modern process control systems are capable of a high degree of automation. Increasingly the function of man is that of dealing with malfunctions. He is the component which the designer includes to make the control system self-repairing.

It should be an objective of design so to integrate man into the control system that the situation is avoided where, on failure of automatic control, there has to be reversion to totally manual control. As far as possible, man is incorporated in such a way as to impart to the whole system something of his ability to degrade gracefully. This is well illustrated by the role of man in the Apollo moonshots and Skylab flights.

Allocation of function should take place at function level, not at some higher level. Thus, for example, if the administration of malfunction is considered, the three functions of detection, diagnosis and correction should be considered separately, as de Jong (20) has emphasized, and the strengths of man and machine in each of these functions assessed.

Task Synthesis

Task analysis and synthesis involve the detailed description of the task, for the cases respectively where the task already exists and where it is being created. This is an activity fundamental to the design process and provides a basis for the other stages such as job specification, selection and training, and interface design.

The early work of Crossman (8,9) emphasized the cognitive aspects of task analysis, as opposed to the physical aspects investigated in traditional time and motion study. A methodology for task analysis applicable to process control has been developed by Annett, Duncan and coworkers (3,11), who have analyzed the task of the process operator on an acid purification plant. The task is broken down into a hierarchy of subtasks, which are described in detail in tabular form with a listing of possible input, feedback or action problems. This process highlights such potential difficulties as identification of equipment, lengthy sequential procedures or information overload and generates suggestions for job aids and training.

An account of task analysis in the specific context of reliability has been given by Swain (39), who emphasizes the identification of error-likely situations. The analysis aims to discover the errors which may occur, the possibility of retrieving these and the effect they are likely to have if not corrected.

Job Specification

Job specification involves combining the individual control tasks into a rounded job in which there is a balanced work load and job satisfaction.

Work load is another area where there are great differences between processes. The point which may be emphasized is that the objective of job design should be to optimize rather than minimize the work load. Investigations of work load may be made using simulation. Siegel and Wolf (37) have described extensive simulations of work load in military systems. Similar but less complex simulations for chemical plants have been carried out by Munro, Martin and Roberts (26).

As the degree of automation increases there is a tendency for the operator's role to be reduced to that of monitoring the automatic controls. This is not a desirable situation, since it does not accord well with either man's abilities or his motivation, but it is one which it is often difficult to avoid.

An obvious approach is to seek to give the operator a secondary task. One of the commonest such tasks in process control is that of logging the process instrument readings. With the advent of process computers, which have powerful data logging facilities, much of this work becomes redundant, though most plants do have a significant number of local gauges which do not transmit signals to the computer, but which may carry quite important readings.

A better alternative may be involve the operator in more active monitoring of the state of health of the process equipment. This is a task which accords more closely with his real purpose in the system, which is to make it more reliable. This is discussed further below.

Selection

Developments in selection and training for process control have long been hindered by the lack of a proper job specification for the process operator and by the lack of objective means of assessing his performance.

There exist selection tests for process operators, such as those described by Hiscock (18). These tests attempt to select men who possess qualities which it is reasonable to suppose are required for process control. However, recent work by Davies (10) has shown a low degree of correlation between such tests and supervisors' assessments of performance.

Studies of the process operator conducted over the last decade provide a background of knowledge for a possible attempt to revise such selection tests. It is possible to state some of the qualities which appear to be desirable in a process operator. The individual's body rhythms are adapted to shift work. His response to the task is one of mild stress; he is neither highly nervous nor totally phlegmatic. He is able to divide his attention between tasks. He has certain signal processing capabilities, such as the ability to estimate averages from a noisy signal. He is able to develop effective strategies and to continue learning. He does not jump to conclusions on the basis of inadequate initial evidence, but revises his opinions.

However, this approach remains a priori in the absence of a methodology for assessing operator effectiveness in process control. Kitchin and Graham (23) have studied some 50 process operators and have devised a method of assessing the work load and difficulty of a process control job from supervisor assessments, but work is still required to develop methods of evaluating performance.

The contribution made by selection to human reliability lies to a considerable extent in the elimination of individuals who are almost totally deficient in some of the needed qualities. If such qualities are assumed to be normally distributed throughout the population, there must be a proportion of individuals who are largely deficient in them. It is known, for example, that a small number of people are quite unable to adapt to shift work or to integrate a noisy signal.

Although the objective of selection is to choose those individuals who are best fitted to the actual process control task rather than the most intelligent, the point may nevertheless be made that in more highly sophisticated systems, where reliability considerations may be relatively more important, abilities such as analytical reasoning and clear expression do become important if the operator is to play his proper part as a member of the process team.

Training

Basic principles of training are well established. These include relevance of material; format and rate of presentation of material; motivation of the trainee; feedback of results to the trainee; follow-up on the job. It is particularly important to train for the job which is actually to be done: a classic example is the training of electronics maintenance technicians which now emphasizes diagnostic techniques rather than electronics theory.

In the context of reliability several points may be stressed. One is that the operator must have a clear understanding of system goals and priorities. Studies of process operators by Engelstadt (15) and by Attwood (6) show that this is not always the case. This is likely to be especially important where the operator has two potentially conflicting tasks, such as keeping the plant running if he can but shutting it down if he must. Where certain operating sequences are crucial these must be identified by the task analysis and overlearned by the operator. This is particularly necessary for sequences which, though important, are performed infrequently.

Training is not a substitute for good design. While it is possible to train a man to use poorly-designed equipment, it is precisely under the stress of emergency conditions that this is likely to fail. Errors arising due to violation of stereotypes in equipment design are well understood.

Interface Design

The traditional interface in process control is a control panel containing quite large numbers of indicators and dials, chart recorders, three-term controllers, switches and pushbuttons, and alarm and status lamps. There may also be a mimic panel in which measurements and statuses are displayed at appropriate points on the process flow diagram.

On many computer systems much of this conventional instrumentation is retained, but in others the sole interface provided is the operator's control panel on the computer. Early panels consisted mainly of digital **display** windows with rotary switch or pushbutton controls. While most panels are still of this type the cathode ray tube is now coming into widespread use. This is an obvious equipment to use in the interface, but its utilization in process control has been held back by economic considerations.

Since both hardware and software costs are high for one-off systems some degree of standardization is necessary in operator's control panels. The manufacturer's approach is to offer a standard panel system, rather than a standard panel. The user thus retains some freedom in the selection of displays and controls.

There are a number of design principles which can greatly increase the reliability of utilization of the interface. The choice between general-purpose and dedicated displays and controls should be carefully made. The retrieval of error may be enhanced by the display of data insertions prior to entry into the program. Software guards may be created against incorrect insertions in certain cases.

Once alternative interface designs have been proposed, they may be assessed for reliability. Rooney and Jacobs (35) have described a comparative assessment of operator's control panels in which the panels were used to carry out control functions on a computer-controlled pilot plant and were assessed by analyzing the execution of particular subtasks through the panel both using flow diagrams and experimentally. A somewhat similar exercise is described by Huters (19).

Prototype Production

The production and testing of a prototype is regarded as an essential stage in the design of many sophisticated man-machine systems. Such prototypes are not normally built for process plants and their control systems. Pilot plants are often used as an intermediate step in the scale-up from laboratory to full scale, but the scale, control system and works environment of such plants usually differ greatly from those of the final installation.

The nearest approach to a prototype is the combination of control interface mockup and process simulator which is used in the design of many electrical power generation systems, conventional and nuclear, but even this is the exception rather than the rule in the design of process control systems generally.

It is desirable, however, in the context of reliability that the process industries should at least be aware of the crucial role played by the prototype in the design of other types of system and should consider its use in appropriate cases.

SOME PROBLEM AREAS

Information Display

The traditional control room instrumentation was described above. Such a control interface offers simultaneous display and easy sampling of large quantities of information. Other important features include spatial and other forms of coding, display in analogue form and trend displays.

In many instances computer control has tended to bring with it rather drastic changes in the interface, which in the extreme case is reduced to an operator's control panel only, without there being much ergonomic justification for this. In particular, information is no longer available at a glance, positive effort is required to obtain it and the amount of information available simultaneously is reduced. Spatial coding of controls and displays is lost. Displays are digital only rather than analogue or trend.

The cathode ray tube offers the possibility of some restitution of what has been lost. It also provides a basis for the development of facilities which the operator has previously never had in process control. Alarm systems and sequential operations are areas which can be expected to benefit particularly.

There is a tendency among instrument designers to assume that what is required from an indicating instrument is a quantitative reading. Results obtained by Murrell (27) suggest, however, that check reading or check reading and resetting are more important industrial uses.

The trend records available on chart recorders are extremely valuable to the operator. Thus de Jong and Köster (21) have described their use by the operator in information sampling, Anyakora and Lees (4) their use in instrument malfunction detection and Clark (7) their use in handling process fault conditions.

The amount of plant which the operator has to supervise tends to increase as control is centralized in large control rooms. The advent of the process computer has enabled this trend to continue, the computer being used to replace large numbers of instruments. There is, however, a danger here. In principle, the computer is an enormously powerful tool for data reduction and display which may indeed enable the operator to cope with larger quantities of information than would otherwise be possible. In practice, however, the actual data reduction and display facilities offered by the computer often do not match the increased information load and do not compensate for the loss of the conventional displays.

A rather neglected aspect is the degree of confidence which the operator has in the information displayed. This will be more credible if there is available other information against which it can be checked, in other words, if there is information redundancy. Belief that a reading is due to instrument error is particularly common. The redundancy which allows this to be checked may be a duplicate instrument, but more often it is some other related measurement or even the instrument's own past signal as shown by a trend record. If the operator doubts the validity of the information displayed, he may delay taking necessary action until it is too late.

Alarm Scanning and Analysis

A function of the control system which is fundamental to reliability is the monitoring of the process for alarm conditions. As plants become more automated such monitoring becomes an increasingly important part of the task of both computer and operator, and the interaction between the two in carrying out this function requires increasing attention. Yet this is a very neglected area of process control.

Process plants are provided with alarm systems which indicate when a process variable has gone out of limits or when an equipment is in an incorrect state. An alarm is given by the lighting of an alarm lamp and by an audible signal.

Such conventional alarm systems are unsatisfactory in many respects. The worst fault is that usually too many alarms are generated. Often these are simply unnecessary alarms or alarms with unduly tight limits. Mixture of alarms and statuses creates further confusion. It may be desirable to indicate the status of items of equipment, such as equipments not in use, but these statuses should be separate from genuine alarms. Another source of trouble is sequential operations which tend to generate signals which in continuous operation would constitute alarms, but which are normal during certain phases of the sequence. Sometimes the operator depends on such signals in order to execute the sequence.

The main impact of process computers on alarm systems has been to increase the number of signals which can conveniently be scanned for alarms and the types of alarm which can be used. The computer usually takes in a large number of analogue and digital inputs all of which can be scanned. 'Indirect measurements', calculated from a number of instrument readings, may be scanned also. In addition to alarms on absolute values of variables, other types of alarm may be used, such as rate of change or deviation from setpoint alarms. Typically all these alarms are printed out on a typewriter log. Once again the main fault is the generation of unnecessary alarms.

The process computer has great potential for the rationalization of the alarm system, but this is at present largely unexploited. The computer can be used to create a more flexible alarm system which adapts to the plant state, to separate alarms and statuses, to analyze alarms and to display alarms in a more assimilable form.

On some processes, notably nuclear reactors, there may be several thousand computer inputs, and as many alarms. Means must therefore be devised to protect the operator from information overload. The method used is alarm analysis, as described for the Oldbury reactor by Kay and Heywood (22) and by Patterson (28) and for that at Wylfa by Welbourne (40). The analysis is based on relating groups of alarms which form a tree. The actual cause of the alarms may or may not be diagnosed. While this technique is becoming established in nuclear installations in the U.K., it involves much engineering effort and has not so far spread to the process industries.

Such work needs to be balanced by an equal attention to the human factors aspects of the alarm system, but much less work has been done on questions such as the appropriate format of alarm displays or the action actually taken by the operator on receipt of an alarm signal.

Again the question of the credibility of the information arises. In diagnostic tasks men and computers tend to operate differently. The computer usually follows a logical algorithm which takes it down both high and low probability paths systematically, whereas man jumps around following high probability paths.

As Rasmussen (29) points out, if the investigation has not exhausted the high probability possibilities, the operator is not likely to believe a low probability cause indicated by the computer.

There is need for design of alarm scanning systems in a more formal way, so that they signal necessary alarms, but not unnecessary ones. The problem is akin to the design of protective systems, from which something might be learnt.

If facilities more sophisticated than simple scanning are considered, then a clear design philosophy is necessary. It has been emphasized, for example, by Rasmussen (29) that an alarm analysis should result either in automatic fault correction or a command to the operator, but not in mere advice.

Work Load, Monitoring and Boredom

There are wide variations in the work load of operators. Underloading is already a problem in some systems. Other work is increasingly automated and the monitoring content of the operator's task tends to grow. In so far as the designer seeks to achieve greater reliability through increased automation, it may be expected that this problem will be particularly severe on processes where high reliability is desired.

Man is not generally regarded as particularly well suited to monitoring tasks. This is certainly true in so far as monitoring involves the routine scanning and comparison of large numbers of formal signals. His relative effectiveness is much greater if it becomes a question of detecting unexpected, ill-defined and complex signals coded in many different forms, of recognizing patterns and of applying a knowledge of probabilities. The designer will no doubt remain reluctant to dispense with such a versatile component.

It is therefore probable that man will continue to be included in control systems where he may have little to do except monitor and is bound to experience boredom. Under these conditions both the man himself and his effectiveness as a system component suffer.

Work on human factors in process control has brought out the importance of the updating by the operator of his knowledge of the state of the process. He needs to do this continually if he is to be prepared to intervene at any time to handle an unforeseen or emergency situation. This is assisted if the operator is in some way continually involved in the control of the process.

It seems probable, however, that this will be a persistent problem area for the foreseeable future and, if there is a solution, it is not yet in view.

One possible approach may be mentioned, however. This is the greater involvement of the operator in detection of incipient malfunction on the plant. This is in a sense a secondary task but one which is more integrated than some others with the operator's role in a highly automated system, which is largely that of making the system self-repairing.

Malfunction Detection

A normal alarm system is activated when a signal violates an alarm limit. In so far as such alarms occur due to equipment malfunction rather than to incoming disturbances they tend to arise from severe failures.

If possible it is desirable to detect failures before they occur. This reduces the risk of disrupting the process or damaging the equipment and facilitates repair. The value of detecting incipient malfunction is attested by the widespread use of techniques such as vibration analysis.

Most malfunction detection on processes, however, is carried out by the process operator. A detailed account of the detection of malfunction in instruments by the operator has been given by Anyakora and Lees (4). These authors (5) have also discussed the whole problem of malfunction detection by the operator and the computer and the development of man-computer interactive malfunction detection systems. These exploit the capabilities of the computer in areas such as data scanning, reduction, comparison and display and those of the operator in areas such as pattern recognition and diagnostic testing.

Thus it is possible to envisage the creation, alongside the usual alarm system, of a malfunction detection system. This requires that means be developed of distinguishing healthy behaviour from incipient failure in plant equipments and instruments and that these be used to check current signals. As suggested above, malfunction detection may provide the operator with a relevant active task which reduces the risk of boredom and contributes to reliability.

Emergency Situations

Although process plants are provided to a greater or lesser

degree with protective systems, there normally remain potential emergency situations which must be handled by the operator.

A distinction may be drawn between an emergency which has been foreseen and for which a well-defined operating procedure exists and one which is quite unexpected and for which a strategy must be devised on the spot.

It is legitimate to include the operator as a component whose versatility is exploited to handle emergencies, but it is necessary to respect the limitations of this component. The operator must be given both the time and the information necessary for him to form an assessment of the situation. He must have sufficient confidence in this assessment to act on it. His action should not be inhibited by confusion as to his priorities, such as avoiding serious hazard or avoiding unnecessary shutdown. If an operating procedure exists, thorough training will increase the probability of his performing it correctly. If it does not, he will presumably have to develop a strategy. Here man's strength lies less in the rapidity of his decision processes than in their versatility, so he needs further time for thinking.

ASSESSMENT OF DESIGN

Once a system design exists it can be assessed for reliability. It was suggested in the previous paper (24) that this might be done using the following taxonomy:
- (1)   Simple tasks
- (2)   Complex tasks
- (3)   Vigilance tasks
- (4)   Control tasks
- (5)   Emergency behaviour
- (6)   Isolated acts
- (7)   Operator incapacity

Simple tasks are discrete tasks comprising a well-defined sequence of operations. Reliabilities of execution of such tasks may in principle be calculated using data for standard task elements. Data banks for human error data exist: the American Institute for Research data bank is described by Altman (2) and by Meister (25) and that at Sandia Laboratories by Rigby (30) and by Swain (38). Simple reliability relations are then used to estimate the reliability of the task from that of the elements, as outlined by Rook (33) and by Swain (39). The application of this approach to a process control task is exemplified by Ablitt's work (1) on the execution of a schedule of trip and warning tests. Ranking techniques may also be used as described by Rook (34).

The distinction between simple and complex tasks is somewhat arbitrary, but broadly the latter are more complex, involve greater decision-making and greater variability. In most process control systems there are a number of tasks in these two categories which can be isolated and for which a demand rate can be estimated.

The residue of the overall process control task is classified as vigilance and control tasks. The vigilance task certainly includes monitoring of formal instrument and alarm signals, such as figure in much experimental work on vigilance. But it involves in addition responding to a wide range of other signals, which may be complex with ill-defined out-of-limit conditions, weak and noisy, infrequent and unexpected.

Green (17) and Ablitt (1) have described work on vigilance in process control in which the object was to assess the reliability of the human operator responding to formal visual and/or audible signals in actual reactor control rooms. Such work has relevance to the reliability of the operator as a last line of defence in a series of protective systems.

Studies of the vigilance problem in process control suggest that factors such as the vigilance effect are less influential in determining the operator's response than factors such as his doubts about the validity of the information presented or about the need for a control intervention by himself.

The control task which remains is rather ill-defined. The operator continually updates his knowledge of the state of the process and decides whether a control action is necessary.

In some cases it may be possible to isolate other tasks involved in process control, such as tracking, scheduling or communication. Thus there have been investigations in other fields of errors in communication.

It is obviously difficult to estimate demand rates for vigilance and control tasks in process control. While it is possible to conceive of an eventual estimate of demand rate for critical vigilance signals, a demand rate estimate for the control task does not seem practical, since it depends so much on the operator's judgment as to whether he should intervene.

Work on human behaviour in emergencies, mainly in flying, has been described by Swain (39) and by Ronan (32) and the use of ranking techniques to scale such emergencies is outlined by Rigby and Edelman (31). Such work would appear to be of fairly direct applicability in process control.

The process operator may also commit other, isolated acts which affect reliability. An example is sabotage. Ablitt (1) describes an incident in which attempted suicide by nuclear reactor explosion was suspected.

The preceding categories define human errors. For critical functions it may also be necessary to take into account the possibility of the operator's being totally incapacitated through some cause such as a heart attack or sleep.

These aspects of the quantification of the reliability of a man-machine system in process control are discussed in greater detail in the previous paper (24).

SOME NEEDED DEVELOPMENTS

Criteria for Control System Reliability

A pre-requisite of design for reliability is the existence of suitable criteria of system reliability. The formulation of reliability requirements for the overall process control system is not well developed, and often is not attempted at all. This is in contrast to the situation with protective systems, where reliability criteria are properly defined.

Criteria of Human Control Intervention

It is also desirable that criteria be developed for human control intervention in automatic systems in process control. At present there are diametrically opposed practices to be found. At one extreme the operator is not permitted even to alter controller setpoints, while at the other he may disarm the protective systems.

Methodology for Reliability Assessment

A methodology for the assessment of the reliability of the overall process control system needs to be worked out on the basis of actual case studies. The operator's task must be analyzed using a suitable taxonomy of error and existing knowledge applied where possible.

Integrated Data Collection

There are numerous schemes for the collection of data more or less relevant to reliability on process plants. Most of these are partial and obtain data for particular purposes, such as maintenance. They are often intermittent, depending on the enthusiasm of individuals. Few yield usable data on human error. In the previous paper (24) some of the sources of reliability data in process firms were listed and an integrated scheme was outlined for the collection of reliability data, including data on human error, in a systematic way, so that the relationships between the different types of data are known. Such a scheme in effect constitutes a reliability information system.

The U.K. Atomic Energy Authority has done much to encourage the setting up of data collection schemes and the interchange, through its data banks, of the information obtained, but there are as yet in the process industries few fully integrated schemes which collect the full spectrum of reliability information.

CONCLUSIONS

The process industries are showing an increasing interest in operator reliability and the background to this has been considered. It is desirable, however, to put the emphasis on the reliability of the man-machine system as a whole and on the work situation of the operator rather than on human error. Among the factors which affect the operator's work situation are the process and control system characteristics. These exhibit wide variations which also determine the degree of reliability required.

The reliability of the control system is particularly affected by certain features of the design process and by the way in which certain design problems are solved. Consideration has been given to these aspects which not only are guidelines for system design but are essential factors to consider in system assessment and operator modelling. Quantitative assessment for reliability has also been discussed together with the problems of a taxonomy for and of application of existing work on human error. Mention has also been made of some needed developments in design criteria, assessment methodology and data collection.

REFERENCES

1   J.F. Ablitt, "A quantitative approach to the evaluation of the safety function of operators on nuclear reactors", U.K. Atomic Energy Authority, Risley, England, Rep. AHSB(S) R 160, 1969.

2 J.W. Altman, "A central store of human performance data", in Symp. Quantification of Human Performance, Albuquerque, N. Mex., 1964.
3 J. Annett, K.D. Duncan, R.B. Stammers and M.J. Gray, Task Analysis, 1971 (London: H.M. Stationery Office).
4 S.N. Anyakora and F.P. Lees, "Detection of instrument malfunction by the process operator", Chem. Engr., Lond., 1972, vol. 264, 304.
5 S.N. Anyakora and F.P. Lees, "Principles of the detection of malfunction using a process control computer", Decision, Design and the Computer, 1972 (London: Institution of Chemical Engineers), p.6.7.
6 D. Attwood, "The interaction between human and automatic control", in F. Bolam (ed.), Papermaking Systems and their Control, 1970 (London: British Paper and Board Makers Association), p.69.
7 J.A. Clark, "Display for the chemical plant operator", M.Sc. Thesis, University of Manchester Inst. Science and Technology, 1972.
8 E.R.F.W. Crossman, "Perception study - a complement to motion study", Manager, 1956, vol. 24, 141.
9 E.R.F.W. Crossman, Automation and Skill, 1960 (London: H.M. Stationery Office).
10 D.G. Davies, "A psycho-physiological investigation of process control skill", M.Sc. Thesis, Univ. of Aston in Birmingham, 1967.
11 K.D. Duncan, "Strategies for analysis of the task", in J. Hartley (ed.), Strategies for Programmed Instruction: An Educational Technology, 1972 (London: Butterworth).
12 Elwyn Edwards and F.P. Lees, "The influence of the process characteristics on the role of the human operator in process control", Data Reduction, Communication and Presentation for Process Operation, 1971 (London: Institute of Measurement and Control), p.1.
13 Elwyn Edwards and F.P. Lees, "The development of the role of the human operator in process control", in T.J. Williams (ed.), Interfaces with the Process Control Computer - The Operator, Engineer and Management, 1971 (Pittsburgh, Pa.: Instrument Society of America), p.138.
14 Elwyn Edwards and F.P. Lees, Man and Computer in Process Control, 1973 (London: Institution of Chemical Engineers).
15 P.H. Engelstadt, "Socio-technical approach to problems of process control", in F. Bolam (ed.), Papermaking Systems and their Control, 1970 (London: British Paper and Board Makers Association), p.91.
16 P.M. Fitts, et al. "Human engineering for an effective air navigation and control system", National Research Council, Washington, D.C., 1951.
17 A.E. Green, "Safety assessment of automatic and manual protective systems for reactors", U.K. Atomic Energy Authority,

Risley, England, Rep. AHSB(S) R 172, 1969.
18. W.G. Hiscock, "Selection tests for chemical process workers", Occup. Psychol., 1938, vol. 12, 178.
19. W.A. Huters, "Process control system planning and analysis", Chem. Engng. Prog., 1968, vol. 64(4), 47.
20. J.J. de Jong, "Basic philosophy of computer control in the processing industry", in W.E. Miller (ed.), Digital Computer Applications to Process Control, 1964 (New York: Plenum Press).
21. J.J. de Jong and E.P. Köster, "The human operator in the computer controlled refinery", Proc. World Petroleum Cong., 1971.
22. P.C.M. Kay and P.W. Heywood, "Alarm analysis and indication at Oldbury nuclear power station", Automatic Control in Electricity Supply, 1966 (London: Institution of Electrical Engineers), p.295.
23. J.B. Kitchin and A. Graham, "Mental loading of process operators: an attempt to devise a method of analysis and assessment", Ergonomics, 1961, vol. 4, 1.
24. F.P. Lees, "Quantification of man-machine system reliability in process control", I.E.E.E. Trans. Reliab., 1973, in press.
25. D. Meister, "Methods of predicting human reliability in man-machine systems", Hum. Factors, 1964, vol. 6, 621.
26. H.P. Munro, F.W. Martin and M.C. Roberts, "How to use simulation techniques to determine optimum manning levels for continuous process plants", Chem. Engr., Lond., 1968, vol. 222, 355.
27. K.F.H. Murrell, Ergonomics: Man in his Working Environment, 1965 (London: Chapman and Hall).
28. D. Patterson, "Application of a computerised alarm analysis system to a nuclear power station", Proc. Inst. Elec. Engrs., 1968, vol. 115, 1858.
29. J. Rasmussen, "Man as information receiver in diagnostic tasks", Displays, 1971 (London: Institution of Electrical Engineers).
30. L.V. Rigby, "The Sandia Human Error Rate Bank (SHERB)", Sandia Laboratories, Albuquerque, N. Mex., Rep. SC-R-67-1150, 1967.
31. L.V. Rigby and D.A. Edelman, "A predictive scale of aircraft emergencies", Hum. Factors, 1968, vol.10, 475.
32. W.W. Ronan, "Training for emergency procedures in multiengine aircraft", American Institute for Research, Pittsburgh, Pa., Rep. AIR-153-53-FR-44, 1953.
33. L.W. Rook, "Reduction of human error in industrial production", Sandia Laboratories, Albuquerque, N. Mex., Rep. SCTM-93-62(14), 1962.
34. L.W. Rook, "Evaluation of system performance from rank order data", Hum. Factors, 1964, vol. 6, 533.

35 T.R. Rooney and H.H. Jacobs, "Evaluation of consoles for process operation", in T.J. Williams (ed.), *Interfaces with the Process Control Computer - The Operator, Engineer and Management,* 1971 (Pittsburgh, Pa.: Instrument Society of America), p.103.

36 H. Sackman, *Computers, System Science and Evolving Society,* 1967 (New York: John Wiley).

37 A.I. Siegel and J.J. Wolf, *Man-Machine Simulation Models,* 1969 (New York: John Wiley).

38 A.D. Swain, "Development of a human error rate data bank", Sandia Laboratories, Albuquerque, N. Mex., Rep. SC-R-70-4286, 1970.

39 A.D. Swain, *Design Techniques for Improving Human Performance in Production,* 1972 (London: Industrial and Commercial Techniques).

40 D. Welbourne, "Alarm analysis and display at Wylfa nuclear power station", Proc. Inst. Elec. Engrs., 1968, vol. 115, 1726.

RISK INDICES FOR EVALUATING THE RELIABILITY OF AN
ELECTRICAL SYSTEM.*

G. MANZONI - P. L. NOFERI - L. PARIS - M. VALTORTA

ENEL - ITALY

0. ABSTRACT

In order to have a quantitative evaluation of the reliability of an electrical system, it is necessary to refer to a number, known as risk index, that measures the security of the system itself.
A definition is given here of some risk indices, each of which takes into account one of the various ways in which the system shows itself to be inadequate to perform its tasks. These are risk indices that characterize the state of permanent, temporary, and dynamic insufficiency.

1. THE RELIABILITY OF ELECTRICAL SYSTEMS

When we speak of the "reliability" of an electrical system in general, we do not refer to the <u>numerical quantity</u> deriving from the classic definition: "probability that a given apparatus will have to perform its functions for a preestablished period under given conditions." Rather do we refer to the <u>concept of the potential ability</u> of the system to perform its function.
It may be assumed that an electrical system performs its functions when it is able to meet the load within the required frequency and voltage limits at all the points at which it is designed to supply it. Given the vastness and complexity of the system, which has to feed an extremely broken-up load, and the indefinite period of operation, it is obvious that the probability of performing its functions completely is, at best, nil, even if the system as a whole has a "reliability" that is, generally speaking,

---
*The full length article is available from the authors.

acceptable from a practical point of view, inasmuch as inadequacies of big magnitude are very rare, while more numerous inadequacies are smaller;
In any case, even if, for an electrical system, the classic definition of reliability cannot be applied, it is necessary to single out one or more values (expressed in probabilistic terms) that may measure the potential ability of the system to perform its functions.
From the practical point of view, it is preferable to make the measurement with reference to situation in which that ability falls short. The values selected to this end are called "risk indices"; in general, they measure, in terms of mathematical expectation, the average breakdown characteristics (e. g., frequency, duration, magnitude) affecting the load as a consequence of any insufficiency that may occur.

2. "STATIC", "TEMPORARY" AND "DYNAMIC" RISK INDICES

The risk indices that can be arrived at quickest, and that are therefore most often used, are those deriving from a comparison, at given instants, in probabilistic terms, between the power (availability) that the system is able to supply under normal conditions, with the components at that moment available, and the power of the load (demand).
These indices are called static (see Section 3) precisely because they do not take into consideration the transient phenomena that occur in an electrical system from the moment in which a forced outage in a component takes place until a new steady state situation is reached.
Obviously, as a consequence of the transient phenomena, the disconnected load may even be greater than the difference between availability and the demand in the new state of the system. To measure the ability of the system to overcome these transients, indices that we shall call dynamic must be used (see Section 5).
It should be remembered that static risk indices take into consideration the availability of the system that emerges when considering faults of relatively long duration, an hour, or even a day, since these require repair and are, therefore, the faults commonly defined as permanent. These indices do not therefore indicate the faults known as transients, lasting fractions of a second up to a few minutes, caused, for example, by the temporary breakdown of air insulations characteristics of overhead lines. In order to take into account these faults too, it is neces

sary to introduce a further class of indices, which we shall call temporary

3. STATIC RISK INDICES

The main among these risk indices, which, as has already been said, are the most often used, may be classified into two main categories. The first are indices expressed by only one parameter (monoparametric) and the second, indices expressed by two parameters (biparametric).

3.1. - Monoparametric Risk Indices

The risk indices belonging to this category are those that measure the reliability of a system by means of the expected value of a suitable quantity representing the breakdown, assumed as the risk parameter, within a given period of time (for example, a year).

In order to evaluate these indices, the behaviour of the various system components is defined by means of the average forced outage rate without stating the duration and frequency characteristics of the outages.

These indices, therefore, provide a measure of the average behaviour of the system during the period under investigation, without stating whether the risk situation of the system itself corresponds to a single long event, or to a certain number of short events.

Of the indices belonging to this category, the following may be considered, (see also Fig. 1).

3.1.1. - The probability of not meeting the annual load peak

In this case, judgement as to the reliability of the system is based simply on the probability of being able, or not, to supply the maximum annual power demand. The risk index is evaluated by comparing the probabilistic distribution of the maximum annual load with that of the system's power availability in the period of the year in which the peak occurs.

Obviously, a judgement that is based only upon this index can be sufficient only if, with that index, a certain degree of confidence has been acquired; and this presupposes a stable relationship between the peak value and the values for the rest of the year. This index was originally applied in order to evaluate the reliability of generation systems; therefore, it is applied, generally, only to bus-bar systems.

3.1.2. - <u>Expected number of days in the year in which maximum daily demand (peak) is not met.</u>
　　This risk index is obtained by adding together the probabilities, calculated for every day in the year, of not being able to meet daily load peak.
It is therefore much more significant than the preceding index, because it takes into account situations that occur throughout the year, albeit only those that correspond to daily peaks.
In this case, too, as in the preceding one, the index takes into consideration only the occurrance of the event "inability to meet maximum load value", without evaluating the extent of the corresponding insufficiency; therefore, for example, a lack of 1 kW is judged in the same way as a lack of some hundreds of MW.

3.1.3. - <u>Expected Annual Value of power shortage and daily load peak.</u>
　　In order to take into account, in some way, the extent of the insufficiency, it is possible to assume as risk index the sum of the average expected values of the power shortage at the peak of the various days in the year.
This index may be expressed directly in power units (MW), or in a value relating to the sum of the daily peaks during the year (expected values).

3.1.4. - <u>Annual expected value of the square (or other power) of the power shortage at daily load peak.</u>
　　In order to stress the importance of the extent of the inefficiency, the risk index assumed may be the sum of the expected values of the square (or other power) of the power shortage at the peak on the various days in the year.
This index may be expressed directly as an absolute value, in MW^2 or in MW^n, or else as a value relating to the sum of the squares, or of the nth powers, of the daily peaks during the year (expected values).

3.1.5. - <u>Expected number of days in the year on which pre-established fractions of maximum daily demand are not met.</u>
　　This index too tends to take account of the extent, and not only of the circumstance of insufficient availability. It is a question of evaluating not only the risk (expressed in days per annum) of not meeting the peak on various days of the year (3.1.2.), but also the risk of not meeting, on various days in the year, some fractions $(1-\alpha)$ of the peak power. The risks,

related to different values of α, are then weighted by means of a suitable linear combination.

Levels α and weighting coefficients may be chosen depending on the types of measure that have to be taken at various levels: lowering of voltage and frequency, disconnecttion of less or more important loads.

3.1.6. - Expected value of energy not supplied

This is the sum of the average expected values in respect of power shortage at all hours during the year; assuming that the load diagram is expressed in hourly intervals, the index represents the expected value for energy that cannot be supplied to the load.

This index may be expressed directly in energy units (MWh) or in a value relating to the energy (average expected value) required annually by the load.

The risk index in terms of expected value for energy not supplied is particularly suitable to dealing with systems with several supply points; indeed, whatever may be the form of the load diagrams at the individual points, the index can be defined in the same way for each of them, and the overall index, if expressed as an absolute value (MWh), is only the sum of the various local indices.

3.1.7. - Combinations of monoparametric indices

Sometimes, in order to judge the reliability of a system, particularly with a view to making an economic evaluation of it, use is made of a linear combination of the aforementioned indices, each of which takes into consideration a specific aspect of the inconvenience caused by the breakdown.

For example, an expression may be used that weights not only the number of days in the year in which breakdown occurs, which in itself causes inconvenience, but also the energy not supplied that measures the extent of the breakdown.

3.1.8. - Links between the monoparametric indices.

The full-lenght article shows that the various monoparametric indices are interlinked. Some examples are reported to illustrate these links and their dependence on the structure of the electrical system.

3.2. - Biparametric risk indices

The risk indices belonging to this category are those that measure the reliability of a system not only by means of the annual expected average value of the quantity representing the breakdown (duration or extent), but also by means of the frequency (number of times per annum) with which the breakdown occurs.

Obviously, besides the frequency of the occurrence, it is possible to calculate, as opposed to the annual expected average value, the expected average value for each single occurrence by dividing the annual expected average value by the annual frequency of the occurrence.

In order to evaluate these indices, the behaviour of the various components, as far as their unavailability is concerned, must be defined not only by the value of the average forced outage rate, but also by the average frequency, (or duration) of individual non-availability.

3.2.1. - The expected number of consecutive hours in which available power is less than the load peak value (duration of occurrence) and expected average frequency of occurrence.

This index is of concrete significance only if the load is constantly equal to its peak value, or if the availability is expressed as on-off (for example, a load connected to a radial system).

3.2.2. - Expected number of consecutive hours in which the load is not met (duration of occurrence) and expected average frequency of the occurrence.

This index, which is analagous to the preceding one, however, takes into account the actual behaviour of the load over a period of time, and is therefore significant.

3.2.3. - Expected number of consecutive days in which the load peak is not met (duration of occurrence) and expected average frequency of the occurrence.

This index may be considered an extension of the monoparametric index dealt with in Section 3.1.2.

3.2.4. - Expected value of the energy not supplied for each occurrence (magnitude of occurrence) and its expected average frequency.

This index may be considered an extension of the monoparametric index dealt with in Section 3.1.5.

4. TEMPORARY RISK INDICES

These indices take into consideration the reliability of a system from the point of view of those unavailabilities that in volve temporary shortage of supply for intervals of time ranging between fractions of a second and many minutes (example switching operation).

In general, the temporary interruptions can be classified according to their duration (classes of duration); for example a) interruption of a duration not greater than a second b) interruptions lasting a few minutes (corresponding to reclosing, or fast or slow automatic switching) c) interruptions of dozens of minutes (due to the manual operations needed for eliminating the faulty component and for possibly restoring supply by other means).

The gravity of a temporary interruption depends on the class of the duration and on the type of load curtailed (for certain types of loads even one interruption of few dozen seconds, may involve interrupting productive activity for a number of minutes. A typical diagram relating to the consumption trend after a very short temporary interruption in supply (three-phase opening and reclosing) is shown in Fig. 2.

With regard to different load behaviour with different classes of interruption duration, different risk indices are generally required.

4.1. - Annual expected value of the temporarirly interrupted power for a given duration class.

This index, within the framework of the respective duration classes, does not take into account the actual duration of the interruption.

4.2. - Annual expected value of the energy temporarily interrupted due to a given class of duration.

This index is obtained by multiplying the power interrupted by the corresponding interruption duration, and it does not take into account the energy curtailed while taking resumption of the load after the supply has been restored (depending on the characteristics of the load and not on those of the system).

5. DYNAMIC RISK INDICES

It is obvious that static risk indices do not provide complete information on the reliability of an electrical system; indeed, they completely ignore the ways in which the reduction in load, caused by possible non-availability of the system, is actually made.

As is known, reduction of the load caused by non-availability may occur either through the dispatcher, according to program me, on the basis of short or long-term forecast, or else suddendly, at the very moment in which there takes place in the system a change in availability, due to the effect of automatic devices (under frequency relays), or because the loads are separated from the generation, or, lastly, due to instability in the operation of the system.

Obviously, in these cases, the disconnected load may be considerably greater than the load corresponding to the reduction of the static capacity of the system (see Fig. 3a) and might also reach the whole load of the system (black-out-see Fig. 3b).

Let us consider for example two systems, both operated according to the same criteria, consisting of units of very different size but with the reserve laid down for each system in such a way that the two systems have the same static risk index. It might therefore be said that the systems are equally reliable. However, it should be noticed that the first system characterized by larger units is not, in practice, as secure as the other. In fact, the dynamic consequences of forced outage to generator are far more serious in the first system than in the second. The following dynamic risk indices may be considered:

5.1. Probability of Frequency during a transient being below the critical Level.

Since, below a certain frequency value (critical level), the system cannot operate, in particular due to the inability of the thermo-electric stations to operate, the corresponding probability coincides with the probability of total breakdown or black-out.

This index takes into consideration only the more serious and rarer consequence of the breakdown of a component of the system, and therefore does not give sufficient information for a fairly secure system, in which less serious but more probable occurrences may be of greater interest.

5.2. - Average Expected Value for Suddenly Disconnected Power.

This index also measures slighter inconveniences, apart from total breakdown, caused by a fault in a component of the system.

Suddenly disconnected power (indicated by S in Fig. 3) consists of:

- Power required by load that are isolated from generation
- Power disconnected from the under frequency relay
- Power required by the loads of the whole system, (or parts

there of that have remained separated) when the frequency drops below the critical value.

6. USE OF RISK INDICES FOR SYSTEMS PLANNING

The full length-article explains how the risk-indices can be used for system planning: e.g. for developing systems having the same level of risk and for giving a penalty to systems having different reliability, by means of the "cost of risk".

REFERENCES

The full-length article reports a list of 15 references.

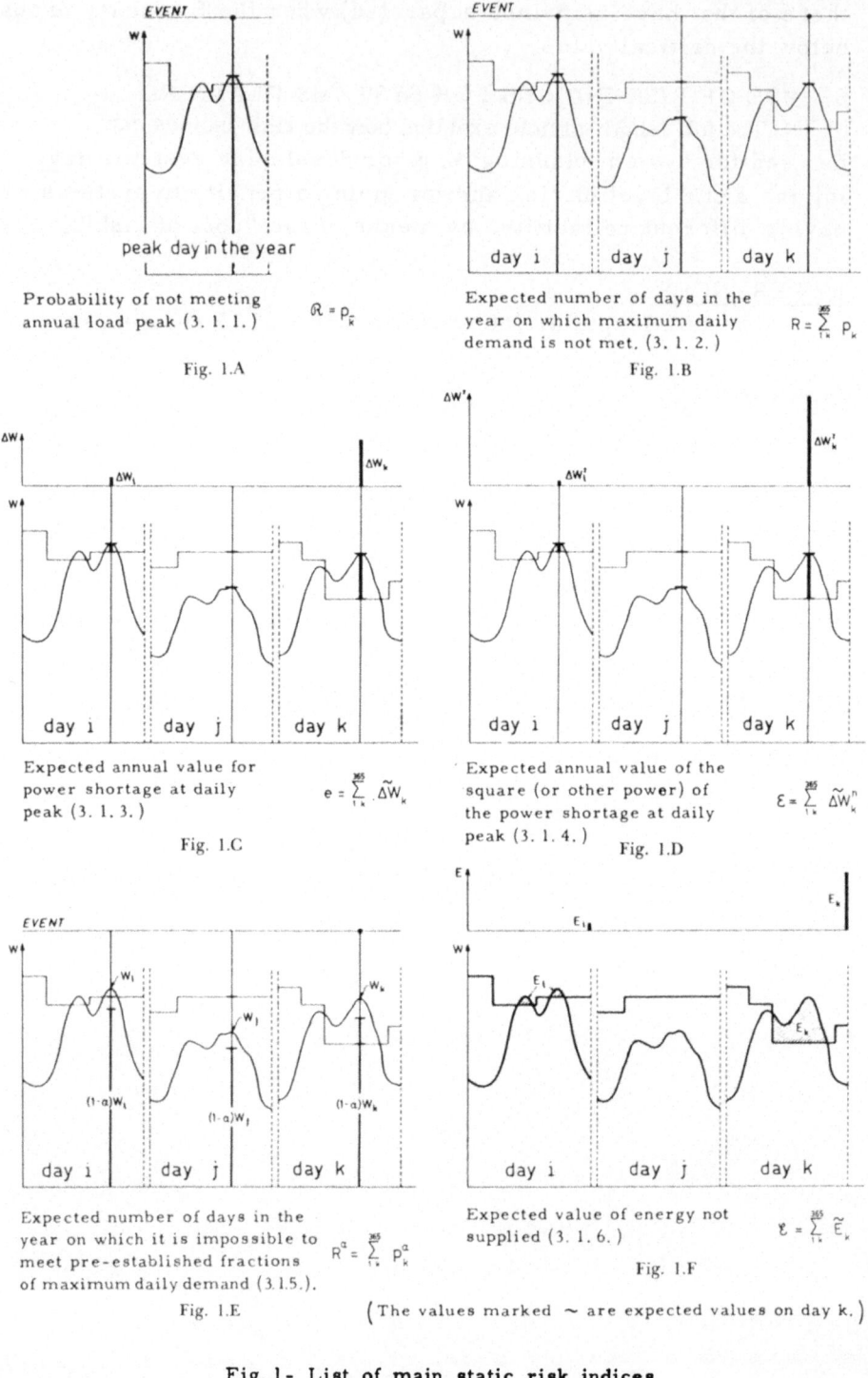

Fig. 1 - List of main static risk indices.

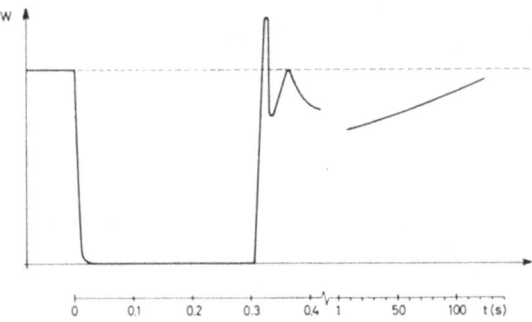

Fig. 2 - An example of the active power consumption trend, following a very short temporary interruption (0.3s) in supply (three-phase opening and reclosing).

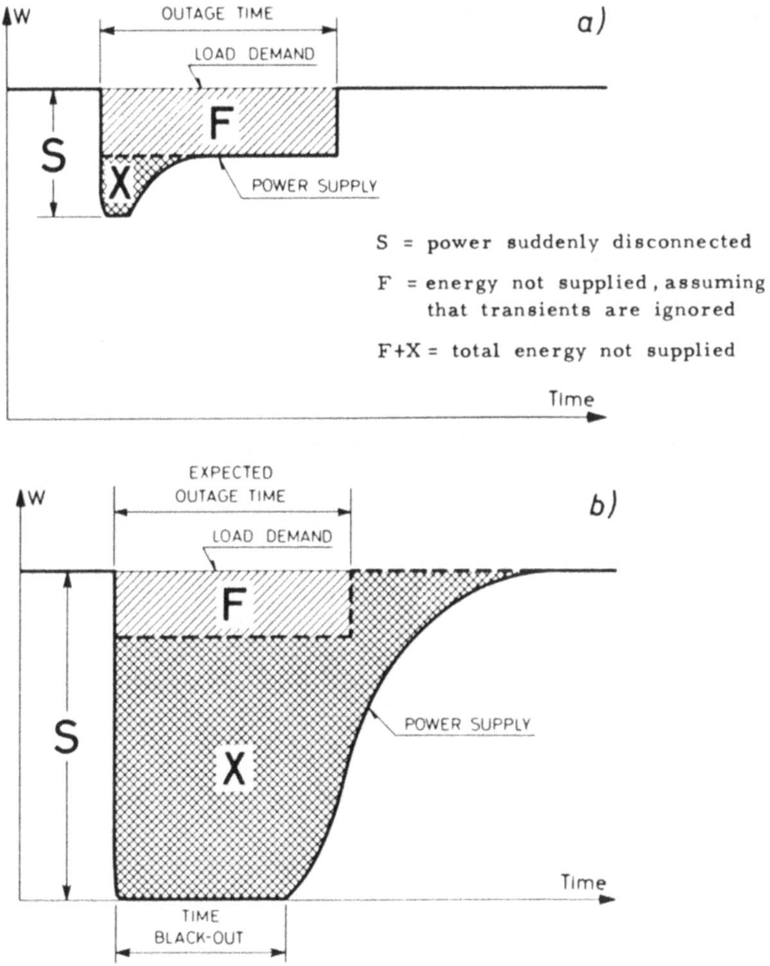

Fig. 3 - Behaviour of power supply following forced outage of a component.

S = power suddenly disconnected

F = energy not supplied, assuming that transients are ignored

F+X = total energy not supplied

RELIABILITY IN MECHANICAL DESIGN AND PRODUCTION

P. MARTIN

THE UNIVERSITY OF LIVERPOOL

1. RELIABILITY OF MECHANICAL EQUIPMENT

The concept of reliability quantification in probabilistic terms is based essentially on the statistical analysis of the times to failure of components operating in actual or simulated service conditions. This concept is difficult to apply to one off items of mechanical equipment. With a view to identifying a systematic approach to this problem, the steps involved in the design and construction of mechanical equipment are briefly examined.

2. MECHANICAL ENGINEERING DESIGN AND PRODUCTION

The information available for the design of engineering systems has evolved from the study of basic categories of engineering elements which may be regarded as:-

1. Structural
2. Kinematic
3. Fluid
4. Thermodynamic
5. Material conversion
6. Electric current
7. Electromagnetic
8. Electronic
9. Nuclear
10. Computer software.

The basic activity involved in the design of mechanical equipment consists essentially of combining elements of the first five of these categories in such a way that their resulting behaviour will satisfy a defined requirement. Interactions between these elements can be difficult to define and this leads to lack of precision in design analysis. The method utilised for storing product design information is usually confined to the use of schemes and detailed drawings backed by technical reports. This information is difficult to check. The many production methods available modify the material properties of components in such a way that laboratory testing does not closely indicate the true strength of manufactured components. These difficulties give rise to low reliability.

In view of this situation it is desirable to build up statistical information on the behaviour of manufactured components in service conditions to provide a basis for quantitative design predictions. The following concept provides a basis for the organisation of such information for mechanical components.

3. SYSTEM FAILURE MODES

Each category of the mechanical engineering elements could be regarded as having a number of failure modes which typify functional failure in equipment. These are:-

1. Structural system failure modes - Fracture, excessive deflection.
2. Kinematic system failure modes - slide or bearing siezure, reduced accuracy of relative movement.
3. Fluid system failure modes - leakage, wrong flow conditions or distorted flow.
4. Thermodynamic system failure modes - Overheating, reduction of efficiency.
5. Material conversion failure modes - Incorrect material properties, incorrect component geometry.

The following significant stages in the design procedure for a particular piece of equipment are given to indicate how this system failure mode concept could be used for checking system interactions during design and also provide a basis for mechanical system failure analysis compatible with existing probabilistic approaches.

4. DESIGN STUDY

Safety considerations related to the operating characteristics in certain modern power stations resulted in a requirement for a new design of actuating system for an electrical relay. Experience with the type of relay indicated the following requirements would ensure satisfactory latching:-

1. The closing mechanism would have to be capable of introducing a defined amount of energy.
2. The relay arm contact velocity at contact touch must be controlled within close limits.
3. The relay arm movement time must be closely controlled.

A design evaluation of possible actuating systems indicated that a low pressure pneumatic system would be likely to satisfy the required design specification. The essential elements of the proposed system could be regarded as three basic subsystems comprising:-

1. The main circuit components and associated insulated components.
2. A mechanical system consisting of a four bar linkage operating with an additional auxiliary slide bar link in the form of a pneumatic ram.
3. A pneumatic system able to supply sufficient energy in the form of compressed air to the pneumatic ram.

Operation of the relay is achieved by operating the solenoid of a spool valve thus allowing air from the reservoir to expand into the operating cylinder.

4.1 System Synthesis

To synthesise a system capable of introducing the required amount of energy and to provide the end of travel piston force dictated by the electrical performance requirements involved the analysis of the transient conditions during closure.

4.2 Detail Design

The detail design of the component parts of the relay was largely dictated by reliability requirements and the necessity to minimise cost. Detailed design features were included to compensate for corrosion, fretting and misalignment. The location of these reliability features was aided by the preparation of a system function diagram as depicted in Figure I. This diagram represents the mechanical system constituents of the mechanism and the pneumatic system. A fluid system F, a kinematic system K and a structural system S, with the salient points of interaction is defined. The information resulting from a large volume of detailed graphical information is thus simply coded for the examination of interactions.

5. SYSTEM INTERACTIONS

The concept of the system function diagram may be taken a stage further. The non redundant interaction points between the three systems are depicted in Figure 2.

6. GENERAL APPLICATION OF SYSTEM CONCEPT

The above concept could be used as a basis for the functional and system interaction analysis of a variety of mechanical systems by providing a basic approach for such systems.

The performance conditions at each of the node points labelled $\{F_i \ (i = 1, 2, \ldots n), S_i \ (i = 1, 2, \ldots n) \text{ etc.}\}$ as shown in Figure 2, at the system boundaries may be quantified. In this way the equivalent of the electrical circuit may be derived for mechanical equipment while it is being designed.

7. CONCLUSIONS

Development of the above concept would lead to

1. A simple systematic basis for checking detail designs.
2. A method of coding information on interactions to provide the basis for fault tree analysis.
3. A means of combining an analysis of a variety of mechanical systems with information resulting from data banks enabling common interacting elements to be located in a variety of equipment types. In this way statistically significant numbers of common elements could be located even in equipment types manufactured in small numbers.

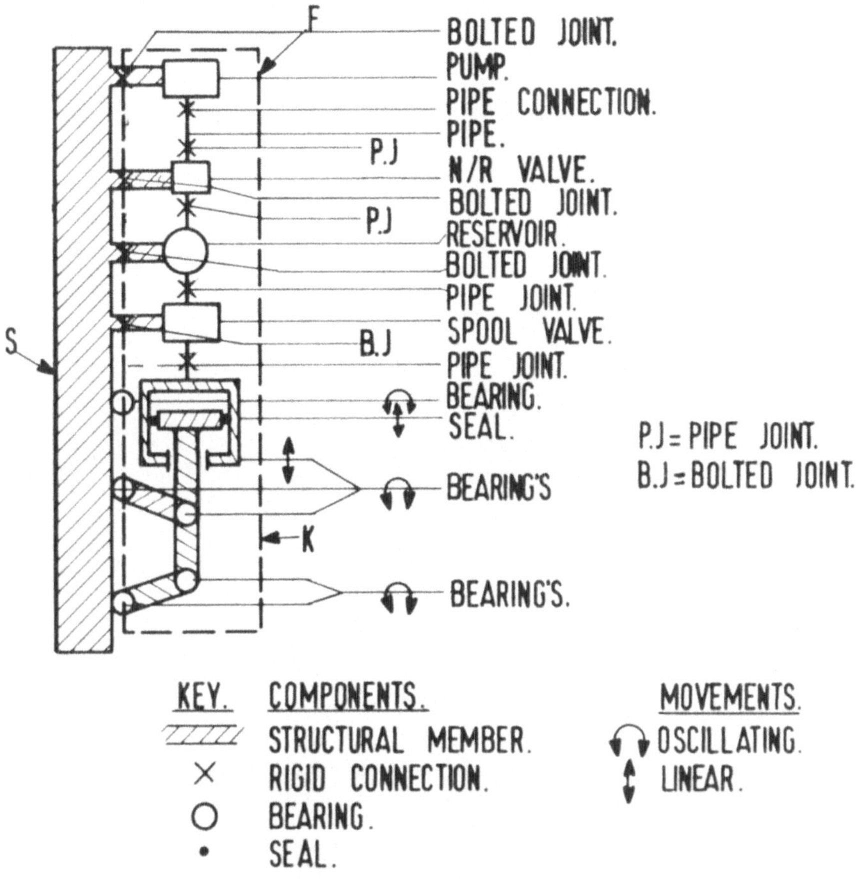

271

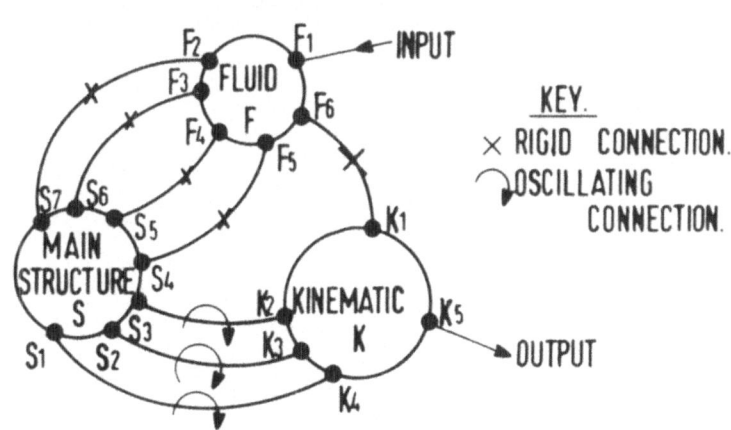

FIG.1. SYSTEM FUNCTION DIAGRAM.

FIG.2. SYSTEM INTERRACTIONS.

THE ANALYSIS AND SYNTHESIS OF SAFE AND SERVICEABLE STRUCTURES

JOHN MUNRO

CIVIL ENGINEERING DEPARTMENT,
IMPERIAL COLLEGE OF SCIENCE AND TECHNOLOGY,
LONDON.

INTRODUCTION

The purpose of this paper is to present a brief conspectus of some aspects of a systems approach to structural design. Important differences separate the study of structural reliability from the main-stream of reliability engineering - notably in the complexity of structural elements and their interactions (Lovelace, 1972). The random nature of many loadings - wind and seismic loading, for example - is combined with the random nature of the strength and stiffness properties. Indeed the synthesis of structural systems in the face of uncertainty presents such difficulty that most practical design procedures have replaced the essentially stochastic and dynamic nature of the problem by mathematical models which are deterministic and quasi-static. Frequently statistically based characteristic strengths and loads are specified and the designer confines his attention to devising trial designs and, by stress, deformation and strength analyses, ensuring that the structure performs adequately with respect to the specified criteria or limit states. These code requirements can be considered in two convenient groups. First the serviceability conditions ensure satisfactory performance at the specified working load and secondly the ultimate load conditions ensure that the structure is safe in the sense that it will not collapse, or malfunction in some defined way, before some specified overload level. Thus the design process is reduced in this first consideration to the successive re-analysis of deterministic models until a feasible solution is generated.

More recently much work has been devoted to the development of direct design procedures using mathematical programming. Such

synthetic methods are clearly more efficient since they can
optimise the design with respect to any desired criterion. However
they tend to produce multiple failure modes and therefore tend to
produce diminished failure probabilities when viewed stochastically.
This is in part due to the use of deterministic codes which are
based on experience acquired in a non-optimisation climate. Thus
for completeness the mathematical programming methods which seek
the most economical design should incorporate the probability of
failure and also the associated cost of failure. Thus the structural synthesis problem becomes one of stochastic programming.

The selection of material to be included in such a brief
summary is necessarily subjective. The present paper emphasises
the role of mathematical programming in structural synthesis. The
deterministic analysis and synthesis for both elastic and plastic
materials is presented fairly fully and, in the later portions of
the paper, some aspects of stochastic analysis and synthesis are
discussed more briefly. The discussion will be restricted to
frame structures.

DETERMINISTIC STRUCTURAL ANALYSIS

Introduction

A final design is frequently achieved only after successive
analyses of conjectured trial designs. Thus for any stage of this
design process the structural geometry and loadings are completely
known. The problem is to evaluate the response of the proposed
structure to the specified loadings and to assess that performance
against specified constraints and criteria. The two main types of
constitutive relations used in such analyses are (i) linearly
elastic and (ii) perfectly plastic. In the elastic case the main
strength constraint is that the frame will support the specified
(working) loadings without exceeding the specified limiting
elastic stresses whilst in the plastic case the main strength
constraint is that the frame will support loadings corresponding
to the working loadings increased by a specified load factor (or
factors) without collapsing plastically. The latter considerations
are facilitated by the "plastic hinge" idealisation and by the
association of plastic collapse with the formation of sufficient
plastic hinges to form a mechanism.

The methods of elastic and plastic analysis will be briefly
described in later portions of this section but first it is convenient to introduce a fundamental concept of structural mechanics
which will be used as the basis of the subsequent discussion.

Structural duality

In the present consideration, the vectorial nature of the fundamental entities plays an important role. The state of stress across an element of the frame can be identified through the simplifying assumptions of engineering theory by a set of generalised forces which will be termed stress-resultants $(\underline{x})$. When a set of loads $(\underline{\lambda})$ are applied to a determinate frame the stress-resultant distributions can be derived directly from statics after making the usual assumption of small displacements.

$$\underline{x} = \underline{Q}\underline{\lambda} \qquad (2.1)$$

Similarly the deformations across a small element (ds) of the frame can be represented by a set of strain-resultants $(\underline{du})$ and if at discrete locations the displacements $(\underline{\Delta\delta})$ due to the strain-resultants $(\underline{du})$ at some element are expressed in the form

$$\underline{\Delta\delta} = \underline{R}\,\underline{du} \qquad (2.2)$$

then it can be shown that the transformations are contragredient.

$$\underline{R} = \underline{Q}^T \qquad (2.3)$$

However, in general, frames are indeterminate and the indeterminacy number (α) (Henderson and Bickley 1955) is given by : -

$$\alpha = ac - r \qquad (2.4)$$

where "a" is six for the general case and three for the planar case ; "r" is the number of stress-resultant releases (hinges, slides, etc.) built into the frame and "c" is the cyclomatic number (Berge, 1962) of the graph of the frame. The static significance of this number is that it gives the number of independent self-equilibrating stress-resultant distributions for the frame and the kinematic significance is that it gives the number of suitable releases which must be introduced so that the deformed shape of the frame, associated with the introduction of any strain-resultant at any element of the frame, can be identified from purely kinematic (geometric) arguments and without recourse to the constitutive relations.

For any indeterminate frame the statics can be associated with a particular stress-resultant distribution $(\underline{x}_o)$ which is in equilibrium with the loads $(\underline{\lambda})$ and a complementary stress-resultant distribution which has α components.

$$\underline{x} = \underline{x}_o + \underline{x}_c \qquad (2.5)$$

where
$$\underset{\sim}{x}_o = \underset{\sim}{Q}\underset{\sim}{\lambda} \qquad (2.6)$$

and
$$\underset{\sim}{x}_c = \underset{\sim}{H}\underset{\sim}{p} \qquad (2.7)$$

The elements of the functional matrices $\underset{\sim}{Q}$ and $\underset{\sim}{H}$ are known influence functions and the vector $\underset{\sim}{p}$ represents the so far unknown indeterminacies.

Now strain-resultants ($\underset{\sim}{du}$) are associated with the stress-resultants and displacements are associated with the loads. Also kinematic discontinuities ($\Delta\underset{\sim}{v}$) are associated with the indeterminacies ($\underset{\sim}{p}$). These discontinuities could, for example, be angular discontinuities across simulated hinges and the indeterminacies could be bending moments at the corresponding locations. The basic algebraic structure of the related statics and kinematics can be represented in the following diagrams.

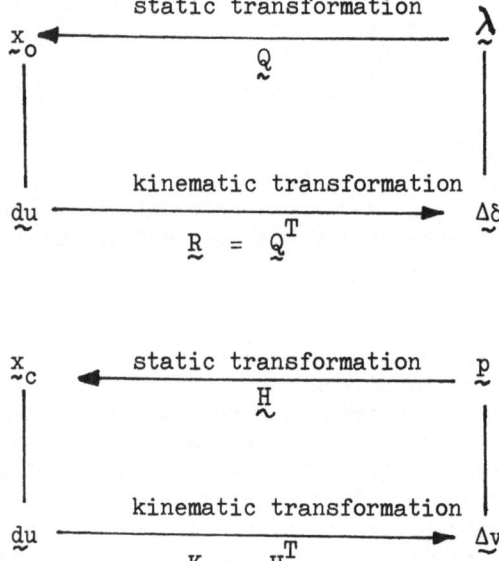

The totality of such relationships is dubbed "static-kinematic duality" and using this concept a primal problem which is purely static can be transformed to a dual kinematic problem and similarly a primal problem which is purely kinematic can be transformed to a dual static problem. The primal-dual nature of the fundamental (vectorial) variables of the mechanics of frames will later be shown to be directly related to the duality of the mathematical programs of structural synthesis.

Elastic analysis

The previous sub-section outlined some fundamental relations in the statics and kinematics of frames. The solution of indeterminate frame problems requires the specification of constitutive relations which mathematically model the material properties. The simplest such relationship is that of a linearly elastic material where the generalised strains ($\underset{\sim}{e}$) are linearly linked to the stress-resultants ($\underset{\sim}{x}$) and are independent of temporal or thermodynamical variables.

$$\underset{\sim}{e} = \underset{\sim}{F}\,\underset{\sim}{x} \qquad (2.8)$$

where $\underset{\sim}{F}$ is the elemental flexibility matrix. Engineering theory uncouples these constitutive relations by reducing this flexibility matrix to a diagonal form.

Assuming small deformations, the generalised strains across an element (ds) are related to the strain-resultants ($\underset{\sim}{du}$) in the following linear way :-

$$\underset{\sim}{e}\,ds = \underset{\sim}{du} \qquad (2.9)$$

Combining equations (2.8) and (2.9) yields

$$\underset{\sim}{du} = \underset{\sim}{F}\,\underset{\sim}{x}\,ds \qquad (2.10)$$

Recalling the kinematic transformation,

$$\underset{\sim}{\Delta v} = \underset{\sim}{K}\,\underset{\sim}{du} \qquad (2.11)$$

then,

$$\underset{\sim}{\Delta v} = \underset{\sim}{K}\,\underset{\sim}{F}\,\underset{\sim}{x}\,ds \qquad (2.12)$$

The discontinuities at discrete locations have now been expressed in terms of the stresses at a particular element. Assuming small displacements, the discontinuities due to the stresses at all the elements can be obtained by superposition

$$\underset{\sim}{v} = \sum \underset{\sim}{\Delta v} = \oint \underset{\sim}{K}\,\underset{\sim}{F}\,\underset{\sim}{x}\,ds \qquad (2.13)$$

However the stress-resultants can be expressed in the following superposed form :-

$$\underset{\sim}{x} = \underset{\sim}{x}_c + \underset{\sim}{x}_o = \underset{\sim}{H}\,\underset{\sim}{p} + \underset{\sim}{Q}\,\underset{\sim}{\lambda} \qquad (2.14)$$

From equations (2.13) and (2.14)

$$\underset{\sim}{v} = \underset{\sim}{G}\,\underset{\sim}{p} + \underset{\sim}{v}_o$$

where
$$\underline{G} = \oint \underline{K}\,\underline{F}\,\underline{H}\,ds$$

and
$$\underline{v}_o = \oint \underline{K}\,\underline{F}\,\underline{x}_o\,ds$$

The discontinuities ($\underline{v}$) are generally zero and so the elastic compatibility equation becomes : -

$$\underline{G}\,\underline{p} + \underline{v}_o = \underline{0} \qquad (2.16)$$

and the solution to this equation can be symbolically represented as : -

$$\underline{p} = -\underline{G}^{-1}\,\underline{v}_o \qquad (2.17)$$

and the stress-resultants become :

$$\underline{x} = \underline{Q}\lambda - \underline{H}\left[\oint \underline{K}\,\underline{F}\,\underline{H}\,ds\right]^{-1}\left[\oint \underline{K}\,\underline{F}\,\underline{x}_o\right] \qquad (2.18)$$

where as previously described

$$\underline{K} = \underline{H}^T$$

Having obtained the stress-resultants from equation (2.18) above, the displacements ($\underline{\delta}$) can now be obtained from

$$\underline{\delta} = \oint \underline{R}\,\underline{F}\,\underline{x}\,ds \qquad (2.19)$$

When the above intrinsic elastic theory of skeletal structures is used as the method of analysis it is frequently referred to as the "flexibility method" or "mesh algorithm" and can be considered to be a generalisation of one of the two Kirchhoff network laws in which the potential differences around a mesh of the network are replaced by the kinematic discontinuities. The other Kirchhoff law also has its counterpart in elastic structural analysis ; when the sum of the currents at a node is replaced by the nodal equilibrium relations then the "stiffness method" or "nodal algorithm" results. The essential computational feature of elastic analysis is that the two algorithms are concerned with the solutions of matrix equations.

Plastic analysis

Consideration will be confined herein to proportional loading; that is a load system which is fixed by a single parameter λ. The main problem of plastic limit analysis is to evaluate the load factor (λ_c) at which the frame collapses plastically. The analysis is based on two fundamental propositions. The automated plastic analysis of frames can then be derived in the form of mathematical programs. The following expositive treatment is confined to flexural frames and rolled steel sections.

First the sections at which plastic hinges may form are identified. Let "c" be the number of such critical sections. The equilibrium equation for the ith critical section can be written as : -

$$\underset{\sim}{m}^i = (\underset{\sim}{k}^i)^T \underset{\sim}{p} + \lambda \underset{\sim}{r}_o^i \qquad (2.20)$$

where r_o is the particular solution bending moment per unit value of λ. The (positive) magnitude (m_*^i) of the positive and negative plastic moments of resistance will be known and the yield conditions become : -

$$-m_*^i \leqslant m^i \leqslant +m_*^i \qquad (2.21)$$

If the equilibrium and yield conditions are satisfied for all values of i (= 1, 2, ..., c) then the corresponding value of is termed a safe load parameter (λ_s). The adjective "safe" is justified by the safe theorem which states that the plastic collapse load parameter (λ_c) is the largest λ_s.

$$\lambda_c \geqslant \lambda_s$$

The safe theorem requires the maximisation of λ subject to the yield and equilibrium conditions and so the mathematical program for the safe method of plastic limit analysis becomes : -

$$\text{Max } w = \begin{bmatrix} 1 & \underset{\sim}{0}^T \end{bmatrix} \begin{bmatrix} \lambda \\ \underset{\sim}{p} \end{bmatrix}$$

$$\begin{bmatrix} \underset{\sim}{r}_o & \underset{\sim}{c}^T \\ -\underset{\sim}{r}_o & -\underset{\sim}{c}^T \end{bmatrix} \begin{bmatrix} \lambda \\ \underset{\sim}{p} \end{bmatrix} \begin{matrix} \leqslant \\ \leqslant \end{matrix} \begin{bmatrix} \underset{\sim}{m}_* \\ \underset{\sim}{m}_* \end{bmatrix} \qquad (2.22)$$

It will be seen that the above program is entirely linearised and consequently after transforming the unrestricted variables to non-negative form the solution can be obtained from the simplex algorithm of linear programming (Dantzig, 1963). The optimal values of the static variables give the value of the collapse load parameter (λ_c) and the corresponding indeterminacies.

The unsafe method of plastic analysis is based on the study of potential collapse mechanisms. It can be shown (Neal, 1956) that all possible collapse mechanisms can be generated from a set of "b" independent mechanisms where

$$b = c - \alpha \qquad (2.23)$$

This generation of mechanisms was greatly assisted by the Neal-Symonds technique of combining "elementary mechanisms" (Neal and Symonds, 1952). More recently the adaptations using "basic mechanisms" (Munro, 1965) have facilitated a dual manipulation of statics and kinematics. Thus the safe and unsafe methods can proceed from the same influence functions. If α suitable critical sections are selected for the basis then any additional critical section can be added to the α basic sections and these can be considered to be the $(\alpha + 1)$ plastic hinges of a basic mechanism. From the previous discussion of kinematics the mechanism compatibility conditions can be derived and these will link the mechanism deformations ($\Delta\theta$) at all the critical sections which participate in the considered mechanism. All the critical sections must also satisfy the parity conditions : -

$$m^i \, \Delta\theta^i \geqslant 0 \qquad (2.24)$$

which ensure corresponding sense of the related static and kinematic quantities. Thus if the parity and mechanism compatibility conditions are satisfied, an unsafe load parameter (λ_u) is defined by : -

$$\lambda_u = \frac{m^T \, \Delta\theta}{r_o^T \, \Delta\theta} \qquad (2.25)$$

and the adjective "unsafe" is justified by the unsafe theorem which states that the plastic collapse load parameter (λ_c) is the lowest λ_u.

$$\lambda_c \leqslant \lambda_u \qquad (2.26)$$

The optimality criterion for the unsafe method is therefore minimisation of λ_u which is expressed in equation (2.25) as a linear fractional form. This apparently non-linear objective function can however be linearised by recalling that the kinematic variables ($\Delta\theta$) are related to mechanism deformation modes which are fixed up to an arbitrary scale factor. By normalising the denominator of the linear fractional form the numerator becomes the new objective function. Next the kinematic variables ($\Delta\theta$) are expressed through complementary variables ($\Delta\theta^+, \Delta\theta^-$)

$$\Delta\theta^i = \Delta\theta^{+i} - \Delta\theta^{-i}$$

$$\Delta\theta^{+i} \geqslant 0 \qquad \Delta\theta^{-i} \geqslant 0 \qquad (2.27)$$

$$\Delta\theta^{+i} \, \Delta\theta^{-i} = 0$$

$\Delta\theta^{+i}$ is the (positive) magnitude of the positive plastic hinge mechanism deformation whilst $\Delta\theta^{-i}$ is the (positive) magnitude of the negative plastic hinge mechanism deformation.

Combining all of the above results and using static-kinematic duality, the mathematical program for the unsafe method of plastic analysis (Charnes, Lemke and Zienkiewicz, 1959, Gavarini, 1966, Munro and Smith 1972, Smith and Munro 1972) becomes : -

$$\text{Min } z = \begin{bmatrix} \underline{m}_*^T & \underline{m}_*^T \end{bmatrix} \begin{bmatrix} \Delta\underline{\theta}^+ \\ \Delta\underline{\theta}^- \end{bmatrix}$$

$$\begin{bmatrix} \underline{r}_o^T & -\underline{r}_o^T \\ \underline{C} & -\underline{C} \end{bmatrix} \begin{bmatrix} \Delta\underline{\theta}^+ \\ \Delta\underline{\theta}^- \end{bmatrix} = \begin{bmatrix} 1 \\ \underline{0} \end{bmatrix} \quad (2.28)$$

$$\Delta\underline{\theta}^+ \geqslant \underline{0} \quad \Delta\underline{\theta}^- \geqslant \underline{0}$$

It will be seen that the program is now completely linearised and can be solved using the simplex algorithm. The optimal solutions of the kinematic variables fix the mechanism deformation modes at plastic collapse. Using a more compact and self-evident notation the unsafe and safe programs can be written in the following primal-dual forms (Dantzig, 1963).

UNSAFE PROGRAM

$$\text{Min } z = \underline{c}^T \underline{x}$$
$$\underline{A}\,\underline{x} = \underline{b}$$
$$\underline{x} \geqslant \underline{0}$$

PRIMAL

SAFE PROGRAM

$$\text{Max } w = \underline{b}^T \underline{y}$$
$$\underline{A}^T \underline{y} \leqslant \underline{c}$$
$$\underline{y} \gtrless \underline{0}$$

DUAL

(2.29)

It will be noted that the primal variables and the kinematic variables are the dual variables. The significance of this duality is twofold. First, from the theoretical viewpoint, the encoding of the theory of plastic analysis in a linear programming framework has allowed any known result from programming theory to be immediately applicable to plastic theory. Secondly, from a computational viewpoint, the optimal solution of any one program yields not only the optimal values of the corresponding primal variables but also through the use of simplex multipliers (Dantzig, 1963) the optimal values of the dual variables. Thus the optimal values of the static and kinematic variables are all identified from a single computation.

The theory can be generalised to more complex structural materials such as reinforced concrete (Munro and Smith 1972, Smith and Munro, 1972) and certain types of timber frames (Munro, Booth and Smith, 1973) or more complex loadings (Smith and Munro, 1973).

DETERMINISTIC SYNTHESIS

Introduction

In the previous section the design process was represented as a series of analyses in which the performance of completely known structural configurations was assessed against specified performance criteria. The final design was therefore evolved through a sequence of such analyses. In the present section designs which are optimal in some defined way will be synthesised directly from appropriate mathematical programs. It will be found that the programs of elastic synthesis are non-linear and take a generalised polynomial form whilst the programs of plastic synthesis can be transformed to linear programming form. Static and kinematic plastic synthesis programs can be derived as strict primal-duals in a generalisation of the previously discussed treatment of plastic analysis.

In the case of elastic synthesis it will be shown that many of the non-linear programs can be formulated in a special form which will be dubbed "signomial program" and it will further be shown that such a program can be approximated by a "posynomial program" whose optimal solution yields an upper bound on the minimal value of the original signomial program. Thus sub-optimal solutions can be derived in a monotonically decreasing sequence which converges on the true optimal. The computational procedures of structural synthesis can therefore be presented as a hierarchy of signomial, posynomial and linear programs in which the related duality theory plays an important role.

Since the plastic synthesis programs can be derived directly in linear form, these will first be discussed before turning to elastic synthesis and the section will be closed by a discussion of some hybrid programs which are of value in practical design.

Plastic synthesis

The simplest problem of plastic synthesis is probably the minimum weight design of a proportionally loaded steel frame such that it will not collapse plastically at a load parameter less than a specified value. The weights per unit length of the rolled steel joists are assumed to be linearly related to their plastic moments of resistance ($\underline{m}_*$). The optimality criterion of minimum weight can be stated as : -

$$\text{Min } z = \rho \, \underline{1}^T \, \underline{m}_* \qquad (3.1)$$

where l_i is the length associated with the ith design variable (m_*^i) and ρ is a density factor.

The yield conditions are : -

$$-m_*^i \leqslant m^i \leqslant +m_*^i \qquad i = 1, 2, \ldots, c \quad (3.2)$$

and the equilibrium conditions are : -

$$m^i = (\underline{k}^i)^T \underline{p} + \lambda \, r_o^i \qquad (3.3)$$

The value of the plastic moment of resistance of a critical section is linked to the prismatic member in which the section is located and the totality of the yield conditions can be written as : -

$$\begin{bmatrix} \underline{J} & -\underline{C}^T \\ \underline{J} & \underline{C}^T \end{bmatrix} \begin{bmatrix} \underline{m}_* \\ \underline{p} \end{bmatrix} \geqslant \begin{bmatrix} \lambda \underline{r}_o \\ -\lambda \underline{r}_o \end{bmatrix} \qquad (3.4)$$

where $\underline{J}$ is an incidence matrix linking critical sections to design variables (Munro and Smith, 1972) and, as in the analysis section, the matrix $\underline{C}$ is assembled from the k^i coefficients. Thus the safe linear program of plastic synthesis can be written in the following form : -

$$\begin{aligned} \text{Min } z &= \begin{bmatrix} \underline{1}^T & \underline{0}^T \end{bmatrix} \begin{bmatrix} \underline{m}_* \\ \underline{p} \end{bmatrix} \\ \begin{bmatrix} \underline{J} & -\underline{C}^T \\ \underline{J} & \underline{C}^T \end{bmatrix} \begin{bmatrix} \underline{m}_* \\ \underline{p} \end{bmatrix} &\geqslant \begin{bmatrix} \lambda \underline{r}_o \\ -\lambda \underline{r}_o \end{bmatrix} \\ \underline{m}_* \geqslant \underline{0} \qquad \underline{p} &\gtrless \underline{0} \end{aligned} \qquad (3.5)$$

The unsafe program of plastic synthesis. The equation (2.25) will be recalled.

$$\lambda_u = \frac{\underline{m}^T \Delta\underline{\theta}}{\underline{r}_o^T \Delta\underline{\theta}} \qquad (2.25)$$

and the unsafe theorem suggested the optimality criterion of plastic analysis that λ_u should be minimised and that the minimal value of λ_u was λ_c. This theorem and its implied optimality criterion must now be rearranged so that it can continue to be stated in terms of kinematic variables but must additionally imply that at optimality the weight is minimised. It can be shown (Chan 1969, Munro and Smith 1972) that this can be achieved by maximising ($\lambda_c \underset{\sim}{r}_o \underset{\sim}{\Delta\theta}$) and notionally scaling $\underset{\sim}{m}^T \underset{\sim}{\Delta\theta}$ so that it equals the minimum weight. Combining these conditions with the mechanism compatibility conditions yields the following unsafe linear program of plastic synthesis.

$$\text{Max } w = \begin{bmatrix} (\lambda \underset{\sim}{r}_o^T) & (-\lambda \underset{\sim}{r}_o^T) \end{bmatrix} \begin{bmatrix} \underset{\sim}{\Delta\theta}^+ \\ \underset{\sim}{\Delta\theta}^- \end{bmatrix}$$

$$\begin{bmatrix} \underset{\sim}{J}^T & \underset{\sim}{J}^T \\ -\underset{\sim}{C} & \underset{\sim}{C} \end{bmatrix} \begin{bmatrix} \underset{\sim}{\Delta\theta}^+ \\ \underset{\sim}{\Delta\theta}^- \end{bmatrix} \begin{matrix} \leqslant \\ = \end{matrix} \begin{bmatrix} \underset{\sim}{1} \\ \underset{\sim}{0} \end{bmatrix} \quad (3.6)$$

$$\underset{\sim}{\Delta\theta}^+ \geqslant \underset{\sim}{0} \qquad \underset{\sim}{\Delta\theta}^- \geqslant \underset{\sim}{0}$$

The above program is seen to be the dual of the safe program of plastic synthesis. These primal-dual programs can also be generalised to more complex structural materials such as reinforced concrete (Munro and Smith, 1972) and certain types of timber frames (Munro, Booth and Smith, 1973) or more complex loadings (Smith and Munro, 1973).

Elastic synthesis

The elastic synthesis of frames generally involves the solution of non-linear mathematical programs. Geometric programming (Duffin, Peterson and Zener, 1967) is a recently developed technique for handling a special class of non-linear programs which appears particularly promising for many structural problems. The subject of geometric programming has been greatly generalised since publication of the above-cited text and the term "posynomial programming" will be used herein to refer to the original form. A standard posynomial program is one which can be formulated in the following way : -

$$\begin{aligned} \text{Min } z &= g_o(\underset{\sim}{x}) \\ g_r(\underset{\sim}{x}) &\leqslant 1 \qquad r = 1, 2, \ldots, p \quad (3.7) \\ \underset{\sim}{x} &\geqslant \underset{\sim}{0} \end{aligned}$$

where the objective function (g_o) and the constraint functions (g_r) are posynomials in the primal variables ($\underline{x}$). The primal posynomial program is therefore generally non-linear with respect to its objective and constraint functions. The theory of posynomial programming develops the dual program which is non-linear only in its objective function and the linear nature of the constraints greatly facilitates the computational algorithm (Frank, 1966). From the duality theory of posynomial programming, the constrained maximum of the dual program equals the constrained minimum of the primal program. If this optimal solution is acceptable the optimal values of the primal variables can be obtained from the optimal values of the dual variables and the design is then fixed. The applications of posynomial programming to structural design has received comparatively little attention but some applications have been made to steel designs (Templeman 1970 and 1972, Morris 1972) and timber designs (Schmidt 1971, Munro and Booth 1973). However only a limited number of elastic synthesis problems occur naturally as standard posynomial programs. The constraint functions may take a signomial form and, even if a constraint function remains a posynomial, the constraint may be a reversed inequality.

$$g_r(\underline{x}) \geqslant 1 \qquad (3.8)$$

A signomial (s_1) can be written as the difference of two posynomials (G_1, G_2) and so a standard signomial constraint can be written as

$$s_1(\underline{x}) = G_1(\underline{x}) - G_2(\underline{x}) \leqslant 1 \qquad (3.9)$$

Introducing a new non-negative variable (x_{m+1}) the standard signomial constraint becomes : -

$$G_1 \leqslant x_{m+1} \leqslant G_2 + 1 \qquad (3.10)$$

or $$x_{m+1}^{-1} G_1 = g_1 \leqslant 1 \qquad (3.11)$$

and $$x_{m+1}^{-1}(G_2 + 1) = g_2 \geqslant 1$$

where g_1 and g_2 are posynomials. Thus the standard signomial constraint can be replaced by a pair of posynomial constraints - one standard and the other reversed - for the augmented variables. A reversed signomial constraint can be treated similarly and the original set of signomial constraints can be replaced by a mixed set of standard and reversed posynomial constraints. The new problem is termed a "reversed posynomial program". Duffin and Peterson (1972) have used the harmonic inequality to replace a reversed posynomial program by generating an approximating standard posynomial program from an arbitrary set of harmonic weights. This new program is termed a "harmonic program". If $\underline{x}'$ is the optimal solution to the harmonic program and $\underline{x}_*$ is the optimal solution to the reversed posynomial program then

$$g_o(\underline{x}') \geqslant g_o(\underline{x}_*) \qquad (3.12)$$

Thus from any starting set of harmonic weights an upper bound to objective function is obtained. New and improved weights can be generated using this solution and a second harmonic program can be generated. If the optimal solution of this second harmonic program is $\underline{x}''$ then

$$g_o(\underline{x}') \geqslant g_o(\underline{x}'') \geqslant g_o(\underline{x}_*) \qquad (3.13)$$

Under certain mild restrictions, this sequence of harmonic programs yields a monotonically decreasing sequence of sub-optimal solutions converging from above on the minimal solution of the reversed posynomial program.

Occasionally a standard posynomial program occurs naturally in terms of an objective function and constraint functions which each contain only a single posynomial term (Munro and Booth, 1973). Such "monomial programs" can be transformed logarithmically (Duffin, Peterson and Zener, 1967) and these programs can be solved using the simplex algorithm. This property of monomial programs can also be used to demonstrate the connection between the duality theories of posynomial and linear programs. Duffin (1970) has also shown that other posynomial programs may be approximated by linear programs using the geometric inequality. Thus a large class of synthesis problems can be formulated in a hierarchy of mathematical programs extending from signomial programs through reversed and standard posynomial programs to linear programs.

Hybrid design programs

It has been noted that the problems of plastic synthesis can be formulated as linear programs and that a broad class of elastic synthesis problems can be formulated as signomial programs. However in general the problems of structural synthesis do not fit conveniently into either class since both elastic and plastic constraints must be satisfied. Additional, and more complex, constraints such as those due to the limited ductility of reinforced concrete (Baker, 1956) may also have to be included in the design process. Hybrid programs of this type have been suggested (Cohn and Grierson 1968, Grierson and Cohn 1970, Munro, Krishnamoorthy and Yu 1972) for the optimal design of reinforced concrete frames. Solutions have generally been obtained by simplex computations after piece-wise linearisations of the design process and significant economies in the volume of reinforcement have been reported.

STOCHASTIC ANALYSIS AND SYNTHESIS

It has been pointed out earlier that the use of deterministic optimisation may cause a reduction in the survival probability by causing a multiplicity of failure modes to be active simultaneously at the specified overload level. It therefore seems logical to consider the failure probabilities of proposed designs and to attempt to ensure consistent reliability levels. Reliability analysis is made difficult by the complex interactions between the many members of a typical structure, by the large number of alternative loadings and by the multiplicity of potential failure modes.

The fundamental case of a single member of random strength R and a single random load S can be readily studied. The random variable Z can be defined as : -

$$Z = S - R \qquad (4.1)$$

and the failure probability (P_f) can be defined as : -

$$P_f = \Pr\left[Z > 0\right] \qquad (4.2)$$

and this probability (Freudenthal 1956, Garrelts and Shinozuka 1966) can be computed from : -

$$P_f = \int_0^\infty \left[F_R(t)\right] f_S(t) dt = 1 - \int_0^\infty \left[F_S(t)\right] f_R(t) dt \qquad (4.3)$$

where $F(t)$ denotes a probability distribution and $f(t)$ denotes a probability density and t is the state variable. The failure probability is greatly influenced by the tails of the distribution and, in general, great uncertainty exists regarding these tails in most structural situations. Thus the failure probability for the fundamental case displays considerable sensitivity to the input statistical parameters. However this fundamental case merely demonstrates the concepts for a single member and single load and it is instructive to consider some of the more complex structural types in a reliability context. "Weakest-link" structures are those which fail if any single member (or critical section) fails. Determinate trusses are of this type. In some indeterminate structures of brittle material a member may fail and cease to carry any load and the structure may continue to support the load. Such structures are termed "brittle-fail-safe" and the evaluation of failure probability must consider the order in which the members fail. The type of indeterminate frame with ductile material which was considered previously in relation to plastic collapse clearly does not necessarily fail when a critical section reaches its plastic moment of resistance. The subsequent re-distribution takes place whilst the critical section maintains its moment. This is an example of the "ductile-fail-safe" type. The random variable Z for this type can be considered in a particular mechanism deformation mode ($\Delta\underset{\sim}{\theta}^i$) for a flexural frame : -

$$Z_i = \lambda \underset{\sim}{r}_o^T \Delta\underset{\sim}{\theta}^i - \underset{\sim}{m}^T \Delta\underset{\sim}{\theta}^i \qquad (4.4)$$

and the overall failure probability can be written as : -

$$P_f = \Pr\left[Z_1 \geqslant 0\right] + \Pr\left[Z_2 \geqslant 0, Z_1 < 0\right] + \Pr\left[\ldots \quad (4.5)\right.$$

A theory has been presented (Stevenson 1967, Stevenson and Moses 1970) for the evaluation of the probability associated with each collapse mode and for the incorporation of the statistical correlation between collapse modes in the evaluation of the frame failure probability.

These analyses can be used as the starting point of a reliability based optimal design process (Stevenson 1967, Moses and Stevenson 1970). The deterministic objective function of minimum cost could ideally be modified to a total cost function which is the sum of the initial cost and a second cost function which is obtained by weighting the cost of failure by the probability of failure. More frequently however the optimality criterion is the minimisation of structural weight, subject to the probability of failure being less than some specified value. Thus the optimisation procedure may be conceptually simpler than in the deterministic case since the many deterministic constraints are replaced by a single constraint associated with the failure probability. This apparent simplification is however balanced by the analytical complexity of evaluating that failure probability.

Some recent attempts (Gavarini 1969, Gavarini and Veneziano 1970) have been made to modify the safe and unsafe programs of plastic analysis to stochastic programming form in which, for example, the random strength characteristics ($\underset{\sim}{m}_*$) are expressed through a vector of random variables. The frame statistical strength (λ) can then be calculated from the known material strength statistical parameters from the theory of stochastical programming. If the random nature of the loads is known then the failure probability can be evaluated. More recently Gavarini (1973) has described some attempts to extend these concepts to the development of stochastic synthesis programs.

CLOSURE

The paper has presented a conspectus of the generic techniques of automated structural analysis and synthesis with particular emphasis on the applications of mathematical programming. Attention has been drawn to the important role played by different forms of duality and their connecting relationships have been indicated. The relevance of linear, posynomial, signomial and stochastic programming to the design of safe and serviceable structures has been briefly discussed.

REFERENCES

BAKER, A.L.L. (1956) The ultimate load theory applied to the design of reinforced concrete and prestressed concrete frames.
Concrete Publications.

BERGE, C. (1962) The theory of graphs and its applications.
Methuen.

CHAN, H.S.Y. (1969) On Foulkes mechanism in portal fram design for alternative loads.
Journ. Appl. Mech. 73, March.

CHARNES, A. (1959) Virtual work, linear programming and plastic limit analysis of structures.
Proc. Roy. Soc. London, A. 251, 110.

COHN, M.Z. and D.E. GRIERSON (1968) Optimal design of reinforced concrete beams and frames.
8th IABSE Congress.

DANTZIG, G.B. (1963) Linear programming and extensions.
Princeton.

DUFFIN, R.J. (1970) Linearized geometric programs.
SIAM Rev. 12.

DUFFIN, R.J. and E.L. PETERSON (1972) Reversed geometric programs treated by harmonic means.
Indiana University Mathematics Journal, 22, 6, Dec.

DUFFIN, R.J., E.L. PETERSON and C.M. ZENER (1967) Geometric programming.
Wiley, New York.

FRANK, C.J. (1966) An algorithm for geometric programming in "Recent Advances in Optimization Techniques".
Edited by Lavi and Vogl, Wiley.

FREUDENTHAL, A.M. (1956) Safety and probability of structural failure.
Trans. A.S.C.E. 121.

FREUDENTHAL, A.M., J.M. GARRELTS and SHINOZUKA, M. (1966) The analysis of structural safety.
J. Struct. Div., A.S.C.E. 92 No. ST1.

GAVARINI, C. (1966) I teoremi fondamentali del calcolo a rottura e la dualita in programmazione lineare.
Ingegneria Civile 18.

GAVARINI, C. (1969) Concezione probilistica del calcolo a rottura.
Giornale del Genio Civile, Agosto.

GAVARINI, C. (1973) Stochastic programming in "Optimum Structural Design - Theory and Applications" Edited by Gallacher and Zienkiewicz, Wiley.

GAVARINI, C. and D. VENEZIANO (1970) Calcolo a rottura e programmazione stochastica - problemi con una variable casuale.
Giornale del Genio Civile No. 4.

GRIERSON, D.E. and M.Z. COHN (1970) A general formulation of the optimal frame problem.
Trans. ASME, Journal of Appl. Mech., June.

HENDERSON, J.C.deC. and W.G. BICKLEY (1955) Statical indeterminacy of a structure.
Aircraft Engineer, December.

LOVELACE, A.M. (1972) Keynote address. International Conference on structural safety and reliability (Washington, 1969).
Pergamon.

MOSES, F. and J.D. STEVENSON (1970) Reliability based structural design.
J. Struct. Div. A.S.C.E., 96, No ST2.

MORRIS, A.J. (1972) Structural optimization by geometric programming.
Int. Journ. Solids Structures, 8.

MUNRO, J. (1965) The elastic and limit analysis of planar skeletal structures.
Civil Engineering and P.W.R.

MUNRO, J. and L.G. BOOTH (1973) On the application of posynomial programming to the optimated design of timber beam components.
International Union of Forestry Research Organisations, University of Stellenbosch.

MUNRO, J. L.G. BOOTH and D.L. SMITH (1973) Collapse design of timber portal frames.
International Union of Forestry Research Organisations, University of Stellenbosch.

MUNRO, J., KRISHNAMOORTHY, C.S., and C.W. YU (1972) Optimal design of reinforced concrete frames.
The Structural Engineer, 50, 7. July.

MUNRO, J. and D.L. SMITH (1972) Linear programming duality in plastic analysis and synthesis.
International Symposium on computer-aided structural design, University of Warwick.

NEAL, B.G. (1956) The plastic methods of structural analysis.
Chapman and Hall, London.

NEAL, B.G. and P.S. SYMONDS (1952) The rapid calculation of the plastic collapse load for a framed structure.
Proc. Inst. Civ. Engrs. Vol. 1 Part III No. 1.

SCHMIDT, L.C. (1971) Discussion.
Procs. Inst. Civ. Engrs. 48, April.

SMITH, D.L. and J. MUNRO (1972) Primal-dual programs of plastic analysis.
Report SAM 72/2 Civil Engineering Dept., Imperial College, London.

SMITH, D.L. and J. MUNRO (1973) Plastic analysis and synthesis of frames subjected to multiple loadings.
International Symposium on Optimisation in Civil Engineering. University of Liverpool.

STEVENSON, J.D. (1967) Reliability analysis and optimum design of structural systems with applications to rigid frames.
Case Western Reserve University.

STEVENSON, J.D. and F. MOSES (1970) Reliability analysis of frame structures.
J. Struct. Div., A.S.C.E. 96, No. ST1.

TEMPLEMAN, A.B. (1970) Structural design for minimum cost using the method of geometric programming.
Procs. Inst. Civ. Engrs. 46,

TEMPLEMAN, A.B. (1972) Geometric programming with examples of the optimum design of floor and roof systems.
International Symposium on Computer-aided Structural Design. University of Warwick.

FUNDAMENTAL PROBABILITY RELATIONS FOR REPAIRABLE ITEMS*

J. D. Murchland

Planning & Transport Research & Computation Co. Ltd.
40, Grosvenor Gardens, London S.W.1., England

ABSTRACT

The relations described apply in a very general way to an "item" of interest to reliability analysis, whether the item be a component, sub-system or the whole system. The item can only be in two states: failed or good. Its history begins at time 0.

Subsequent behaviour can be characterized by several functions, defined as averages over all realizations under the conditions prescribed.

$P(t)$ probability that the item is failed at time t
$W(t)$ expected number of failures to time t
$w(t) = W'(t)$ failure rate at time t
$X(t)$ expected number of repairs to time t
$x(t) = X'(t)$ repair rate at time t
$U(s,t)$ probability that the item is failed at time s or fails in the interval $(s, s+t]$

The integral of P over an interval is the expected time spent failed there.

The first relation is

$$P = W - X, \quad \text{for all } t \geq 0,$$

---

*A detailed, twenty-three page manuscript is available from the author or editors.

provided that the initial values of W and X are chosen so that
$P(o) = W(o)-X(o)$. No restrictions on the item's alternation
between good and failed states are necessary for this "net transition" relation (save that W be finite, of course).

If an item is not just a component, but has its state
entirely determined by the states of a number of statistically
independent predecessor items, P for the item will be a polynomial
in the predecessor P's (call them P_i). Such polynomials may be
indirectly specified by a fault-tree, structure network or in
other ways appropriate to the problem in hand. Assume that P is
amenable to calculation, if nothing better offers by exploiting
its linearity in each P_i separately ("expansion").

w for the item can be obtained from the predecessor w_i by
use of the failure rate relation

$$w = \sum_i w_i \frac{\partial P}{\partial P_i}, \quad \text{all } t \geq o.$$

The relation needs the conditions that w exists for all t, "strong
independence" of predecessors and positivity in each predecessor.
The partial derivative amounts merely to picking out those terms
in P which contain P_i, numerically if need be.

Note that the x_i are conveniently absent. The same relation
holds for x, and dP/dt. From w, W can be found for each time of
interest by numerical integration.

The calculation of U is usually intractable, but fortunately
it can be bounded:

$$\underset{s \leq a \leq s+t}{\text{Max}} \{P(a)\} \leq U(s,t) \leq \min \{P(s)+W(s+t)-W(s),1\}.$$

The lower bound is approached by an item whose good intervals are
either very short or else very long, and the upper attained by an
item whose good intervals have a constant duration.

In summary, the reliability behaviour of a two-state system
can be characterized by $P(t)$, $W(t)$, $X(t)$ and bounds on $U(s,t)$ for
various times, infinity often being of chief concern. In favourable
cases these quantities can be readily calculated for each item of
the system by using the above relations sequentially, starting
from values of $P(t)$ and $w(t)$ for each component. The latter have
to be found by an expedient method of analysis, renewal theory for
instance, depending on the type of component. For the usual types
the results are known.

In some problems, such as electrical connection or computer
communication networks, it is natural for more than two states to
arise, perhaps representing distinct degrees of failure. This requires the appropriate elaboration of these quantities and the
three relations - an extension made without difficulty.

UNAVAILABILITY STATISTICAL MODELS FOR POWER SYSTEM
RELIABILITY STUDIES

Panichelli S., Salvaderi L., Scalcino S.

ENEL - Centro Nazionale Studie Progetti (CNSP) - Roma (Italy)

SUMMARY

The reliability parameters of the components of a power system are considered in the frame of probabilistic models.

In particular, failure bunching in severe weather conditions is discussed with its reflection on reliability.

1. GENERAL REMARKS ON UNAVAILABILITY CAUSES

Various papers deal with the quantitative evaluation of reliability of a power system by making use of different indices, (1)

It is obvious that apart from the data relating to the generation power , it is necessary to have data describing completely permanent and transient unavailabilities.

These unavailabilities are:

a) Permanent unavailabilities, which cause a need for repairs or substitutions. These in turn comprise:
 - unavailabilites caused by faults or by a maintenance work which cannot be deferred;
 - unavailabilities caused by occasional maintenance work which can be deferred:
 - unavailabilities due to causes external to the component itself and which cannot be deferred:
 - unavailabilities due to external causes and which can be deferred;
 - unavailabilities due to planned maintenance.
b) Transient unavailabilities, in which a re-closing operation is sufficient to put into service again the component concerned.

The quantities which are to be looked for, in respect to each component, in order to describe the unavailability are in general the following:
- unavailability probability = U (%); average duration of an unavailability = r:
- average number (frequency) of the faults, both permanent and transient, referred to time unit

$$\lambda_{tot} = \lambda_p + \lambda_t$$

2. THE EXPONENTIAL MODEL FOR THE STATISTICAL EXAMINATION OF THE UNAVAILABILITIES

A well known model suitable to describe the behaviour of each element subjected to internal faults, permanent and transient, is that corresponding to the state diagram illustrated in figure 1a [1]. Effecting, successively, a separation between permanent and transient events, we obtain the state diagrams of the figures 1b and 1c, where m_i = "time to fault"
r_i = duration of the fault or "time to repair";
t_i = "time between faults" which is identical to m_i in the case of the transient faults.

The latter are random and the introduction of further power models for their study is necessary in order to design a system whose behaviour is, naturally, random.

We assume that the events happen independently in random fashion (2)

The statistical model which corresponds to that assumption is the exponential one.

For "time to failure" $(\alpha = \lambda)$

S = probability that the "time to failure" be $t_1 \leq t \leq t_2$
P (t) = probability that the time to failure be $\leq t$, that is failure within the time t

$\overline{P}(t) = 1 - P(t)$ = probability of survival to time t.

An estimate $\hat{\lambda}$ of λ (rate of failure) can be obtained from the experimental data through the evaluation, during a certain observation period T_{ob}

The estimated value $\hat{\lambda}$ may then be expressed in terms of T_{ob} by the equation

---

[1] - This model adheres very well to the real condition for lines but is less faithful in the case of thermal generators for which derated states may exist

$$\hat{\lambda} = \frac{\text{number of forced internal faults of the components under examination during } T_{ob}}{\text{total duration of the operating time } T_o \text{ of all the components during } T_{ob}}$$

In the same way:

$$\hat{\mu} = \frac{\text{number of repairs due to forced outages during } T_{ob}}{\text{total duration of repair time due to forced outages } (T_r) \text{ during } T_{ob}}$$

$T_r = T_{ob}$ - (maintenance period + operating time T_o + period of unavailability due to other causes).

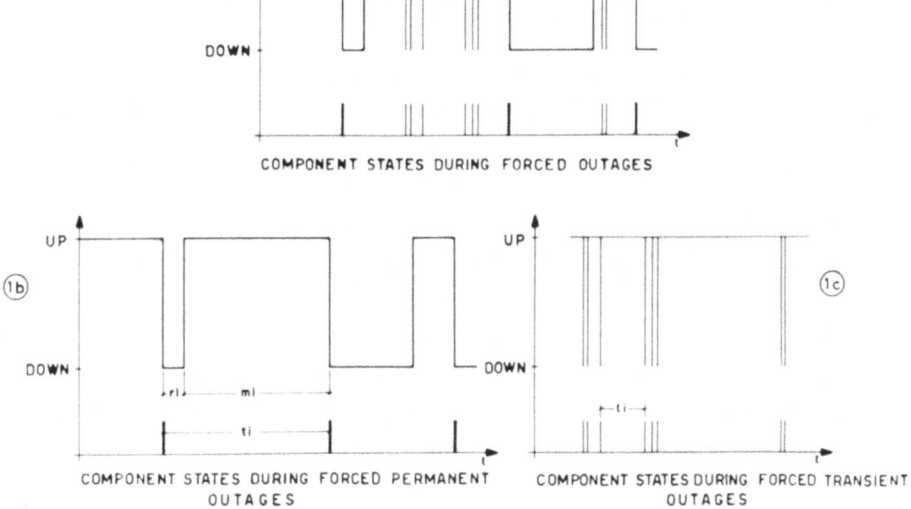

Figs. 1a, 1b, 1c

State diagram for components of electric systems which are subjected to forced permanent and transient faults.

The application of the exponential model to the unavailabilities which are illustrated in the figures 1b and 1c leads to

a) **Permanent faults**
- The mean duration of an out-of-service condition, that is the mean repair time (mean of the distribution) has the value: $r = 1/\mu$
- the mean duration of the operating time, i.e. the "mean time to failure" has the value: $m = 1/\lambda_p$

The behaviour of the element is, therefore, expressed by the mean quantities m and r; their sum $T = (m + r)$, "mean time between the faults", represents the period of the phenomenon (3).

From the above mentioned definitions the "forced unavailability" of an element is derived,

$$U = r/(m+r) = \frac{\lambda_p}{\mu + \lambda_p} = r/T$$

and, since this is a probability, U is dimensionless.

It is possible to achieve a simplification in which the mean repair time is very small in comparison with the mean time to failure:

$$U = r/(m+r) \cong r/m = \lambda_p T$$

Consequently the permanent unavailability is usually defined:
- for the generation system: by means of parameters U and r;
- for the components of transmission and distribution systems: by means of parameters λ_p and r, and it is assumed that $U \cong r \lambda_p$

b) **Transient faults**

Durations of "out of service" conditions: as the "out of service" time durations coincide with those of the "reclosing operation" which are practically null, there is no meaning in extracting their distribution P.

- The mean duration of the operating time, i.e. the "mean time to failure = mean time between failures" (mean value for the distribution) has the value $T_t = 1/\lambda_t$

3. APPLICATION OF THE EXPONENTIAL METHOD TO THE OVERHEAD LINES: IMPORTANCE OF THE FAILURE BUNCHING ON THE RELIABILITY OF A POWER SYSTEM

The constant "predisposition" model (corresponding to the exponential distribution) is really not always applicable to the whole life of a component of a power system; it must in fact be remembered that a typical "hazard function" of several physical phenomena has the well known "bath-tub" shape. In order to include also the initial period the artifice of assuming several values of rate of failure in the first years of life (immature rates) is sometimes used, values larger than those which occur after the starting period (mature rates).

It must be remarked that overhead lines suffer from atmospheric conditions and we have periods, during which (4) the fault frequencies are very different from one another. The influence of the phenomenon on the reliability of the system is very remarkable.

While the yearly average per hour unavailability of a line is $U_m = h_m/8760$ (h_m is the mean yearly total hours of unavailability of the line) the per-hour probability of unavailability due to adverse atmospheric conditions is $U_s = h_s/S_{tot}$ (h_s = hours of unavailability due to storms; S_{tot} = total of hours of storms in a year) this probability can be in practice much higher than the first one.

The composit probability that both lines are out of service, is given by the squared evaluation of the probability relative to one line: if, then, in the risk evaluation one would take into account the mean probabilities only, some phenomena that really turn up would not be caught, with heavy evaluation error (5).

For a system composed of parallel lines 30 km long (with conditions equal as to the mean yearly probability and duration of normal and stormy weather) the error curve increases rapidly with the ratio U_s/U_m and, for ratios U_s/U_m between 20 and 80, as occurs in reality H_{eff}/H_{med} is of the order of 5 to 75 where H_{eff} is actual hours out of service and H_{med} is mean probability based on U_s/U_m. As for the physical phenomenon, the exponential model with constant failure rate can be thought to interpret this by the following assumptions:
- a geographic area subject to uniform meteorological conditions is considered
- during the year, the fluctuations, in this zone, of the different weather conditions having similar characteristics are grouped together
- each group of so determined conditions, "normal", "average" or "stormy" will show a constant value of the "predisposition to fault" and then <u>inside each group</u> the proposed model will be applicable.

In reality a power system can be extended to zones having different meteorologic characteristics: it can then be thought to be subdivided into various geographically limited areas, each of which is subject to uniform meteorologic characteristics.

We think it is realistic to proceed by making use of the "density" of the transient faults by assuming that the low, medium, high intensity of transient faults is usually accompanied by meteorologic normal, average, stormy weather.

The distributions of the durations of the normal, average and stormy situations being random are not far from the exponential with a mean value that we shall call N, M and S for each condition of normal, average and stormy weather. Then

$$\lambda_n = \frac{\text{number of faults occurred during normal weather conditions}}{\text{yearly percentage of normal weather}} = \frac{N_n}{N/(N+S)}$$

$$\lambda_s = \frac{\text{number of faults occurred during stormy weather conditions}}{\text{yearly percentage of storm duration}} = \frac{N_s}{S/(N+S)}$$

N_n and N_s are the faults occurring in the year in the two examined situations, permanent or transient: the fault frequencies are conventionally expressed in "number of faults/100 km.year".

We explicitly remark that what has been said on fault frequency is not true for the durations r_i of the permanent faults (for the transient ones it has already been said that the problem does not exist). As we can assume that the repair times are roughly independent of the atmospheric conditions; the distribution of the time to repair r_i will be unique for all conditions.

4. SOME PROCESSING EXAMPLES RELATING TO ITALIAN OVER-HEAD LINES

In order to verify the theoretical hypotheses and the described methods of examination of the statistics, a survey has been effected on the permanent and transient faults of an area of the Italian network which is supposed to present uniform weather conditions and corresponds to the line layout of the Compartment of Milan. The analysis has been extended to a rather limited number of years (1968 - 1969 - 1970) we have confined ourselves, indeed, to the 132 kV - lines, whose total length is of about 3,200 km and which contains approximately 190 lines.

4.1 Weather condition distribution

The "grouping" of the different weather conditions has been effected, as aforesaid, through the density of the transient faults. The obtained results are interesting and confirm our hypotheses.

The respective total yearly durations of the various weather conditions have been the following: normal weather conditions 90,5% average weather conditions 8,1% stormy weather conditions 1,4%

the percentage being referred to the total number of hours contained in one year; the distribution of the durations shows a good agreement with the exponential distribution.

4.2 Distribution of the time between the transient faults

Duration: it has been said that, for the transient faults, the duration of the interruptions, (transient time to repair) is conventionally assumed to be equal to zero.

Frequencies: we have tried to identify, in the distributions of the "time to fault" (m_i) (which, in the case of transient faults, are identical to the time between faults (t_i)) the existence of three components presenting a constant "predisposition" to the fault, each of the components corresponding to the different weather conditions which have been previously examined. In order to do that, we have decomposed the experimental distribution into the sum of three exponential curves. The results are shown in fig. 2.

The transient unavailability rates which can be derived from the curves are: for normal weather condition λ_{tn} = 3 faults/100 km. year; for average weather condition λ_{tm} = 39 faults/100 km. year; for stormy weather condition λ_{ts} = 370 faults/100 km.year whereas the corresponding annual mean value results:

$$\lambda_t = 10 \text{ faults}/100 \text{ km.year}$$

Ratios between each value and the mean value are 0,3 - 3,9 - 37 respectively.

4.3 Distributions of the durations for the permanent outages

We refer now to the forced outages which cannot be deferred. As indicated, we have separated those due to "internal causes" and "external causes".

a) Internal causes:

Frequency: the span of three years of available data provided too limited a number of permanent faults to allow a distribution of the "time to fault" (m_i).

However, mean values for the different weather conditions have been determined, by utilising the total number of faults which take place in each of the weather conditions which have been described.

The values of the permanent failure rates, conventionally referred to 100 km/year are for normal weather conditions

$$\lambda_{pn} = 0,5 \text{ faults}/100 \text{ km. year; for average weather conditions}$$

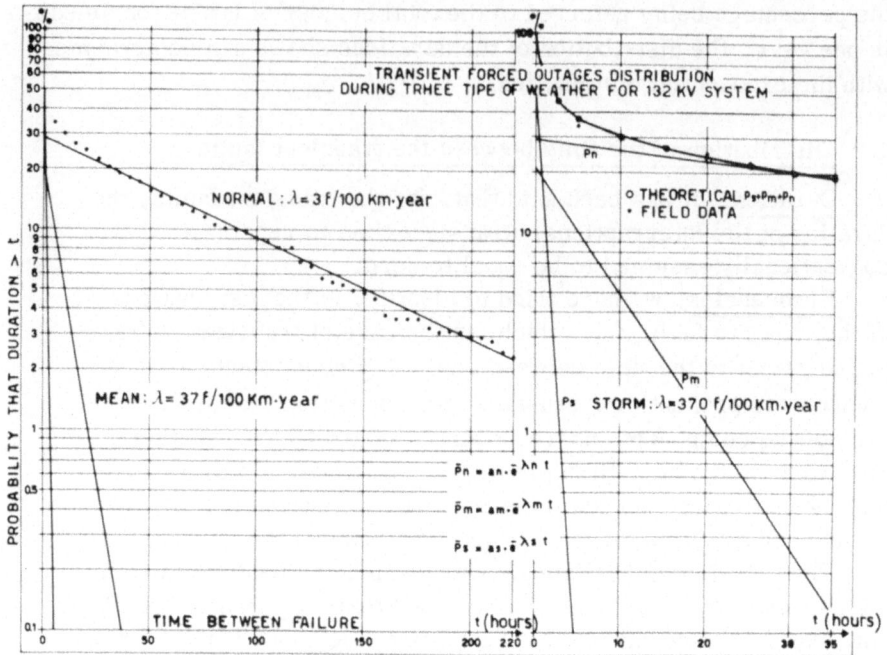

Fig. 2 Decomposition of the cumulated distribution curve of the time between transient faults in three exponential distributions for the different weather conditions for a 132 KV network.

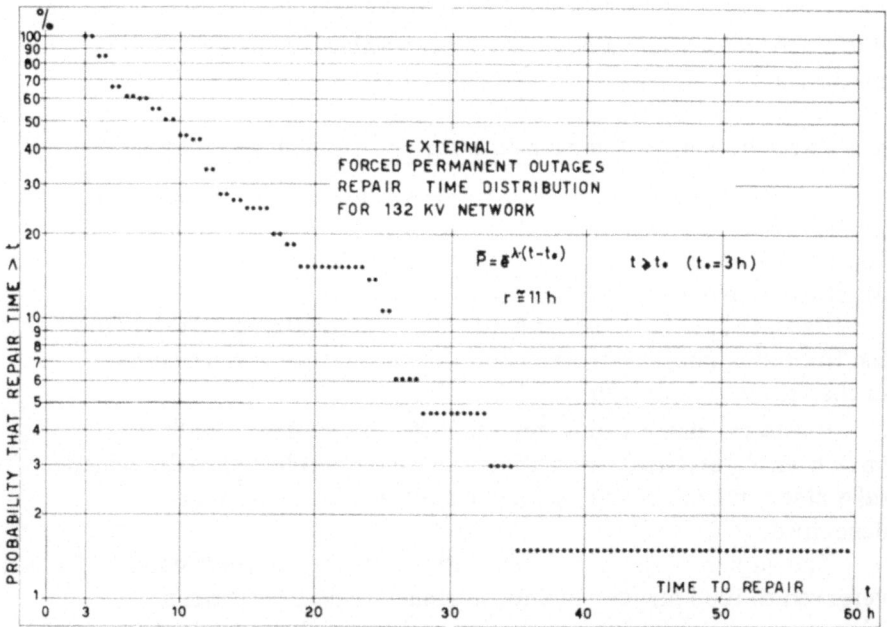

Fig. 3 Distribution of the external fault durations for the 132 KV lines of the Compartment of Milan.

$\lambda_{pm} = 1,5$ faults/100 km.year- for stormy weather conditions

$\lambda_{ps} = 17$ faults/100 km. year

A search for the annual mean value would have provided, instead:

$\lambda_p = 0,86$ faults/100 km. year

Ratios between each value and mean value are 0,58 - 1,75 - 20 respectively.

If we consider the transient failure rates presented in paragraph 4.2 and the actual permanent failure rates for the above mentioned weather conditions, we observe that both increase as the weather conditions become worse. This indicates that between permanent and transient faults a concomitance does exist, with the consequence upon the dynamic risk index of the overall system, as already explained.

Duration: the distribution of r_i time durations has been derived and in this case, too, the result has confirmed the hypotheses, in the sense that the distribution of the repair times is of an exponential nature

The mean time to repair was found to be: $r = 12$ h

b) External causes: [2]

Frequency: given the low number of available data it has been impossible to affect a verification of the dependence on the weather situations. There is an indication that this dependence does not exist. The annual mean failure rate is $\lambda_{pest} = 0.11$ faults/line. year.

Duration: The distribution is illustrated in fig. 3, from which it can be seen that the shape is exponential, with a mean duration: $r_{est} \cong 11.3$ h

4.4 Unavailability due to external causes, which can be deferred

We point out the unavailability due to these causes in order to demonstrate their importance under the profile of the reliability evaluation, particularly in the case of the 132 kV network.

As already said, since (by definition) the relative interruptions can be deferred, a differentiation in the different weather conditions is devoid of any meaning. The mean values that, altogether describe them are the following: mean number of out-of-service

---

[2] - Among the external causes an important one is represented by the unavailability of station switches (breakers). ENEL-DPT statistics give, as percentage of the fault causes for switches, due to weather agents, 16, 5% for 1968.

conditions: $\lambda = 1.5$ times/line. year; average duration of an out-of-service condition : $r \cong 28.5$ h.

The corresponding number of mean unavailability hours is $h = 44$ h/line. year : recalling that the forced unavailability hours (mean annual value) for each 100 km of line are about 10 h, it is possible to understand - even if the relative interruptions can be carried out in the most favourable periods - the importance of the operational difficulties introduced by the phenomenon, and therefore the validity of our previous statement.

5. CONCLUSIONS

5.1 - The general fitness of the exponential model to describe the parameters relating to the unavailability of each component of an electric system has been justified. The parameters are usually provided in terms of "frequency" and "duration" for the transmission and distribution systems, and in terms of "annual unavailability" and "duration" for the generating system.

5.2 - The same model has been identified for the particular case of the overhead lines, showing that the effect of fluctuating environments on the lines themselves implies the necessity of establishing the model for uniform weather conditions.

5.3 - The expedients have been indicated which have to be followed when affecting the statistical surveys, in such a way that they provide data which are liable to be directly employed in the study of reliability for power systems, in the designing phase, with particular reference to the need to take into account the failure bunching phenomenon, for the determination of which we have suggested the transient fault "density"

5.4 - Some numerical examples show the applications of the concept to some real situations.

REFERENCES

(1) MANZONI - NOFERI - PARIS - VALTORTA: Risk indices for evaluating the reliability of an electric system. Conference on "Generic techniques in system reliability assessment, Liverpool 1973".
(2) HAHN - SHAPIRO: Statistical models in engineering. H. Chestnut - Ed. New York 1967.
(3) IEEE TUTORIAL COURSE: Probability analysis of power system reliability. IEEE ed. 1971.
(4) PATTON: Determination and analysis of data for reliability studies. IEEE Trans. PAS, Vol 87, 1968/1.
(5) NOFERI - PARIS: Quantitative evaluation of power system reliability in planning studies. IEEE Winter Meeting, New York 1971.

Computer-Aided Synthesis of Fault Trees for Complex Processing Systems

Gary J. Powers and Frederick C. Tompkins, Jr.

Department of Chemical Engineering, Massachusetts Institute of Technology, Cambridge, Massachusetts, USA

Introduction

This paper describes a procedure for generating fault trees for complex chemical processing systems. With some modifications the procedure could be extended to other types of systems including nuclear power plants. The procedure and associated computer programs have been developed so that complete and detailed fault trees for complex chemical plants could be rapidly generated and evaluated.

The logic flowsheet for the safety analysis procedure is shown in Figure 1.0. The basic steps in the procedure are:
1. Identify the hazards which exist.
2. Estimate the consequences of each hazard.
3. Determine the probability of occurance of each hazard by constructing fault trees for the system.

The first piece of required information is a description of the system equivalent to a detailed equipment and instrument flowsheet. Since computers aren't very good at reading flowsheets, an equivalent description in terms of linked data lists is automatically constructed. The data structure gives the connectivity of the flowsheet as well as information on each unit in the process.

Secondly, property data on all the species in and around the process is requested. The kinds of data required are: flammability, explosivity, toxicity, corrosivity, and the reactivity of all species combinations. If the data does not exist it is

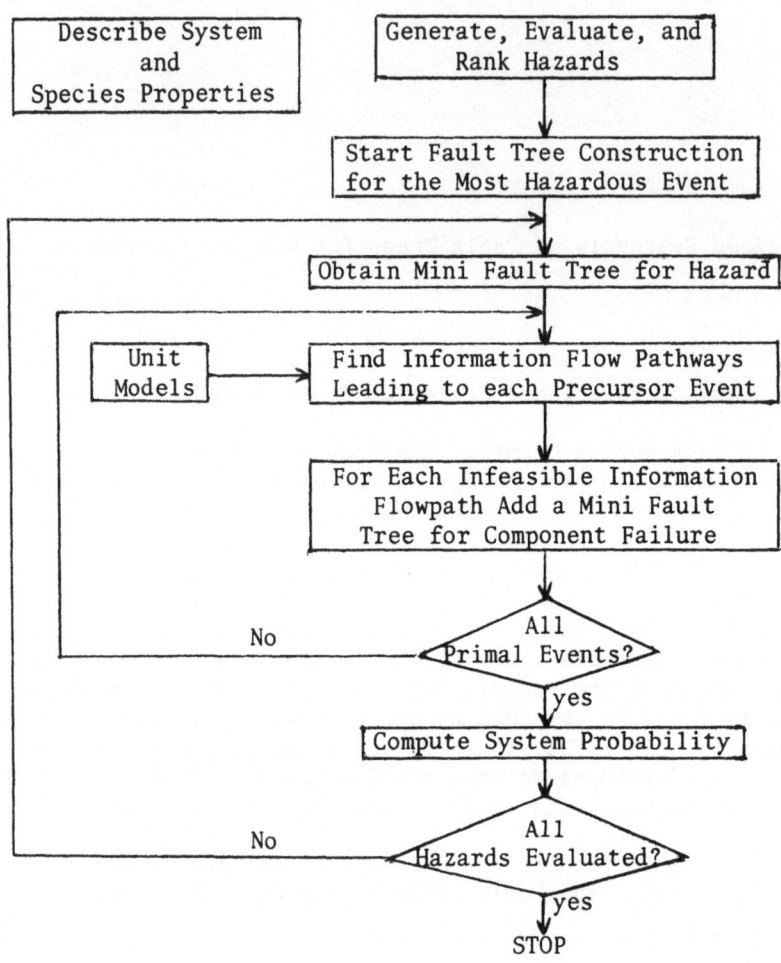

Figure 1.0 Safety Analysis Logic

predicted from molecular characteristics by several correlations built into the program. Great care has been taken in developing the input section of this program to insure that all pertinent species, including trace components at concentrations down to the limit of analytical capabilities, and their hazardous properties have been included. Once the system is described and the properties of the species are known a list of hazards is generated. The hazards are related to either the properties of the species (explosivity, flammability, toxicity, corrosion, reactivity) or the characteristics of the equipment in the process (pressure ratings, corrosion ratings, temperature ratings, loading limits, etc.). Hazard matrices which indicate the species combinations and location of each hazard are then generated.

Each of the hazards is then evaluated. In the current version of the program very simple evaluation methods are used. These methods are patterned after those used in the fire insurance industry. They involve a ranking index for each species for each class of possible hazard. The indices are normalized on a scale from 0 to 10; with 10 being extremely hazardous. A total hazard index is computed for each hazard. The total index involves the properties of the species, the amount of the species present, the location of the equipment in question, the potential business interruption expense, potential human injury, etc. A more detailed analysis of each hazard is possible, of course, if additional information on the system is available. However, the principle objective of this program is to reveal potential hazards rather than to analyze their consequences in detail.

The result of the hazard evaluation is a ranked list of hazards for the system. The fault tree generation starts with the most hazardous event. The probability of the hazard occuring is the desired information. Unfortunately, most systems are too complicated to directly assess the probability of occurance of the final hazard event. It is necessary to combine probability of failure data for components within the system (valves, pumps, sensors, controllers, heat exchangers, operators, etc.) with knowledge of the system structure and behavior to predict the probability of the final hazard. The construction of a fault tree for the system supplies the bridge between component failure data and the final hazard. A fault tree consists of a logic diagram which identifies all event sequences which could lead to a specified failure event.

The first step in generating the fault tree is to obtain a mini-fault tree for the hazard in question. These fault trees have been defined for a wide range of possible hazards and are stored in the program's library. The fault tree for one type of explosion is shown in Figure 2.0. The fault tree indicates that for this type of explosion to occur, at the location in question, the correct species, concentrations, temperatures, pressures, and ignition source must be present. This version of the fault tree for explosion is simplified and more complicated versions can be utilized.

The top of the complete fault tree is hence defined by the logic required for the occurance of the final hazard. The fault tree branches to precursor events defined by the hazard mini-fault tree. It is then necessary to determine the probability of occurance for each precursor event. For example, with the explosion fault tree one must determine whether the exploding species are

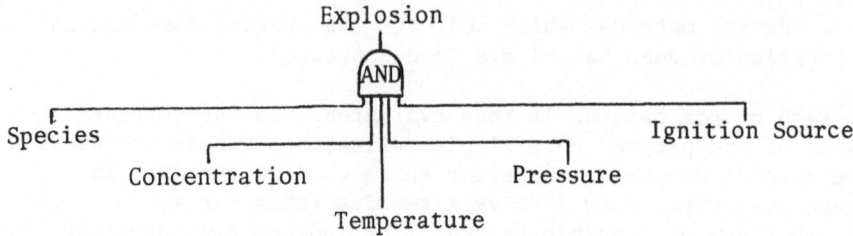

Figure 2.0 Simplified Hazard Fault Tree for an Explosion

normally at the location in question. If they are, then the probability is set to 1.0 and the other precursor events investigated. If the exploding species are not present then it is necessary to continue the fault tree generation to predict the probability of their being present. Similar procedures can be used for other precursor events such as concentration, pressure, temperature, etc.

The general method used to generate fault trees for these precursor events is to:

1. Identify all feasible information flow paths in the system which could lead to each event.

2. Use mini-fault trees for components when the failure of these components is required for the completion of information flow paths.

Information flow paths are defined by the equations which describe the mass, energy and momentum flow within the system. Equations which describe the control structure of the system also form part of the information flow structure. These equations can be derived from a fundamental analysis of the behavior of units within the system. In the present version of the program, a modular approach has been taken. Steady-state mass, energy, and momentum balance equations for units commonly found in chemical processing systems have been formulated. Heat exchangers, valves, pumps, reactors, distillation columns, etc. have been included in the models. A simplified means has been developed for representing the non-linear algebraic equations which describe the performance of a given unit. A steady-state gain matrix which gives the change in each dependent variable as a function of changes in the independent variables is formulated. Figure 3.0 illustrates this matrix for a simple counter-current heat exchanger. These matrices (one for each unit in the system) define the pathways along which information can flow in the system. For example, the explosion fault tree defined above indicates that a certain temperature (for example, higher than the normal temperature) is necessary for an explosion to occur. Assume that the temperature corresponds to one of the streams leaving the heat exchanger in Figure 3.0. The gain matrix would then indicate which of the variables associated

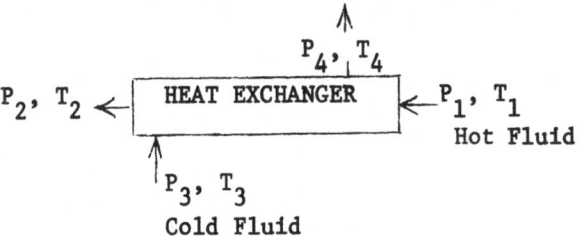

Dependent Variables

	$\dot{m}_1$	$\dot{m}_3$	T_2	T_4	$\bar{\rho}_1$	$\bar{\rho}_3$	$\bar{\mu}_1$	$\bar{\mu}_3$	U	P_2	P_4
T_1	o	o	+	+	−	−	−	−	o	+	+
T_3	o	o	+	+	−	−	−	−	o	+	+
U	o	o	−	+	+	−	+	−	×	−	+
A	o	o	−	+	+	−	+	−	o	−	+
P_1	+	o	+	+	−	−	−	−	+	−	+
P_2	−	o	−	−	+	+	+	+	−	×	−
P_3	o	+	−	−	+	+	+	+	+	−	−
P_4	o	−	+	+	−	−	−	−	−	+	×

(Independent Variables on the left axis)

Figure 3.0 A partial steady state gain matrix for a counter-current heat exchanger. Only the sign of the gain is given. The sign of the gain is for a positive change in the independent variable.

with the heat exchanger could cause the outlet temperature to
increase. These variables could then be further investigated to
determine how they could have reached the indicated state. This
procedure is recursively applied to each new set of precursor
variables until primal events are encountered. A primal event is
an event which is defined to be causative. For example, a stream
entering the process from outside the system could be causing a
sequence of events. If we have no "upstream" information on that
stream, the events associated with that stream are considered
primal. Another example is a system failure caused by loss of
electricity. One could define loss of electricity as a primal
event and accept a probability based on past experience. The
other alternative is to further investigate the power system to
determine why it might fail. Hence, the scope of the system
defines what is to be called a primal event.

Information flow paths are found by defining sources and
destinations of information within the system and then searching
for ways to connect them. In the case of a species being present
at a location where it is not normally found, the destination is
the location in question and the sources are all the locations
where the species is normally present. The information flow paths,
which are the same as physical flow paths in this case, are all
the possible pathways from the sources of the species to the
destination. Some of the flow paths will be feasible within the
normal operation of the system. Other paths will require mis-
operation of the equipment; i.e. the wrong valves open or closed,
pumps off when they should be on, etc. These two classes of
flow paths are detected using the gain matrix for each unit in the
flowpath. There may occur flow paths which are not feasible when
checked against the description of the system and the gain matrices
for the components in the system. These flow paths require failure
of the components. One example, for the case of physical flows
relating to the presence of a species, is the counter-current
heat exchanger discussed above. In the normal operation of this
type of exchanger, species in one inlet stream do not mix with
species in the other inlet stream. In order for species to flow
from one of the inlet streams into the other it is necessary to
consider a failure of the equipment. Failure models have been
defined for the units commonly found in chemical processing
systems. The models take the form of mini-fault trees. In the
case of a heat exchanger a fault tree containing the event
"leakage between streams" is part of the failure model. These
failure models can be quite extensive and complex. It is important, of course, to include all conceivable failure modes in the
unit failure models. For a given infeasible information flow path
there may exist several possible failures of the component which
will open the path. Each of these failures may then represent a
new point for continuing the development of the fault tree. The
fault tree given in Figure 4.0 illustrates a simplified model for
a pneumatic controller.

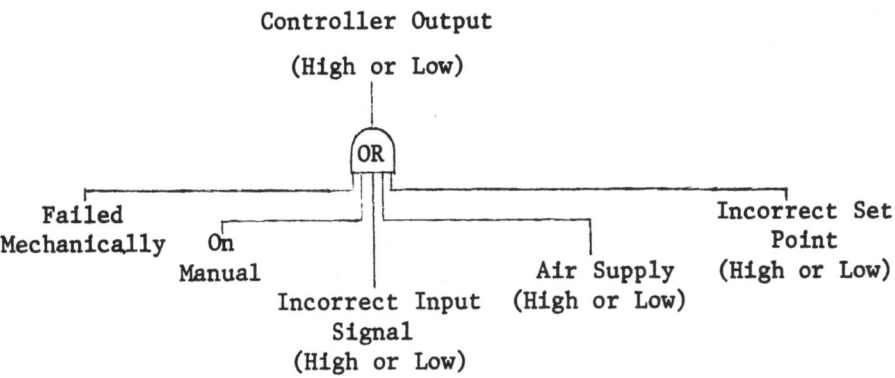

Figure 4.0 Simplified Failure Fault Tree for a Pneumatic Controller

The procedure of tracing out possible information flow paths is continued until only primal events are encountered. The probabilities of the primal events are then used to compute the probability of the final hazard event.

The probability calculation methods have been described elsewhere (Vesely, 1970; Fussell, 1973). With these methods, it is possible to determine both the probability of the final event and the critical path of events in the fault tree. The events along the critical path are the ones which control the system failure probability. Design changes should focus on these events.

After generating the fault tree for the most hazardous event the next most hazardous event is considered. In a similar manner a fault tree is generated for this hazard. When all the hazards have been evaluated and analyzed a complete set of fault trees exists for the system.

Fault trees generated by these means tend to become rather large. Current research is aimed at developing methods for "pruning" low probability pathways out of the tree as it is being generated. Interaction with a safety analyst during computer generation of the fault tree is also being incorporated into the program. A liquified natural gas system containing well over 500 streams and 200 pieces of equipment is being evaluated by this approach.

Conclusion

The methods described in this paper portend a more systematic and rapid approach to the safety analysis of chemical processing systems using fault trees.

References

1. Powers, Gary J., Frederick C. Tompkins, Fr., Computer-Aided Fault Tree Synthesis for Chemical Processes, to be presented at the Philadelphia AIChE Convention, 1973.

2. Fussell, J.B., Synthetic Tree Model - A Formal Methodology for Fault Tree Construction, PhD Thesis, Georgia Institute of Technology, December, 1972. A methodology similar to the one described in this paper. Examples for electrical circuits.

3. Vesely, W.E., and R.E. Narun, "PREP and KITT: Computer Codes for the Automatic Evaluation of a Fault Tree," Idaho Nuclear Report IN-1349 (August, 1970).

4. Recht, J.L., "Systems Safety Analysis: The Fault Tree," National Safety News, April, 1966.

5. Fussell, J. and W.E. Vesely, A New Methodology for Obtaining Cut Sets for Fault Trees, Hydrodynamics and Safety Analysis Methods, Transactions Am. Nuclear Soc., p. 262, 1973.

THE ROLE OF THE MAN-MACHINE INTERFACE IN SYSTEMS RELIABILITY

Jens Rasmussen

Danish Atomic Energy Commission, Research Establishment Risö

ABSTRACT

Human malfunctions in abnormal tasks are an important factor in low probability events, but human behaviour in higher level mental tasks cannot yet be predicted. It is therefore important to verify the limits of use of existing methods of reliability prediction.

THE PROBABILITY/SEVERITY RELATIONSHIP

The probability of an abnormal event can be assumed to be inversely proportional to the related consequence to the system operation. This is in agreement with the frequency/severity plot of injuries in American industry shown by Johnson (1972) and is also reflected in the nuclear safety criterion suggested by Farmer (1967). The importance of low probability events imposing severe risks on the system has to be faced, if reliability prediction should be of any real value. In the safety assessment of nuclear plants for instance, the look-out is for failure probabilities in the range $10^{-5} - 10^{-7}$ per year or less.

Neither the functional analysis of the system to identify the relevant causes and consequences of faults nor the probability analysis itself can cover all possible events. The analysis must be based upon a number of assumptions and approximations, and there is a danger that important, but low probability, fault modes are excluded from the analysis. It should therefore be realized that a quantitative reliability figure only constitutes a minor part of the result from the analysis. A very significant part of the result is given implicitly in the assumptions and approximations, as they very often identify conditions, which may be of low probability but vital to the total reliability. Furthermore, they typically involve conditions, which are dealt with by the plant personnel, and it is therefore important that the assumptions are approximations underlying the analysis are interpreted and documented carefully to facilitate their verifications during plant operation.

THE CAUSE/CONSEQUENCE ANALYSIS

The first important step of the analysis of a process system is a cause/consequence evaluation aiming at an identification of the relevant fault traces through the system. A fault tree analysis based upon typical component faults may not identify low probability, but risky fault traces. A vital part of the analysis will be to trace also the possible but improbable faults and combinations of faults, from a postulated set of consequences. This in itself implies an interface problem between the system and the analyst, as it demands a detailed knowledge of the practical layout of the technical system and of the working conditions and behaviour of the plant personnel.

The prime condition to be fulfilled by a reliable analysis is of course that all relevant traces are identified. In a complex system the analysis cannot cover all physically possible faults and their combinations, and it is therefore important to have systematic heuristic methods to identify relevant traces. Such a method should support the creative or inventive powers of the analyst, and we have briefly considered the "morphological" method suggested by Zwicky (1967), which may be a fruitful approach.

Johnson (1972) has recently published a comprehensive work on systematic evaluation of accidents using a similar approach. Johnson traces the possible causes starting from a rather high level of abstraction and controling the tracing of faults systematically though several levels of detail, such as:

"an accident is
- an unwanted transfer of energy,
- because of lack of barriers and/or controls,
- producing injury to persons, property or process
- preceded by sequences of planning and operational errors, which failed to adjust to changes in physical or human factors, and produced unsafe conditions and/or unsafe ads,
- arising out of the risk in an activity,
- and interrupting or degrading the activity"

We find it very important to develop systematic methods for cause/consequence tracing with tight coupling to appropriate models to facilitate the complete analysis (Nielsen 1971).

A clear systematic approach to the identification of relevant fault traces furthermore facilitates adequate documentation of the analysed mechanisms considerably. This documentation is a vital part of the man-machine interface. A trivial, but important condition of a reliable analysis is of course that it deals with the system actually operating. The system, however, may be subject to changes. Equipment can be modified and improved according to operational experience, as well as working procedures and instructions will be changed-planned or unnoticed. A considerable risk therefore exists that the conditions of plant reliability will be unintentionally violated. To avoid this the analysis must be documented in a systematic form, which can be readily interpreted and used by the operational staff.

THE HUMAN FACTOR IN SYSTEM RELIABILITY

Our attention was directed towards the human element in the system by a review (Rasmussen 68) of reported major incidents and accidents. Its purpose was to enable us to judge whether our methods for reliability evaluation also included such cases - and we found they did not.

Among the cases reviewed are 30 cases reported in USAEC Nuclear Safety Bulletin. In 70 - 80% of these cases, the incidents were initiated from human maloperation in the system. Furthermore, the maloperations did not take place during normal tasks, but overwhelmingly during abnormal or special tasks under abnormal plant conditions, such as modifications, repairs or cleaning and calibration operations; typically operations which are difficult to predict and analyse, and therefore normally covered by suitable assumptions in the analysis.

This is quite reasonable from the traditional reliability point of view, as this type of faults normally account for a small fraction of the total number. In a British fault record 8,000 cases from nuclear installations including trivial technical and human faults, the human faults amount to only 10% of the total (Ablitt 1969). But in our context it is most unfortunate that the source of severe incidents is very likely found in a class of faults, which are normally excluded from the analysis by proper assumptions. A few examples will illustrate this point of view.

The reliability of a system very often depends heavily upon an assumption of mutual independence of fault mechanisms. Physical sources of common mode faults such as flooding by water, rupture by missiles or trucks, etc., may be indentified by a morphological search. But coupling due to people moving around in the plant? If an abnormal condition in the plant, for instance due to a technical fault in a subsystem, calls for manual intervention, there is a probability that an operator misinterprets the situation, and manipulates another part of the system. The result is a coincidence of two faults, which are physically independent and as such difficult to predict at an office desk, although it may be likely to happen, judged from the actual working conditions. The problem is that although it may be possible to predict the probability of operator failure to execute the required function, it may be almost impossible to predict what he does instead.

In redundant systems the assumption of independence can lead to extreme reliability figures - but the actual figure may very likely be controlled by the probability of a faulty repair - which is repeated in more units.

Probability modelling of a complex system is often simplified substantially if proper function of equipment is assumed verified at certain intervals and after repair. This assumption is vulnerable and sometimes unrealistic, partly because repair and test in itself can be faulty, but also due to technical difficulties in testing the equipment without putting it into operation. Furthermore pressure of work during plant shut down can be great to regain plant operation in due time to avoid the operational consequences from process cool down or processes like xenon poisoning in nuclear plants. Therefore, test and calibration procedures may be postponed to the restart phase. This may be critical, as faults introduced during repair and modification work may leave the plant in an abnormal state, which is not covered by the protection of the normal safety system. Although such periods are normally relatively short, they can in our experience contribute significantly to the total risk of the plant.

In other words, in evaluating the reliability of a system like process plants, the role of the human functions in the system should be considered not only to include primary human functions in the reliability analysis itself, but also to verify the assumptions of the analysis, as the assumptions are ultimately administered by plant personnel.

PREDICTION OF HUMAN RELIABILITY

Several important approaches have been made towards the development of methods for predicting human reliability. Such methods have recently been reviewed by Meister (1972) and are the subject of other lectures of this meeting.

The basic assumptions of these methods are typically:
- The task is well defined and the procedure followed can be formulated in detail,
- The procedure can be broken down into a sequence of behavioral units, i.e. subtasks or task elements,
- Data on the reliability of the individual subtask are available together with the parameters characterizing the relevant task situation.

Typically these assumptions do not fit the work procedures found in process plant environments. The work procedures may be known in detail under task conditions, where the physical environment paces the man, and thus forces him to use a known sequence of subtasks, as is the case in e.g. manual assembly processes.

In modern automated process plants, however, the human function is typically higher level mental data processing and decision making, and the human work procedures are constrained by the physical environment to a much lesser degree.

Consequently, the normal practice is to try to control the work procedure in critical tasks by work instructions which take into account the possible deviation from normal working conditions that have been identified during system design.

However, in the analysis of accidents it is frequently seen that such safe work procedures have been operationally "improved" to fit the normal work situation in a way that does not take account of the predicted risk.

In reliability assesment this tendency has to be faced in a realistic way. As long as the prime cause to have the people in the plant is the human ability to adapt to the operational needs of the plant and to improvise in all plant conditions not foreseen by the system designer, it is not reasonable to expect them to follow work procedures which are troublesome in the normal work, just for the sake of conditions they possibly never meet.

The concluding remarks in reports investigating accidents, which are due to inappropriate procedures, frequently prescribe "tighter administrative control" of work instructions. A more realistic approach is the situational one, as advocated by Rigby and Swain, who argue that a work situation can only be reliable if properly fitted to normal variability of human behaviour. Human actions due to normal psychological mechanisms should not be classified as operator errors, even if they do lead to system faults. Rather the work situations have not been designed in a way resulting in predictable procedures.

Unfortunately, very few studies have been made to describe the procedures evolving in higher level mental tasks in real life working conditions and to relate them to controlling factors in the work situation. Consequently we have recently initiated such studies. We have examined mental procedures in control room environment and in an electronic work shop, and the prelininary analysis has identified features which we find illustrating in the present discussion.

We have found that the creativity and adaptability of man often result in the evolution of several basically different mental procedures for the same type of task, all capable of ending up with the same result. The procedures may differ in several basic aspects, such as the amount of data or observations which are needed; the complexity of the mental data processing which is implied; the depth of functional knowledge regarding the system anatomy and function which is used; and finally the time spent on the task (Rasmussen and Jensen 1973). A fault in an electronic system, for instance, may be located from a minimum number of observations if a careful deduction is used, based upon a detailed knowledge of system anatomy and internal functioning. However, the fault may also be located by a rapid sequence of observations or measurements and simple checks against normal values in a diagram without considering the functioning of the system. In this way different procedures can be available to a human operator for a specific type of task, procedures which fit the different working conditions in which the task is met.

The choice between the different procedures depends upon the performance criterium adopted by the man in the actual working situation. The rapid stream of simple decisions may be valued in some cases, due to the low cognitive strain implied, in other cases the complex reasoning may be chosen due to informational economy. The important point is that the performance criterium of the designer and the real life operator most probably are different, and the designer very likely will not predict the actual procedures used by trained personnel, unless he is very familiar with the actual task conditions from studies on site.

A further prerequisite to be able to use the classical reliability methods for evaluation of human behaviour is the breakdown of the procedures used into a sequence of typical and generally used units. This can be done for a task in which the elements of the sequence on the work steps are defined and cued by the environment as for manual tasks in production. But again it is not the case for higher level data processing tasks in plant environments.

Newell and Simon (1958) have argued that mental processes underlying human data processing and decision making can be decomposed into a sequence of elementary units, and as such simulated by a digital computer program. But in our experience this is not the whole truth, and Dreyfus (1965) has criticized the assumption and stressed the role of holistic, intuitive processing, which cannot be decomposed into elementary units.

Discussing the decisions of chess playing, he argues: playing chess "... may involve noticing that "here something interesting seems to be going on", " he looks week over here" etc. Only after the player has zeroed in on an area does he begin to count out, to test, what can be done from there.-"

In other words the first, important step in the mental work sequence, the identification of the task situation and the appropriate goal, may be based upon wholistic processes, pattern recognition, intuition and "feelings", and based hereupon a sequental processing may take place.

From our reviews of reported accidents we have found the identification of the proper task in abnormal plant conditions to be very critical, and our conclusions from preliminary analysis in control room environments tend to support the view of Dreyfus.

The operator seems to have a "process feeling", some sort of internal dynamic model of the environment, which all the time keeps him prepared for the normal tasks to come. This means that he may only be prepared to look for very little information at the actual time of a task, and it is not possible to predict whether the information actually underlying his decisions is properly updated. Furthermore the source of information chosen may be convenient sources during the normal working condition, such as noise from the system, e.g. relay clicks, rather than information planned by the designer to be task defining and therefore displayed to the operator and considered in a prediction.

Fundamentally this effect of the "process-feeling" links the elements of a task sequence together and they cannot be treated individually. A control room operator typically does not perform isolated actions on well specified bits of information, he is an integrated part of a dynamic situation. This causes difficulties which are hard to predict when abnormal plant conditions suddenly demand the operator to switch to other tasks and performance criteria. The reactions to abnormal plant conditions can only be treated in the light of the normal working conditions prior to the event and setting the process-feeling and thus the expectations of the operator. The reactions can only be treated in isolation, if the man-machine interface can be designed to break the routines of the operator and to set the initial conditions of his data processing in a predictable way at the start of a task.

It is worth noting that the basic aspects of the procedures adopted by man for a task normally will depend upon the frequency of the task. The very frequent tasks are met by procedures based upon pattern recognitions and trained, partly subconscious routines; less frequent tasks by procedures based upon plans or instructions, whereas the unique, very infrequent task may call for improvisation and complex, deductive reasoning related to understanding plant anatomy and functioning.

Again the frequency/risk relationship intrudes our problem. In a reasonably well designed system an inverse relationship can be expected between the frequency of an event calling for manual intervention and the risk implied in the event, and again the frequent events are easier to predict and analyse, whereas the infrequent, but critical events are of major importance to the system user. As discussed above, the familiar tasks set the stage for the unexpected, new events and consequently the operator tends to approach a new task by the most probable hypothesis, although most safety regulations tend to force the man to consider first the hypothesis covering the most critical cause.

To see how far we can get in planning a man-machine interface that will cause personnel to adopt predictable procedures, it is very important to have methods for prediction of human reliability verified by field tests and to have a clear identification of the characteristics of those work procedures and working situations they can be used to analyse, and to have more studies to identify the procedures evolving during process plant operation under different typical working conditions.

CONCLUDING REMARKS

The methods of probabilistic reliability evaluation today are efficient tools internally in the design offices for process plant equipment. To reach the state, where the methods can be used to a quantitative evaluation of the reliability of a complete operating process plant and an assesment of plant safety, it is imperative to create a closer relation to the realities of process plant operation. This implies an interdisciplinary cooperation between the fields of reliability engineering, human factors engineering and plant operation, and a careful verification of the methods including an explicit statement of the limits of their appropriate use.

REFERENCES

Ablitt, J. F., Private Communication. Safeguards Division, Authority Health and Safety Branch, U.K.A.E.A., Risley, 1969.

Dreyfus, H. L., What Computers Can't Do (Harper and Row, New York, 1972) 259 pp.

Evans, R.A., Editorial. Ask a Silly Question .... IEEE Trans Reliab R-21 (1972) 129.

Farmer, F.R., Reactor Safety and Siting: A Proposed Risk Criterion Nucl. Safety 8 (1967) 539-548.

Green, A.E. and Bourne, A.J., Reliability Technology (Wiley-Inter-science, London, 1972) 636 pp.

Johnson, W.G., The Management Oversight and Risk Tree-MORT. SAN 821-2 (1972) No pagination.

Meister, D., Comparative Analysis of Human Reliability Models. AD-734 432 (1971) 481 pp.

Newell, A., Shaw, I.C., and Simon, A., Elements of a Theory of Human Problem Solving. Psychol Rev 65 (1958) 151-166.

Newell, A., Simon, H.A., Human Problem Solving (Prentice-Hall, Erslewood Cliffs, N.J., 1972) 920 pp.

Nielsen, D.S., The Cause/Consequence Diagram Method as a Basis for Quantitative Accident Analysis. Risø-M-1374 (1971) 27 pp.

Rigby, L.V., The Nature of Human Error. SC-DC-69-2062 (1969) 12 pp.

Rasmussen, J., Man-Machine Communication in the Light of Accident Records. International Symposium on Man-Machine Systems, Cambridge, 8-12 September 1969. IEEE Conference Record No. 69 (58-MMS. Vol. 3)

Rasmussen, J. and Jensen, Aa., A Study of Mental Procedures in Electronic Trouble Shooting. Risø-M-1582 (1973) 71 pp.

Swain, A.D., Design Techniques for Improving Human Performance in Production (Industrial and Commercial Techniques Ltd., London, 1973) 140 pp.

Zwicky, F., The Morphological Approach to Discovery, Invention, Research and Construction In: New Methods of Thought and Procedure. Edited by F. Zwicky, and A.G. Wilson, (Springer, Berlin, 1967) 314-333.

HUMAN PERFORMANCE RELIABILITY MODELING IN TIME CONTINUOUS DOMAIN

Thaddeus L. Regulinski
Department of Electrical Engineering
Air Force Institute of Technology
Wright-Patterson Air Force Base, Ohio
USA

INTRODUCTION

Literature dealing with mathematical modeling of human performance can generally be divided along the lines drawn by the principles which underlie model formulation and the techniques adopted by the modeler. One of the most discernible divisions is formed by the principles entailed in the ATOMISTIC and the SYSTEMS approaches. The atomistic approach attempts to break down the human performance to its most fundamental molar and micro behavioral tasks. In contradistinction the system approach deals with the human performance in its totality, and if elements of the performance are distinguishable, they are treated only as a function of the totality.

Crisscrossing the line between the atomistic and systems approaches are numerous human performance modeling techniques which generally reflect the disciplinary background of the modeler. Case in point are the techniques generally associated with system scientists, cyberneticists, and behavioral scientists. The modeling techniques of the system scientist are most compatable with the optimal control theory; those of cybernetician with identification theory, and those of the behavioral scientist with reliability theory.

The foundation for these and all other techniques is the stochastic modeling process, and it is the purpose of this paper to formalize the modeling of human performance functions, and to delineate stochastic techniques useful in quantifying such functions.

STOCHASTIC MODELING PROCESS

In Figure 1, the stochastic modeling process is illustrated showing the domains of data generation and processing, mathematical model formulation, prediction, and validation.

Appropriate methods for the generation and processing of human performance data are heavily dependent upon the task performance observed, and upon the modeler's goals of data processing.

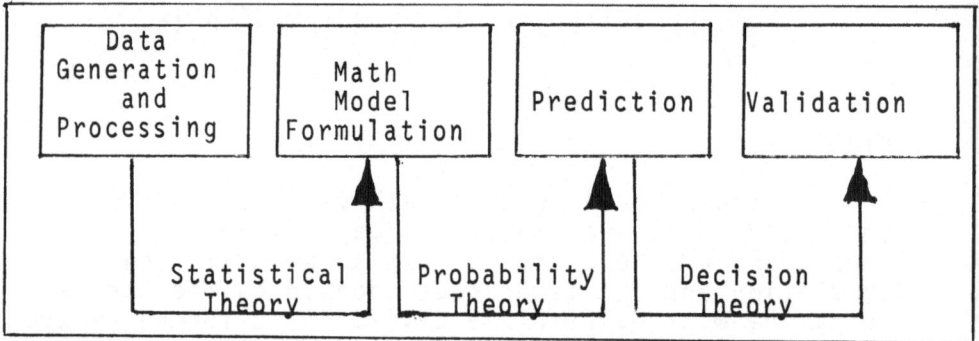

Fig 1: Stochastic Modeling Process

Nevertheless, some methods are fundamental to a large body of human performance data acquisition. These are: data recording which may include storage if data cannot be manifested directly; data conditioning which may involve digitization, quantification, and A/D conversion; data qualification which generally involves editing, formation of individual sample records, formation of ensembles, and ordering; and lastly, stochastic analysis of data for randomness, stationarity and ergodicity.

From the domain of data generation and processing, the modeling process moves to mathematical model formulation domain via statistical theory. The essence of model formulation lies in defining and quantifying functions associated with the random variable under study, and from data statistics, isolating the probability density function (pdf) governing the behavior of the random variable.

Once the governing probability density function is isolated, the modeling process can move to the domain of prediction via probability theory. Involved here are the techniques of estimating the parameters of the isolated pdf, predicting the behavior of the random variable, and establishing suitable criterion against which predictions can be evaluated.

Lastly, the process moves from the domain of prediction to the domain of validation via decision theory. The major activity in this domain is hypothesis testing for validation of the isolated pdf. However, activities in the domain may also include multiple decision and sequential testing, modeling of utility and cost functions, formulating rules for decisions, and testing by the likelihood ratio function. In this domain, as in the two preceeding ones, the governing pdf is fundamental to any mathematical operation which may be undertaken.

ANALYTICAL TECHNIQUES

Human performance functions can be modeled in either the analytic or numerical STOCHASTIC realm, or in the analytic or numerical DETERMINISTIC realm. Whether modeling is to be performed in the stochastic or deterministic realm is dictated principally by randomness or nonrandomness of generated data. A number of methods are available for testing randomness. Two such methods are the Runs Test described below (Ref 1), and the Reverse Arrangement Test (Ref 2).

The Runs Test

The runs test is conducted by finding the data median point, say $\bar{x}$, so that the generated data points are divided into two sets $S_1 = \{X; x_i < \bar{x}\}$ and $S_2 = \{X; x_i > \bar{x}\}$. Thus, if data are listed in order collected, each data point is classified into one of two mutually exclusive categories, and denoted arbitrarily by (-) if data point is less than $\bar{x}$, or by (+) if the data point is greater than $\bar{x}$. A run, denoted by r, is defined as a sequence of identical observations (-) or (+), followed by or preceeded by either a different observation or no observation. Randomness is accepted if

$$r_{N; 1-\frac{\alpha}{2}} \leq r \leq r_{N; \frac{\alpha}{2}} \tag{1}$$

where N denotes the total number of data points, α denotes the level of significance, and the values of $r_{N;\alpha}$ such that $\Pr\{r_N > r_{N;\alpha}\} = \alpha$ can be found in numerous statistical reference books.

Model Formulation - Discrete Tasks

If data randomness can be ascertained modeling in the stochastic realm can be justified. A reasonable first step in model formulation is to define the performance function which is relevant to the task. One such function is the human performance reliability which for the discrete task may be defined by

$$R_h = \Pr\{\text{task performance without error}|\text{stress}\} \quad (2)$$

and for the time-continuous task

$$R_h(t) = \Pr\{\text{task performance without error in } (t_0, t) \text{ time}|\text{stress}\} \quad (3)$$

In these definitions stress is used in its generic sense. It includes the totality of all factors, psychological, physiological, and environmental, having an effect on human performance.

Underlying the definition given by (2) is the assumption that the task random variable generated, call it X, assumes a finite number of values. If such is the case then X is by definition a discrete random variable. Associated with it is a number

$$p(x_i) = \Pr\{X = x_i\} \quad (4)$$

called the probability of x_i,

$$F(x) = \Pr\{X \leq x\} \quad (5)$$

called the Cumulative Distribution Function, CDF, and

$$f(x) = F(x^+) - F(x^-) \quad (6)$$

called the probability mass function (discrete probability density).

Numerous human performance discrete models can be and have been generated from the functions given in expressions (4) through (6). For example, calling upon the limit function of the ratio of successful trials, N_s, to the total number of trials, N, one can model,

within limitations implied, the reliability of some repetitive discrete tasks simply as

$$R_h = N_s/N \tag{7}$$

Another model frequently found in literature is based on the concept that discrete human performance can be approximated from the computation of the difference between unity and the error rate (Ref 3). Formalizing the concept, one can model performance reliability as

$$R_h = 1 - \frac{F(x^+) - F(x^-)}{1 - F(x^-)} \tag{8}$$

Still another model can be derived from the Binomial probability mass function (Ref 4). Denoting N to be the total number of independent trials of a task, and X to be the number of times human error occurred with probability p in the N trials, then

$$R_h = Pr\{X=k\} = \binom{N}{k} p^k (1-p)^{N-k} \tag{9}$$

where k = 0 by definition.

The Binomial random variable is closely connected with the Bernoulli random variable which, at a given task trial, takes on the values of 1 or 0 depending on whether or not an error occurred. It follows then that another viable human performance reliability model may be formulated from the Bernoulli probability mass function (Ref 5). Analogous rationale may be stated for modeling performance reliability from the Poisson probability mass function (Ref 4). The model can be derived from equation of expression (9) by letting p be very small and N very large, so that $Np = \mu$. Letting

$N \to \infty$, and keeping k finite, in the limit there will emerge the Poisson model of the performance reliability function given by

$$R_h = Pr\{X = k\} = \frac{\mu^k e^\mu}{k!} \tag{10}$$

where, again, k = 0 by definition.

Model Formulation - Time-Continuous Tasks

In time-continuous performance situations, as exemplified by vigilance, stabilizing, tracking, or controlling type tasks, the random variable generated, call it T, is continuous over some range of definition. The cumulative distribution function of a continuous random variable is given by

$$F(t) = \int_t f(t)dt \tag{11}$$

where f(t) is the probability density function (pdf). Clearly, it follows from (11) that

$$f(t) = dF(t)/dt \tag{12}$$

The instantaneous error rate, e(t), relates the CDF and the pdf by

$$e(t) = \frac{f(t)}{1-F(t)} \tag{13}$$

where e(t)dt is the probability of error occurrence in (t, t+dt) time interval, given errorless performance up to t time.

Using the expression of equations (11) through (13) it has been shown that the quantification of human performance reliability in time continuous domain is (Ref 6)

$$R(t) = \exp\left\{ - \int_0^t e(t)\, dt \right\} \tag{14}$$

Alternatively, human performance reliability can be expressed in terms of the probability density function governing the random variable, T, time-to-error, by

$$R_h(t) = \int_t^\infty f(t)dt \tag{15}$$

Depending on how time is measured in a specific task performance, the mean time-to-first error, mean time-to-error, mean time between errors, and the such can be determined from

$$E(T) = \int_0^\infty tf(t)dt = \int_0^\infty R_h(t)dt \tag{16}$$

Numerous other human performance measures can similarly be formulated. For example, recognizing the capacity of the human to correct self generated errors, it is germane to model some performance function relevant to error correction. Call such performance measure human performance correctability function (Ref 7). Denote the function by $C_h(t)$ and define it by

$$C_h(t) = \text{Pr (Completion of task error correction in } t \text{ time}|\text{stress)} \qquad (17)$$

It can be shown that the quantified version of expression given by (17) is

$$C_h(t) = 1 - \exp\left\{-\int_0^t c(t)dt\right\} \qquad (18)$$

where $c(t)$ is the instantaneous error correction rate and has a definition analogous to expression of equation (13). Alternatively,

$$C_h(t) = \int_0^t p(t)dt \qquad (19)$$

expresses the correctability function in terms of the probability density, $p(t)$, governing the random variable time-to-error correction.

Still another time-continuous human performance measure may be formulated from the random variables which led to the formulation of the reliability and correctability functions. Specifically, if T_A denotes the random variable time-to-error, and T_B time-to-error correction, a new random variable formed from the ratio T_A/T_B may be of interest as a metric of system or process control. Call such function controllability, denoting it by $Y_h(\tau)$, and defining it by

$$Y_h(\tau) = \text{Pr\{Ratio of time spent in control to time spent out of control equals or exceeds } \tau|\text{stress\}}$$
$$= \text{Pr}\{T_A/T_B \geq \tau|\text{stress}\} \qquad (20)$$

It, therefore, follows that
$$Y_h(\tau) = \int_\tau^\infty f(\tau)d\tau \qquad (21)$$

where τ may represent some critical value in system or process specification. If it may be more relevant to a given system or process to formulate the controllability function from the ratio of $T_A/(T_A+T_B)$ then the results of expressions developed in equations (20) and (21) would, of course, be modified.

Validation Techniques

Close examinations of formulated functions given by equations (15) through (21) shows that their tractability depends on isolation of the underlying probability density functions governing the random variables generated. Numerous methods for the isolation of governing pdf's exist. One such test is the Kolmogorov-Smirnov (K-S) test of validity (Ref 8). The basic K-S algorithm calls for the comparison of a sample cumulative distribution function $S_N(t)$ of a size N sample taken from a population with unknown cumulative distribution function, $G(t)$, to a theoretical distribution, $F(t|\Theta)$, where Θ is the known parameter(s) of F. Integral part of the algorithm is to test the null hypothesis
$$H_0: \quad G(t) = F(t|\Theta) \qquad (22)$$
against the alternate hypothesis
$$H_1: \quad G(t) \neq F(t|\Theta) \qquad (23)$$
and to determine the maximum difference
$$D = \max|F(t_i|\Theta) - S(t_i)| \qquad (24)$$
over the set of generated data $T = \{T_i | i=1,2,\ldots N\}$.

The null hypothesis can be accepted at any level of significance α specified if $D < d_{\alpha(N)}$, where $d_{\alpha(N)}$ is the critical value of the K-S test statistic and can be found in tables of Ref 8.

Since there exists prior evidence of applicability, pdf's to be tested should at least include the Normal, Lognormal, Gamma, Weibull, Rayleigh, Exponential, Beta,

Erlang, smallest extreme value, and largest extreme value. It is a fact, however, that the K-S test may reject all, some, or none of the densities selected for testing. In the event the test fails to reject two or more, use can be made of the Likelihood Ratio Test defined by

$$L(T) = f_1(T|H_1)/f_0(T|H_0) \tag{25}$$

where

$$f_j(T|H_j) = \prod_{i=1}^{N} f_j(t_i|H_j) \tag{26}$$

The null hypothesis

$$H_0: \text{Distribution is } F_0(t|\Theta_0) \tag{27}$$

can then be tested against

$$H_1: \text{Distribution is } F_1(t|\Theta_1) \tag{28}$$

with the threshold of the test η defined by

$$\eta = \{P_0(C_{10}-C_{00})\}/\{P_1(C_{01}-C_{11})\} \tag{29}$$

where P_0 and P_1 are the priori probabilities of H_0 and H_1 respectively, C_{00} is the cost when H_0 is true and H_0 is chosen, C_{10} is the cost when H_0 is true and H_1 is chosen, C_{11} is the cost when H_1 is true and H_1 is chosen, and C_{01} is the cost when H_1 is true and H_0 is chosen. The Likelihood Ratio Test can then be written as

$$L(T) \begin{matrix}>H_1\\ \eta \\ <H_0\end{matrix} \tag{30}$$

which states that if $L(T)$ is less than η, choose H_0, and if greater than η, choose H_1.

The value of the threshold can generally be taken to be unity based on the following rationale. Since, as a rule, the value of the threshold must be estimated and priori probabilities are unknown, it is reasonable to assume that the two distributions under test have equal likelihood of occurrence, i.e., $P_0 = P_i$, unless, of course, specific information is available on which to base a better estimate. Further, it can be assumed that the cost for an incorrect decision is the same, regardless of the decision, i.e., $C_{01} = C_{10}$, and that there is no penalty cost for the correct decision, i.e., $C_{00} = C_{11} = 0$. Clearly, under the assumptions stated, the expression of equation (30) can be written as

$$\ln L(T) \underset{<H_0}{\overset{>H_1}{}} \ln 1 \qquad (31)$$

or

$$\sum_{i=1}^{n} \{\ln f_1(t_i|H_1) - \ln f_0(t_i|H_0)\} \underset{<H_0}{\overset{>H_1}{}} 0 \qquad (32)$$

which forms a computational algorithm useful for computer application. For two or more hypothesis, each pair can be tested against each other, H_0 vs H_1, H_0 vs H_2, and H_1 vs H_2. With unity threshold value, this leads to one hypothesis being selected over any other two.

In testing for stationarity and ergodicity data generated from a task performed by a number of subjects can be considered a random process consisting of the ensemble of sample records produced by the individual subjects. Stationarity in the wide sense requires that some statistical properties of the random process be time invariant. Ergodicity requires that some statistical properties of a single sample record be representative of the entire ensemble. The former can be tested by dividing each sample record into K equal time intervals and computing at these intervals the mean and the mean square values. The mean values set $(\bar{t}_1, \bar{t}_2...\bar{t}_K)$ and the mean square values set $(\bar{t}_1^2, \bar{t}_2^2....\bar{t}_K^2)$ can then each be tested by the runs test for the presence of variations other than those due to sampling variations. In similar way ergodicity can be tested by computing the mean and mean square values for each individual sample record and testing those for variations. Data are considered stationary and ergodic if there existed no significant variations of the mean and mean square values from one time interval to another and from one sample record to another.

SUMMARY

Because there exists a myriad of human performance models and methods of deriving them, it is useful to formalize the modeling process, the modeling of human performance measures, and the techniques functional in quantifying them. In essence, the modeling process consists of formulating from generated data those metrics which are meaningful to human performance task under consideration; quantifying the metrics; isolating the probability density function governing the behavior of the random variable generated by the task, and predicting its behavior; and validating the propounded models.

Literature is replete with applications of discrete human performance models, however, in time-continuous domain performance models and their applications tend to be somewhat scarce. It is believed that the propounded human performance reliability, correctability and controllability functions may be potentially useful in such applications as process or system control, vehicle stabilization, detection, tracking, and other tasks subject to modeling in time-continuous domain.

REFERENCES

1. Siegel, S. Nonparametric Statistics For the Behaviroal Sciences. McGraw-Hill Book Company, New York, 1965.

2. Bendat, J.S. and Pierson, A.G. Measurement and Analysis of Random Data, John Wiley and Sons, Inc., New York, 1966.

3. Jenkins, J.P. "Proceedings of U.S. Navy Human Reliability Workshop", Report NAVSHIPS 0967-412-4010, Washington, D.C., July 1970.

4. Meyer, P.L. Introductory Probability and Statistical Applications, Addison-Wesley Publishing Co., Reading, Mass, 1970.

5. Indgren, B.W. and McElrath, G.W. <u>Introduction to Probability and Statistics</u>, The Macmillan Co., New York, 1959.

6. Regulinski, T.L. and Askre, W.B. "Mathematical Modeling of Human Performance Reliability", <u>IEEE Proceedings of the 1969 Annual Symposium on Reliability</u>, Vol 2-1, January 1969.

7. Regulinski, T.L. and Askren, W.B. "Stochastic Modeling of Human Performance Effectiveness Functions," <u>IEEE Proceedings of the 1972 Annual Symposium on Reliability and Maintainability</u>, Vol 5-1, January 1972.

8. Massey, F.J. "The Kolmogorov-Smirnov Test for Goodness of Fit", <u>ASA Journal</u>, Vol 46, March 1951.

OPTIMIZATION OF REDUNDANT SYSTEMS - A NON-LINEAR
PROGRAMMING APPROACH

H.V.K. Shetty, and D. P. Sen Gupta

Department of Electrical Engineering, Indian
Institute of Science, Bangalore 560012, India.

ABSTRACT

The object of this paper is to optimize reliability of a system provided with active redundancy subject to inequality constraints, by using a non-linear programming (NLP) technique. Although the constraints imposed are linear in the examples considered here, the method used is applicable to non-linear constraints as well. The constrained objective function is converted into a sequence of unconstrained functions by imposing "penalty". The solution is achieved by two methods, namely 'Conjugate Gradient Method'[5] and 'Newton Search Techniques'[4] and the relative merits of these two methods have been discussed.

INTRODUCTION

The problem considered here can be defined as one of optimal allocation of parallel redundancy under multiple linear (non-linear) constraints by non-linear programming techniques. Various authors[7] have adopted different techniques such as integer programming 0-1 type, Lagrange-Multiplier technique, dynamic programming, geometric programming etc., These techniques have their advantages and disadvantages and their effectiveness depends on the objective function to be optimised and the constraints imposed. In particular, dynamic programming techniques would require a large number of iterations for the final solution, especially when the constraints are large. Lagrange multiplier technique has the limitation that it is valid for equality constraints only, and Kühn Tucker optimality conditions have to be satisfied when inequality constraints are considered. Besides, the choice of a proper Lagrange multiplier is difficult and its value affects the answers. So, an attempt has been made to analyse the optimization problem considered in reference 6 with inequality constraints by using penalty function approach. The technique described in this paper has the advantage that it can take care of any number of constraints of any kind.

REVIEW OF THE NLP TECHNIQUE USING PENALTY FUNCTIONS

The principle of transforming a constrained optimization problem into an unconstrained optimization problem using penalty function and proof of its convergence are given in reference 1-2. The following is a brief description of the principles of the slacked unconstrained minimization technique[2] (SLUMT) and sequential unconstrained minimization technique[1] (SUMT).

An optimization problem may be stated as below:

To minimize $f(\bar{x})$ subject to the constraints

$$g_j(\bar{x}) \geqslant 0 \qquad j = 1, 2, \ldots, m \qquad (1)$$

SLUMT in brief

An auxiliary function for equation (1) is defined as

$$P(\bar{x}, r_k, t_j) = f(\bar{x}) + \sum_{j=1}^{m} \left[g_j(\bar{x}) - t_j \right]^2 / r_k \qquad (2)$$

and

$$\begin{aligned} t_j &= g_j(\bar{x}), \text{ if } g_j(\bar{x}) \geqslant 0 \\ &= 0, \qquad \text{if } g_j(\bar{x}) < 0 \end{aligned} \qquad (3)$$

The function $P(\bar{x}, r_k, t_j)$ is minimized for a non-negative decreasing sequence of r_k. As r_k approaches zero, P and f coincide.

SUMT in brief

Here, the auxiliary function for equation (1) is

$$P(\bar{x}, r_k) = f(\bar{x}) + r_k \sum_{j=1}^{m} 1/g_j(\bar{x}) \qquad (4)$$

The solution procedure is identical to the previous method. Although these two methods have many similarities in that a penalty is imposed when one of the constraints is violated, one is not a trivial variation of the other. In SUMT the initial point has to be within the feasible region which need not be the case for SLUMT. Also, the latter method can be used for a combination of equality and inequality constraints and achieves feasibility and optimization simultaneously.

MATHEMATICAL MODEL

System unreliability of an n-stage series-parallel system is

$$Q_s = 1 - \prod_{i=1}^{n} (1 - q_i^{x_i}) \qquad (5)$$

The optimization problem is to minimize

$$Q_s(\bar{x}) = 1 - \prod_{i=1}^{n} (1 - q_i^{x_i}), \text{ subject to cost}$$

and weight constraints

$$b_1 - \sum_{j=1}^{n} c_j x_j \geq 0 \qquad b_2 - \sum_{k=1}^{n} \omega_k x_k \geq 0 \qquad (6)$$

This can be easily converted into the form given in equation (1). For a standard bridge configuration the reliability expression is formulated by minimal tie-set or minimal cut-set procedure. The constraint equations will be the same as equation (6).

INITIAL VALUES AND CONVERGENCE CRITERIA

Fiacco[1] suggested the following expression for evaluating r_1, the initial value of r_k.

$$r_1 = -\nabla f(x^\circ)^T \cdot \nabla p(x^\circ) / |\nabla p(x^\circ)|^2 \text{ where } p(x) = \sum_{j} 1/g_j(x)$$

and x° the initial values of x. (7)

The general convergence criteria to judge the optimum for a constrained optimization problem is by calculating the root mean square of the violated constraints.

$$\delta = \left\{ \sum_{i \in I} \left[\frac{g_i(\bar{x}) - s_i}{s_i} \right]^2 \right\}^{\frac{1}{2}} \qquad (8)$$

where s_i s are the specified value of $g_i(\bar{x})$ are the calculated value and I is a set of violated constraints.

RESULTS

The parameters of the system considered and the results obtained are given in Tables 1 and 2. The computation was carried out on a IBM 360/44 computer.

TABLE - 1
5-stage series-parallel system

	\multicolumn{5}{c}{stages}	
	1	2	3	4	5	Constraints
Cost (Units)	5.0	4.0	9.0	7.0	7.0	100
Weight (Units)	8.0	9.0	6.0	7.0	8.0	104
Reliability	0.9	0.75	0.65	0.8	0.85	
Opt. no. of components	2	3	4	3	2	

Time for convergence using SLUMT and conjugate gradient techniques ...... 6 sec.s

Using SLUMT and Newton Search technique ...... 4 sec.s

maximum reliability 0.984915

TABLE - 2
Bridge network

	1	2	3	4	5	Constraints
Cost (units)	1.2	2.3	3.4	4.5	4.0	60
Weight (units	5.0	4.0	8.0	7.0	10.0	130
Reliability	0.8	0.7	0.75	0.84	0.65	
Opt. no. of components	4	5	6	5	1	

Time for convergence using SLUMT and conjugate gradient technique ...... 5 sec.s

Using SLUMT and Newton Search tech. ...... 3 sec.s

Maximum reliability 0.99999

CONCLUSIONS

The use of penalty function makes the optimization procedure straightforward. The optimum number of components at each stage is obtained by rounding off the computed results to nearest integers. This method does not seem to have the drawbacks presented by other methods. Time taken by SUMT and SLUMT formulation was identical for the problems considered, although the latter is likely to result in faster computation for longer problems. Newton search technique is quicker than the conjugate gradient method but becomes unwieldy for large system if the Hessian matrix containing second order partial derivatives has to be fed into the computer. Conjugate gradient technique also needs first order derivatives but these may be obtained numerically by the computer.

REFERENCES

1. Fiacco, A.V. and McCormick, G.P., 'Computational algorithm for the sequential-unconstrained minimization technique for non-linear programming', Management Science, vol.10, No.2, pp 601-617, 1964.

2. Fiacco, A.V., and McCormick, G.P., 'The slacked unconstrained minimization technique for convex programming', SIAM J., Appl. Math., vol. 15, No.3, May 1967.

3. Shooman, M.L, 'Probabilistic reliability: an engineering approach' - McGraw Hill Book Company.

4. Pierre, D.A., 'Optimization theory with applications', John Wiley and Sons, Inc. 1969.

5. Fletcher, R. and Reeves, C.M. 'Function minimization by conjugate gradients', Computer Journal, 7, pp 149-154, July 1964.

6. Tillman, F.A., 'Integer programming formation of constrained reliability problems', Management Science, vol 13, pp887-897, July 1967.

7. Messinger, M, Shooman, M.L., "Techniques for Optimum Spares allocation: A tutorial review". IEEE Transactions on Reliability, Vol. R-19, No.4, Nov, 1970, p 156-166.

SOFTWARE RELIABILITY: ANALYSIS AND PREDICTION

Martin L. Shooman[*]

Department of Electrical Engineering and Electrophysics
Polytechnic Institute of New York
Brooklyn, New York, U.S.A.

ABSTRACT

With the advent of large sophisticated hardware-software systems developed in the 1960s, the problem of computer system reliability has emerged. The reliability of computer hardware can be modeled in much the same way as other devices using conventional reliability theory; however, computer software errors require a different approach.

The paper begins by describing the types and causes of software errors and provides working definitions of software errors and software reliability. Some of the basic data on frequency of occurrence of errors is then discussed. The paper then summarizes and references some of the software reliability models which have been proposed and concentrates on one developed by the author.

This newly developed probabilistic model predicts reliability based on the initial number of errors in a program, the number removed, and the number remaining is the program. The model constants are calculated from operational test data taken on the software performance.

The calculations result in a decreasing probability of no software errors versus operating time (reliability function). The rate at which the reliability decreases is a function of the man-months of debugging time. Similarly, the mean time to occurrence of operational software errors (MTTF) is obtained. The MTTF increases slowly and then more rapidly as the debugging

---

[*]This work was supported by the Office of Naval Research, Statistics and Probability Program under Contract N00014-67-A-0438-0013.

effort (man-months) increases. The model permits estimation of software reliability before any code is written and allows later updating to improve the accuracy of the parameters when integration or operational tests begin.

1.0 INTRODUCTION

1.1 The age of large computers

The growth in computer hardware can be illustrated in several ways. First of all the total number of computer installations in the United States has grown phenominally (see Table 1-1) reaching about 50,000 in 1970. Sackman, (see Ref. 1, p. 30) quotes an estimate that the European Economic Community will have approximately 10,000 computers by 1970. Most of the above mentioned computers are large general purpose scientific and business computers. There are also many special purpose military and industrial computers. A detailed count is difficult to obtain; however, in Ref. 3, p. 4, an estimate of the current U.S. Air Force annual expenditures for computer hardware is quoted between $300 and $400 million per year. The corresponding estimate of the cost of procuring computer software is $1,000 million to $1,500 million!!

1.2 Hardware vs. software errors

Ref. 3 cites "a software error aboard a French meteorological satellite caused it to 'emergency destruct' half of its force of weather balloons instead of interrogating them." In another case, that of the Apollo spacecraft guidance computer (see Ref. 5) the similarity of program names led to the wrong program being called which destroyed the guidance systems parameters, necessitating a lengthy reinitialization.

The most dramatic story of a software error comes from the popular American science fiction movie, "2001 A Space Odyssey," (see Ref. 4).

Although the above three examples all dealt with space missions, mainly because they make the point so vividly, many similar examples of the problems caused by software failures in military and industrial computer systems can be cited.

1.3 Some computer failure data

We may shed further light on the distribution between hardware and software errors by considering some actual data. The data given in Ref. 7 lists hardware and software failure rates for a typical real-time data acquisition system. The data represents 9 months of operation totaling 1701 hours. Inspection of the data shows that 48% of the failures were due to software. This is a startling figure if one realizes that although most computer projects estimate and predict the hardware reliability, and use redundancy techniques and high reliability parts to improve the hardware reliability, the software is left to the skill and hard work of the programming team with no quantitative assessment of design progress. We may obtain a typical estimate of what type of system reliability can be obtained in practice. Assuming a simple failure model, we may compute the mean-time to failure, MTTF as the reciprocal of the failure rate. Thus, the MTTF $\approx$ 30 hours for the data of Ref. 7. In other words about one failure per day will occur if the equipment is used 24 hours/day and one failure every three days if used 8 hours/day.

1.4 Computer engineering

In this section we try to briefly discuss some of the scant knowledge of software engineering which directly relates to software reliability. One facet inherent in any engineering design is schedule and manpower needs. In Fig. 1-1 we see estimates of the speed at which code is produced measured in lines of machine code per man hour. If we approximate the median between the 10th and 90th percentile in Fig. 1-1, we obtain: in 1955 machine language programming productivity was 200 machine instructions/man month; in 1970 programming in a higher level language has raised this figure to about 700 machine instructions/man month; the projection for 1985 using structured programming techniques (top down, go to free, cf. Ref. 23) is 1000 machine instructions per man month. At the outset we might say that the switch from machine language to higher level language programming (FORTRAN, etc.) was of great help in that it increased productivity by a factor of 3.5. A deeper consideration of these facts makes this progress appear less spectacular. In the compilation process each line of code written in a higher level language expands into between 5 and 10 lines of machine code. Thus, we might say that in going from machine to higher level language we have decreased productivity by a factor of two due to the time to write an instruction, yet increased productivity by a factor of about seven due to the expansion of code by the compiler. This is only part of the story since this only measures code productivity. From Table 1-1 we see that this phase only represents from 15% to 25% of the total effort.

If we use the SAGE data in Table 1-1, coding represents 15% of the effort. Also, using the previously quoted figure that a programming project done in a higher level language proceeds at a rate of about 700 instructions per man month or 8400 machine instructions per man year, yields the following conclusions: A 12,000 word FORTRAN program would yield about a 84,000 word machine language program and would require about 10 man years to code the program. In addition, about 30 man years would be required for checkout and test (debugging). Also, about 27 man years would be required for analysis and design. One must take care in working with these gross rules of thumb to remember that time can only be traded for manpower within reason. The favorite illustration is: it takes one woman 9 months to produce a baby but 9 women can't produce a baby in one month. Suppose two years time were allotted for the above example. A reasonable preliminary schedule might be to start with a design group of 27 working during the first year. After 6 months, the analysis group should have enough work done so that coding can begin, so a group of 10 programmers join the project after 6 months and should be completed with their work at the 18 months mark. After about 9 months a group of 10 programmers could form a test team to start testing and debugging their first sections of code which have been produced. After about 1 year, another group of 17 programmers could join for one or more additional test teams and begin debugging on the new sections of code being written. Notice that at the peak manpower level about 64 people are working on the project!! If we assume that with overhead the salary of an analyst is $40,000 and that of a programmer is $30,000, the overall labor cost of such a piece of software is over $2,000,000.

We now leave the subject of schedules and turn to some of the other measures of system performance which are used to evaluate a program: core size (number of machine language instructions), running time of the program, load factors, ease of change, and portability. When we say core size, we mean memory in general and with the advent of low cost electronic memory and inexpensive disk storage units, this is a less significant constraint than it once was. However, in space and aircraft applications where memory size is fixed by power, weight, and volume considerations, investigations have found (see Ref. 6) that if the programmers have a fixed core size which is too small for their initial design, then the tricks they use to "squeeze" the program into the core they have leads to great problems with errors and debugging. The influence of program run time on reliability has received little attention. It seems only obvious that if there is a very tight maximum on program run time (sometimes correctly and sometimes erroneously specified) that additional program tricks will be required which again contribute to the error and debugging problems. The implication is that not only will the debugging costs increase but that more sophisticated debug resistant errors may remain in the released version of the software to plague one during operation. Many people have observed that the effect of increased load on a system is to cause more frequent software errors. Most people using time-sharing systems feel that the mean-time between system crashes is strongly dependent on the number of users. We might attempt to summarize the above factors in the following equation

$$R \sim \left(\frac{m_u}{m_a} - k_1\right)^{x_1} \left(\frac{t_u}{t_a} - k_2\right)^{x_2} \bigg/ L^{x_3} \quad \text{for } \frac{m_u}{m_a} > k_1 \quad \frac{t_u}{t_a} > k_2 \qquad (1\text{-}1)$$

where

$m_u \equiv$ memory size used

$m_a \equiv$ memory size available

$t_u \equiv$ time used

$t_a \equiv$ time available

$L \equiv$ load factor

$k_1, k_2 \equiv$ constants

$x_1, x_2, x_3 \equiv$ experimentally determined powers.

2.0 DEFINITION OF A SOFTWARE BUG

2.1 Problems in definition

Definitions are always very difficult in the field of reliability, since we are trying to model a diffuse and complex physical situation by a mathematical model. The concepts of how to count multiple, repeated, and transient errors in hardware is most difficult. The author knows of cases where in a military contract, the contractor had to perform a reliability test of h hours on the equipment in question. If x or fewer failures occurred during the duration of the test, then the equipment passed. A board was created composed of government and contractor engineers. This failure board considered the test results and voted on each occurrence of abnormal behavior noted in the log book kept during the test to determine whether the occurrence should count as a failure. The most difficult case to decide on was where a certain logic card type failed m times. If this could be shown to be a simple repetition of the same transient failure, it would only count once, whereas if it was m separate failures, it would count m times. Often voting on the board was along strict "party lines," but often there were honest differences of engineering opinion.

If the definition of failure is difficult in the case of hardware where we have more experience and theoretical guidelines, it is even a more exacting task in the case of less well understood software. The author hopes the definitions proposed in the remainder of this section are a step toward a set of working definitions. At least they raise the sailent issues in the reader's mind if he wishes to formulate his own definitions.

2.2 Definition of a bug

The following definition of a software bug is proposed:
"One or more software bugs exist in a system if a software change is required to correct a single major error or minor error so as to meet specified or implied system performance requirements."

The following definitions are imbedded within the above definition:

Change - Any alteration (addition, deletion, correction) of the program code whether it be a single character or thousands of lines of code. Changes made to improve documentation or satisfy new specifications are important to record and study but are <u>not</u> counted as bugs.

Major error - A catastrophic event which interrupts or could interrupt most or all major system functions, e.g. an an infinite loop, system crash, a major memory overflow or data base corruption, etc.

Minor error - A marginal event which allows or could allow some portions of the system to operate properly while interrupting others, e.g. some missing output, some wrong output, an inaccurate computation, a recoverable transient error, etc.

Specified performance requirements - A written requirement, figure of merit, or parameter which qualitatively or quantatitively defines system performance.

Implied system performance - An unwritten requirement which is understood by the majority of the project team to be essentially equivalent to a written requirement.

Inherent in the above definitions and discussion is the assumption that errors can be and are detected and recorded. The detection of errors can be effected by monitoring the system (or simulated system) performance or by reading the code. Furthermore, it is assumed that each error is sufficiently well investigated so that it can be classified as hardware, software, operator, or unknown, and that the unknown category is small, say less than 20%.

2.3 Multiple bugs and changes

In this section we amplify on the previous definition in order to classify the cases of repetitious errors, multiple errors corrected by one change, and multiple changes to correct one error.

We now introduce the concept of internal and external errors. An external error is a performance error of the system, which is generally detected by executing the code. Theoretically, an external error could also be found by reading the code ("eyeballing"). An internal error is a coding error which is always found by reading the code either by man or machine.

The programmer could have initiated code reading due to one or more of the following factors:

(1) an external error has been detected and he is trying to find the corresponding internal error.

(2) he was reading the code to verify a program change before submitting it to the computer.

(3) he was performing a code reading test to detect errors.

(4) a colleague told him of an actual or potential error.

We may think of internal errors as causes and external errors as effects. Thus, if a single internal error results in an associated single external error, we call it a single bug. If an internal error results in a minor or no detectable external error, then no bug exists. If an external error exists and we are sure it is a software problem, then a bug exists regardless of whether or not we can find the corresponding internal error.

One important class of external errors for which no internal error can be found are transient errors. These exists for too short a time for isolation of the cause. A transient error which repeats itself enough times so that the associated internal error is found is no longer transient. If before the internal error is found, the transient occurs m times, it is still only counted as a single error if the symptoms are the same.

Theoretically, many internal errors can combine to cooperatively cause one external error. We expect that this is an event with a low probability of occurrence. Since only a single external error exists this would be classified as one bug.

A more common multiple error case would be where one internal error causes m external errors. Initially, if the m external errors are known, but no corresponding internal error has been found, then we classify this result as m bugs. At a later time, several days or weeks hence, when the common internal bug is found, we decide to reclassify this result as a single bug. Thus, if we are recording the cumulative number of bugs frequently, say each day, then the above event would first count as an increase of m bugs, and later when the single common internal error was found there would be a decrease of m-1 bugs. Of course, if the data were taken less frequently, the diagnosis of the internal error and the external error would fall within the same time interval and only the net result, i.e. one bug, would be recorded. In any event we could probably treat the number of bugs as defined herein as an upper bound if multiple external errors occur frequently.

2.4 Old or new bugs

It is useful in trying to model the dynamics of the debugging process to know whether a bug is old or new. To be more precise we might use the terminology previously corrected bugs and generated bugs.

A previously corrected bug is one which reoccurs in substantially the same form after the programmer terminated his work on a code change believing that the error was corrected. A conclusive decision that a bug was a previously corrected one can only be made based on an internal error diagnosis.

A generated error is one which did not exist until it was created as a by-product of a code change made to correct some bug. A generated error is usually best diagnosed by finding an internal error. However, it is sometimes possible to base such a classification on an external error, e.g. if a newly created variable appears in the wrong output form. Of course, all the above definitions rely on subjective judgment of the programmers. However, it is hoped that qualified personnel could use these definitions to obtain repeatable results.

3.0 DEFINITION OF SOFTWARE RELIABILITY

3.1 Factors in the definitions

In the early days of hardware reliability there was much major soul searching and thrashing about until a widely accepted definition of reliability was formulated. In the preceding section some working definitions of software bugs were developed. In this section we will attempt to develop a working definition of software reliability.

Our experience in the hardware area has taught us that reliability must be defined as the probability that some event occurs over a period of time (operating time). In our case the event is success of the software, i.e. error free software operation. We are now helped by the definitions of the previous section in defining the event error free software operation. Error free software operation over the operating time interval 0 to t means that no software bug (external software error) occurred over that interval. Clearly, if we are to define an external error, we must define what constitutes successful performance.

Another factor which must be specified is the hardware environment. This suggests certain obvious factors as well as more subtle ones. Obviously, a FORTRAN program written for use on an IBM computer will probably need modification before it can be used on a UNIVAC computer. We find more subtle differences when a computer is being specially designed for a project and the software and hardware are to be mated on the laboratory prototype. Often there are last minute changes which make for significant differences between the prototype and the field installation. Also, as an economy measure, the prototype may be a smaller configuration than the actual field installation. We are now faced with the situation where the final tests on the hardware complex will be carried out on a computer which differs from the operational model.

Finally, we must be concerned about the intended task the system must perform. Although there are always extensive efforts to write comprehensive specifications, problems occur. For example, returning to the previously cited APOLLO program, it was decided that the astronauts would never enter a call for the wrong program during the orbit phase of flight so no error checking for program calls was incorporated in the software. Subsequently, an astronaut while in orbit called for the ground initialization and alignment program. Would you call this a software error?

In many cases a system is originally sized for a particular data input rate. As the system begins to function successfully in the field after elimination at the initial bugs, there is a trend to employ it more widely. The error rate of the system increases as the data input rate increases. A classic example of this effect is the initial deployment of the SABER Airlines Reservation System. Everytime a new group of terminals was added to the system from a new city or group of cities a new crop of bugs was encountered. The system was allowed to grow, in stages, at a controlled rate by establishing an upper failure rate limit and a lower failure rate limit. When debugging removed enough errors so that the system reached the lower failure rate limits, new terminals were added until the upper limit was reached. As soon as the debugging team reduced the failure rate to the lower limit, the cycle was repeated.

3.2 Definition

A definition for software reliability is given below in keeping with the factors discussed above. This definition is a slight modification of one given by Hesse (see Ref. 8):

"Software reliability is defined as the probability that a given software program operates for some time period, without an external software error, on the machine for which it was designed given that it is used within design limits."

Once we have related reliability to a probability, as in the above definition, the mathematical basis of the measure is well founded. Of course, the problems in interpreting terms such as external error and design limits still exists.

3.3 Reliability Theory

The following brief development of the reliability function, hazard function, and mean time to failure is included for those unfamiliar with reliability theory (see Ref. 9). We begin with the standard probability functions

$$R(t) = P(\underline{t} > t) \tag{3-1}$$

$$F(t) = 1 - R(t) \tag{3-2}$$

$$f(t) = \frac{dF(t)}{dt} = -\frac{dR(t)}{dt} \tag{3-3}$$

where: $\underline{t} \equiv$ the random variable time to failure.

$t \equiv$ a particular value of the random variable.

$R(t) \equiv$ the reliability function, which yields the probability of no failure in the interval 0 to t.

$F(t) \equiv$ the cumulative distribution function, which yields the probability of failure in the interval 0 to t.

$P(\underline{t} > t) \equiv$ the probability that the time to failure lies outside the interval 0 to t.

Workers in the field of reliability have found it convenient to define a different conditional probability function called the failure rate or hazard function, $z(t)$. One may define $z(t)$ in a manner analogous to the definition of $f(t)$ in terms of the probability that a failure occurs in the interval t to $t + \Delta t$

$$\left\{\begin{array}{l}\text{Probability of a failure}\\ \text{in the interval t to t}+\Delta t.\end{array}\right\} \equiv P(t < \underset{\sim}{t} < t+\Delta t) = f(t)\,\Delta t \qquad (3\text{-}4)$$

$$\left\{\begin{array}{l}\text{Probability of a failure}\\ \text{in the interval t to t}+\Delta t\\ \text{given the fact that failure}\\ \text{did not occur prior to t}\end{array}\right\} \equiv P(t < \underset{\sim}{t} < t+\Delta t \mid \underset{\sim}{t} > t)$$

$$= z(t)\,\Delta t \qquad (3\text{-}5)$$

From these definitions it can shown that

$$z(t) = \frac{f(t)}{R(t)} = -\frac{1}{R(t)}\frac{dR(t)}{dt} \qquad (3\text{-}6)$$

Solving this differential equation for R(t) subject to the initial condition that the item is initially good, i.e., $R(t=0) = 1$ yields

$$R(t) = e^{-\int_0^t z(x)\,dx} \qquad (3\text{-}7)$$

Another measure which is often used is the mean time to system failure, MTTF. This is simply given as the first moment of the random variable $\underset{\sim}{t}$.

$$\text{MTTF} \equiv \int_0^\infty t f(t)\,dt \qquad (3\text{-}8)$$

It can be shown that Eq. 3-8 can be reduced to the simpler computational form

$$\text{MTTF} = \int_0^\infty R(t)\,dt \qquad (3\text{-}9)$$

For the simple case where z(t) is a constant Eqs. (3-7) and (3-8) yield

$$z(t) = \lambda \qquad (3\text{-}10)$$

$$R(t) = e^{-\lambda t} \qquad (3\text{-}11)$$

$$\text{MTTF} = 1/\lambda \qquad (3\text{-}12)$$

3.4 System Crash Rate

If we refer back to our definitions of Section 2 we would equate an operating system crash to a catastrophic error. This still leaves out marginal errors so if we talk about crash rate, it is a lower bound on the software failure rate. The main reason for talking about crashes is that they are such a dramatic event; there is a much better chance that they will be recorded.

In Table 3-1 are recorded some data on mean time between system crashes. Some of this data was accumulated in private communications, whereas other data is quoted in the literature.

4.0 ERROR DATA AND MODELS

4.1 Introduction

When one attempts to apply probability and statistics to an engineering problem two approaches immediately suggest themselves. The pure statistical approach is to define the variables and performance measure(s) and construct and carry out an experiment. The experimental results are statistically analyzed to determine quantitatively what relationships (if any) exist among the performance measure(s) and the variables. The other approach is to formulate a probabilistic hypothesis about how the variables interact and write a corresponding equation relating the performance measure(s) to the variables. Based on these models, experiments are planned to verify the hypotheses and to determine the constants in the equations. The reader may wish to view these approaches as analogous to the distinct approaches of a theoretical and an experimental physicist. In either case it is appropriate to begin by examining the data available in the literature at the outset.

The first relevant question is what data should we examine. If we ask experienced software managers what quantitative measures they use to gauge the progress of a software program they answer, none or refer to graphs of the cumulative number of errors removed from the software. Those who believe this is a significant measure look for the slope of the curve to approach zero before deciding the software is sufficiently debugged for release. This section of the paper discusses some of the sparse experimental data available in the open literature on the number of errors removed from a computer program. The next section builds upon this data and certain hypotheses to evolve a probabilistic error model.

Also of interest are the error types and frequencies of occurrence. Unfortunately, only a small amount of data has been accumulated in this area, and until the gross models, (such as those discussed above which treat all errors alike), have been verified such refinement in a model is perhaps unjustified.

4.2 Post release data

As is the case with all studies, data is difficult and generally costly to obtain. Good and complete records are not kept in most situations. Record keeping is generally better in the case of large military and space programs and after release of a large commercial operating system. Although a fairly large amount of such data exists, military secrecy and industrial proprietary policies inhibit its publication in many cases. Some of this data which has been published appears in Refs. 8 and 10.

Let us assume that a typical operating system for a large computer is undergoing continual development and that new features and capabilities are being added. The manufacturer's development group deals with a continually changing product, but external versions (generally called releases) are only made available periodically, say every 6 months. Although the manufacturer tries to thoroughly test each release, the exercising of the program by a fair proportion of the large and diverse user community is more comprehensive than any test he can devise. Consequently, soon after release of a new version, the number of errors found per month (error rate) rises rapidly to a peak. As these are diagnosed and corrected, the number of residual errors decreases and the error rate begins to decrease. When a new release is distributed, this behavior is repeated. Such typical behavior is sketched in Fig. 4-1. Note that the vertical axis is normalized by dividing by the total number of machine language instructions. This should allow us to compare both large and small programs to see if there is a behavior pattern independent of size. Detailed data on the normalized number of errors since release for three different supervisory systems (operating systems) is given in Fig. 4-2. Note that the horizontal axis units are months of debugging τ. In this case τ is identical with operating time, t; however, this is not always the case. Note that the shapes depicted in Fig. 4-2 a, b, c vary. If we assume that the number of remaining errors decreases monotonically and that the error discovery rate is proportional to number of remaining errors, exponential decay is obtained. This explains in a gross way the "tail" of the curves. The initial behavior may be due to the fact that initially only a few installations are using the new release and it is not until a few months later that a sizeable proportion of users have instituted this software. Thus, it might be more appropriate to let τ represent a more general resource variable such as user-months, or program-run-months. Unfortunately, no such data was available to serve as a more realistic horizontal scale parameter. (See Ref. 10 for a more detailed discussion of these curve shapes.)

In Fig. 4-3, the error rate curves for four applications programs are presented. In this case the origin $\tau = 0$ represents the start of program integration where all the individual modules of code are put together to form a system. As is well known, at this point, incompatabilities between the modules crop up and a new set of interface errors must be debugged. We may employ the same argument used previously to describe the tails of the curves in Fig. 4-3. Also, if we think of τ as a general resource variable which is a function of man-hours of debugging and computer test hours, this may explain the initial behavior. Again, no data is available to test this hypothesis.

4.3 Error model

Referring to the data discussed in the previous section we see that although the curve shapes differ the vertical and horizontal scales are similar. Based on this result we can proceed to formulate a general error model using the number of machine language instructions as a normalizing factor.

Basically, the error model used in this paper assumes that the total number of errors in the program is fixed and that if we record the cumulative number of errors corrected during debugging, then the difference represents the remaining errors. The following section on reliability models will relate the probability of encountering a software bug to the number of residual bugs.

The normalized error rate is defined as

$$\rho(\tau) = \text{errors/total number of instructions/month of debugging time.} \qquad (4\text{-}1)$$

Thus, Figs. 4-1, 4-2, and 4-3 are plots of $\rho(\tau)$ vs. τ.

Since we are interested in the total number of errors removed, we will define a cumulative error curve, $\varepsilon(\tau)$, which is the area under the $\rho(\tau)$ curve:

$$\varepsilon(\tau) = \int_0^\tau \rho(x)\, dx = \text{cumulative errors/total number of instructions} \qquad (4\text{-}2)$$

and $\rho(\tau)$ is of course the slope of the $\varepsilon(\tau)$ curve:

$$\rho(\tau) = d\varepsilon(\tau)/d\tau \qquad (4\text{-}3)$$

A curve of the cumulative error data for the supervisory system A of Fig. 4-2 is shown in Fig. 4-4. If similar curves for $\varepsilon(\tau)$ were drawn for the other examples of Figs. 4-2 and 4-3, all would start at zero, increase slowly, then more rapidly, and

and finally, more slowly approaching a slowly increasing or zero rate. Because of this similarity in behavior a cumulative curve such as $\varepsilon(\tau)$ is not too useful to depict differences in behaviors; thus, the derivative curve, $\rho(\tau)$, is more useful for this purpose. Both curves are needed for a detailed study.

If we assume that the total number of errors in the program E_T is constant and that the program contains I_T the instructions and that no new errors are added during debugging, then the asymptote which the $\varepsilon(\tau)$ curves approach is E_T/I_T. If we assume that all detected errors are corrected errors, then by inspection of Fig. 4-4 we can write an expression for the number of residual errors:

$$\varepsilon_r(\tau) = (E_T/I_T) - \varepsilon_c(\tau) \tag{4.4}$$

We assume that in any sizeable program it is impossible to remove all errors, so

$$\varepsilon_c(\tau) < E_T/I_T \tag{4.5}$$

$$\varepsilon_r(\tau) > 0 \tag{4.6}$$

Also since we assume that most programs eventually reach a reasonable debugged state we may assume that for large τ $\varepsilon_r(\tau) I_T/E_T$ is small.

In order to test the hypothesis that the normalized behaviors of $\varepsilon_r(\tau)$ and $\rho(\tau)$ hold for a wide variety of program sizes we make the following comparisons with the data in Figs. 4-2 and 4-3: (1) In order to test the hypothesis that the normalized number of errors E_T/I_T is somewhat constant for a variety of programs we compute the ratio and compare the results. (2) An allied hypothesis is that debugging proceeds at a roughly similar average rate ρ_0 over an entire project. The results are given in Table 4-1. The value of E_T/I_T varies about the average by + 48% and - 31% and that of ρ_0 varies about the average by + 75% and - 31%. Note that all these programs are about 1/4 million machine language statements in size. The data is often "dirty" since in some projects only program corrections are counted; whereas, in others specification and improvement changes are lumped in with actual error changes. Furthermore, the applications programs presented debugging information during the program integration phase of software development; whereas, the supervisory programs reported errors after release. It is not unreasonable that the errors found during system integration and after release of a large software package are roughly commenserable. Based on three software programs, it has been shown that the ratio of changes after release to changes during integration and test was about 0.8 (see Ref. 30). If we compare the average value of E_T/I_T for the supervisory programs to that for the application programs in Table 4-1, the ratio is about 0.7. Based on the above factors the data appears to verify the hypothesis that E_T/I_T and ρ_0 are approximately constant for similar size programs. We now present similar data for small programs in Table 4-2. In this case both the values of E_T/I_T and ρ_0 vary about the average by +79% and -36%. The data in Table 4-2 includes data taken during module test as well as during integration testing and as might be expected (because of the two phases being lumped) the average value of E_T/I_T for the small programs data is 2.15 times larger than the large program data whereas, the value of ρ_0 is 1.53 times larger. Drawing these various facts together allows us to state that within a factor of perhaps 2, the values of E_T/I_T and ρ_0 seem to be constant for a wide variety of programs. Furthermore, within a similar factor, the number of bugs per machine instruction found during module test, integration testing, and after release are roughly the same.

One further comment is in order before we leave the subject of error models. Some experienced programmers have challenged the assumption that no new errors are generated during debugging. In Fig. 4-5 three dynamic debugging behaviors are illustrated. In Fig. 4-5a no new errors are added and the situation depicted is just the one which we have been discussing. In Fig. 4-5b errors are added; however, the removal rate exceeds the generation rate and equilibrium is obtained. If the number of errors added is small percentage wise, even cases (a) and (b) are approximately numerically equivalent. Figure 4-5(c) depicts a case where the error generation rate exceeds the error removal rate and the process diverges. A newly devised model, formulated by this author and his co-workers describes error generation in cases (b) and (c), and is being prepared in memorandum form.

5.0 RELIABILITY MODELS

5.1 Introduction

In order to formulate a reliability model one can take a microscopic or a macroscopic approach. In the microscopic approach we would try and identify individual bugs (either deterministically or probabilistically), the type of bug, the path in the program and how frequently the path is traversed. Initial attempts along these lines have convinced this author that such an approach, while necessary in the long run, involves a more detailed knowledge of program structure and bug types than is now available. The macroscopic approach where all bugs are lumped and treated equally will be employed here. The validity of the result depends on considerable 'averaging' occurring in a large program.

5.2 Basic assumptions

We assume that operational software errors occur due to the occasional traversing of a portion of the program in which a hidden software bug is lurking. We begin by writing an expression for the probability that a bug is encountered in the time interval Δt after t successful hours of operation. This must be proportional to the probability that any randomly chosen instruction contains a bug, i.e., the fractional number of remaining bugs $\epsilon(\tau)$.

From a study of basic probability and reliability theory (see Ref.9) we learn that the probability of failure in time interval t to $t+\Delta t$ given that no failures have occurred up till time Δt is proportional to the failure rate (hazard function) $z(t)$.

$$P(t < t_f \leq t+\Delta t \mid t_f > t) = z(t)\,\Delta t = K\,\epsilon_r(\tau)\,\Delta t \qquad (5\text{-}1)$$

where $t_f \equiv$ operating time to failure, (occurrence of a software error)

$P(t < t_f \leq t+\Delta t \mid t_f > t \equiv$ probability of failure in interval Δt, given no previous failure.

K = an arbitrary constant*

Note that in Eq. 5-1 two time variables appear: first there is t the operating time in hours of the system and second there is τ the debugging time in months (or more generally, the debugging resource variable). Once the assumptions in Eq. 5-1 have been made the reliability and mean time to failure functions follow directly.

5.3 Reliability model

By combining Eqs. 5-1 and 3-7 and assuming that K and $\epsilon_r(\tau)$ are independent of operating time t we obtain for the reliability function

---

*In earlier work (see Ref. 10 or 12) an attempt was made to achieve a more micromodel by splitting K into two factors, K' an arbitrary constant, and r_p the instruction processing rate. This elaboration is not included here since, to date, no data has been obtained to define or calculate r_p.

$$R(t) = e^{-\left[K\varepsilon_r(\tau)\right]t} = e^{-\gamma t} \tag{5-2}$$

Basically the above equation states that the probability of successful operation without software bugs is an exponential function of operating time. When the system is first turned on, t = 0 and R(0) = 1. As operating time increases the reliability monotonically decreases as shown in Fig. 5-1. We depict the reliability function for three values of debugging time, $\tau_0 < \tau_1 < \tau_2$. From this curve we may make various predictions about the system reliability. For example, looking along the vertical line $t = 1/\gamma$ we may state:

1. If we spend τ_0 hours of debugging, then $R(1/\gamma) = 0.35$

2. If we spend τ_1 hours of debugging, then $R(1/\gamma) = 0.50$

3. If we spend τ_2 hours of debugging, then $R(1/\gamma) = 0.75$

5.4 MTTF model

A simpler way to summarize the results of the reliability model is to compute the mean time to (software) failure, MTTF by substituting Eq. 5-2 into Eq. 3-9.

$$\text{MTTF} = \frac{1}{K\varepsilon_r(\tau)} \tag{5-3}$$

If we let $\rho(\tau)$ be modeled by a constant rate of error correction ρ_0 (see Ref. 10 for other models) then solution of Eqs. 5-3 and 4-4 yields

$$\text{MTTF} = \frac{1}{K\left[\dfrac{E_T}{I_T} - \rho_0\tau\right]} = \frac{1}{\beta(1-\alpha\tau)} \tag{5-4}$$

where $\beta = \dfrac{E_T}{I_T} K$ and $\alpha = \dfrac{\rho_0 I_T}{E_T}$

In Fig. 5-2, $\beta \times$ MTTF is plotted vs. $\alpha\tau$. We see that the most improvement in MTTF occurs during the last 1/4 of the debugging.

5.5 Other models

Other similar models have been proposed in the literature. Jelinski and Moranda, Ref. 13, propose a hazard function of the form

$$z(\tau) = \phi\left[N - (i-1)\right] \quad (5-5)$$

where: ϕ = Constant of proportionality.

N = Total number of errors present.

i = Number of errors found by debugging time τ_i.

Comparison of Eq. 5-5 with Eqs. 5-1 and 4-4 shows them to be identical if

$$E_T = N \quad (5-6)$$

$$\frac{K}{I_T} = \phi \quad (5-7)$$

$$\varepsilon_c(\tau) = \frac{i-1}{I_T} \quad (5-8)$$

Equations 5-6 and 5-7 are merely notational differences and Eq. 5-8 is nearly the same (i.e., would be identical if $\varepsilon_c(\tau) = i/I_T$).

In another paper Schick and Wolverton, Ref. 14, modify Jelinski and Moranda's model and assume that the failure rate is proportional to the number of remaining errors and increases with operating time t

$$z(t) = \phi\left[N - (i-1)\right]t \quad (5-9)$$

One rationale for postulating an increase in z(t) would be if operation were viewed as a succession of different trials which gradually closes in on the remaining errors (sampling without replacement). However, one could argue to the contrary that z(t) should decrease with t, since the latter errors are the subtle ones which take a long while to encounter in operation. The author believes that in most cases of large, intricate, well tested, real-time systems the hazard will remain constant once the initial field debugging of a new release is finished. The small number of subsequent patches generated should not be significant. Failure should be caused by rare combinations of input data and path traversals, with the time between failures governed by an exponential distribution, yielding a constant hazard. Experimental data is necessary to choose among these hypotheses.

Other related reliability and error models are discussed in Refs. 15, 16 and 17. We now turn in the next two sections to a discussion of how we can experimentally measure reliability and use these measurements in conjunction with the models of this section to determine the unknown model parameters K and E_T.

5.6 Experimental reliability data

If we had just deployed a large hardware-software system for field use we could monitor its reliability by carefully recording the operating time and documenting each failure in detail. Thus, we could obtain the times between failure. Investigation of each failure should allow one to classify all failures as hardware, software, operator, or unknown. If we segregate the software times between failure and plot their average week by week, we will have a quantitative measure of operational software reliability. We would expect the operational MTTF to increase for the first month (year, in some cases) or so as software bugs detected in service are removed, then gradually to level off to a relatively constant value. This is, of course, an after the fact evaluation of the software design and does not allow one to measure progress and/or need for improvement of the software design while it is under development.

The earliest stage at which an entire system can be functionally tested is during system integration using the system exerciser (functional test) program. If this test is performed at the beginning of system integration, the result will be a succession of very short runs and immediate crashes. Most software test personnel would instinctively comment that this is as expected since the system is still in "poor shape" and such a test should be delayed until the end when the system is in "good shape". A bit of reflection leads one to the conclusion that it is just this frequent crashing which leads to a quantitative assessment of the poor initial reliability.

We now focus on the test data and how it should be analyzed. The necessary information which must be recorded for each run of the system test program is how long the test ran, whether an error occurred and if the error is a software error. Sufficient dumps and other documentation must be recorded for subsequent analysis in order to segregate errors into hardware, software, operator, etc. errors. Each of the r successful runs represent $T_1, T_2, \ldots T_r$ hours of success. If there are n total runs then each (n-r) unsuccessful run represents $t_1, t_2, \ldots t_{n-r}$ successful run hours before failure. The total number of successful run hours H is given by

$$H = \sum_{i=1}^{r} T_i + \sum_{i=1}^{n-r} t_i \qquad (5\text{-}10)$$

Assuming that the failure rate is constant, we denote it by λ and compute it as the number of failure per hour

$$\text{Failure Rate} = \lambda = \frac{n-r}{H} \qquad (5\text{-}11)$$

The MTTF for a constant failure rate is the reciprocal (see Eq. 3-12) of the failure rate

$$\text{MTTF} = \frac{1}{\lambda} = \frac{H}{(n-r)} \qquad (5\text{-}12)$$

Now Eqs. 5-11 and 5-12 represent the total system failure rate and MTTF. Since we are mainly interested in software failures, we assume that the outputs as well as dumps are carefully investigated for the $(n-r) = x$ failures. Based on the above analysis the failures are divided into x_h hardware failures, x_s software failures, x_o operator failures, and x_u unknown failures. Hopefully, the unknown ratio x_u/x will be 25% or smaller so that most of the data is classifiable.

Then the software failure rate and MTBF are defined by

$$\lambda_s = \frac{x_s}{H} \qquad (5\text{-}13)$$

$$\text{MTTF}_s = \frac{H}{x_s} \qquad (5\text{-}14)$$

Thus, based on the results of this occasional test we can plot λ_s and MTTF_s vs. τ the debugging time. Such charts should allow a quantitative measure of the progress in improving software quality. Thus, after τ_a hours of debugging we would have a measure of MTTF and $R(t)$ and by _extrapolation_ of the curves we could _predict_ MTTF and $R(t)$ after $\tau_b > \tau_a$ months of debugging. Unless we knew the functional form of the variations in $R(t)$ and MTTF with τ and could determine an appropriate extrapolation scheme, accurate predictions would be limited to small excursions into the future.

5.7 Estimation of model constants

Rather than use the raw experimental data and extrapolation for prediction we can assume an underlying model for λ_s and MTTF_s and use the data to estimate the model parameters. If the hypothesized model is correct, then predictions using this technique should be superior to the extrapolation technque of the previous section.

For the software reliability model defined in Sections 5-2 and 5-3

$$R(t, \tau) = \exp\left[-K\left(\frac{E_T}{I_T} - \varepsilon_c(\tau)\right)t\right] \quad (5\text{-}15)$$

$$\text{MTTF}(\tau) = \frac{1}{K\left[\frac{E_T}{I_T} - \varepsilon_c(\tau)\right]} \quad (5\text{-}16)$$

Note that if we assume a known program size and careful collection of error data, then I_T and $\varepsilon_c(\tau)$ are known values and only the constants K and E_T remain to be determined. These two unknowns K and E_T can be evaluated by running a functional test after two different debugging times, $\tau_1 < \tau_2$ chosen so that $\varepsilon_c(\tau_1) < \varepsilon_c(\tau_2)$. We then equate Eqs. 5-14 and 5-16 at times τ_1 and τ_2

$$\frac{H_1}{x_{s_1}} = \frac{1}{K\left[\frac{E_T}{I_T} - \varepsilon_c(\tau_1)\right]} \quad (5\text{-}17)$$

$$\frac{H_2}{x_{s_2}} = \frac{1}{K\left[\frac{E_T}{I_T} - \varepsilon_c(\tau_2)\right]} \quad (5\text{-}18)$$

Taking the ratio of Eq. 5-17 to 5-18 and using Eq. 5-14 yields

$$\hat{E}_T = \frac{I_T\left[\left(\lambda_{s_2}/\lambda_{s_1}\right)\varepsilon_c(\tau_1) - \varepsilon_c(\tau_2)\right]}{\left(\lambda_{s_2}/\lambda_{s_1}\right) - 1} \quad (5\text{-}19)$$

Once $\hat{E}_T$ has been computed from Eq. 5-19, we obtain $\hat{K}$ by substituting Eq. 5-19 into 5-17 which yields

$$\hat{K} = \lambda_{s_1} / \left[(\hat{E}_T/I_T) - \varepsilon_c(\tau_1)\right]. \quad (5\text{-}20)$$

The "hats" above E_T and K in Eqs. 5-19 and 5-20 denote estimates of the parameter. Note that if there was no debugging between τ_1 and τ_2 so that $\varepsilon_c(\tau_1) = \varepsilon_c(\tau_2)$, the numerator of Eq. 5-19 becomes zero, i.e., Eqs. 5-17 and 5-18 are no longer independent and the estimate fails.

Further discussions of this parameter estimation technique, the more powerful maximum likelihood estimation technique, and accuracy questions appear in Ref. 18.

6.0 ALLIED AREAS

This paper has concentrated on one aspect of software reliability, its measurement theoretically and experimentally. There are many other allied areas relating to this subject: system recovery techniques (Ref. 21), program design for low error content (Refs. 22 and 23), the production of standard computational programs which are very reliable (Ref. 29), etc.

An historical account of the SAGE air defense system, one of the first realtime hardware-software systems is worth reading (Ref. 24). Other material of interest appears in the Record of the 1973 IEEE Symposium on Computer Software Reliability (see Ref. 15), the Proceedings of the Brown Symposium, (see Ref. 10), the NATO Conferences on Software Engineering (see Ref. 28) and the book edited by Rustin (Ref. 25). More material is appearing each month in the computer research journals on this dynamic new field.

REFERENCES

1. Harold Sackman, "Computers, System Science, and Evolving Society," John Wiley and Sons, New York, N.Y., 1969.
2. Terry M. Walker and William W. Cotterman, "An Introduction to Computer Science and Algorithmic Processes."
3. United States Air Force Report, "Information Processing/Data Automation Implications of Air Force Command and Control Requirements in the 1980's"(CCIP-85), Vol. 1, SAMSO/XRS-71-1, April 1972.
4. Arthur C. Clarke, "2001 A Space Odyssey," based on the MGM film screenplay of the same name, Signet Books, New York, N.Y. 1968.
5. M. Hamilton, MIT Charles Stark Draper Labs., Cambridge, Mass., Private Communication.
6. Barry W. Boehm, "Software and Its Impact: A Quantitative Assessment," Datamation Magazine, May 1973.
7. E. Yourdon, "Reliability Measurements for Third Generation Computer Systems," Proceedings of the 1972 Annual Reliability Symposium, IEEE, New York, N.Y.

8. J. Hesse, A. Kientz, J. Dickson and M. Shooman, "Quantitative Analysis of Software Reliability," 1972 Annual Reliability Symposium Proceedings, IEEE, New York, N.Y., January 1972.
9. M. Shooman, "Probabilistic Reliability: An Engineering Approach," McGraw-Hill Book Co., New York, N.Y., 1968.
10. M. Shooman, "Probabilistic Models for Software Reliability Prediction," Conference on Statistical Methods for the Evaluation of Computer System Performance, Brown Univ., November 1971. Published in "Probabilistic Models for Software," Freiberger Editor, Academic Press, New York, N.Y., 1972, pp. 485-502.
11. Fumio Akiyama, "An Example of Software System Debugging," IFIP Congress 71, Ljubljana, Yugoslavia, Aug. 1971.
12. M. Shooman, "Probabilistic Models for Software Reliability Prediction," 1972 International Symposium on Fault-Tolerant Computing, Newton, Mass., June 21, 1972, IEEE Computer Society.
13. J. Jelinski and P.B. Moranda, "Software Reliability Research," same source as Reference 10.
14. G.J. Schick and R.W. Wolverton, "Assessment of Software Reliability," 11th Annual Meeting, German Operations Research Society, Hamburg, Germany, September 1972.
15. E. Girard and J-C. Rault, "A Programming Technique for Software Reliability," Record of the 1973 IEEE Symposium on Computer Software Reliability, New York, N.Y., May 1973, p. 44.
16. B. Littlewood and J.L. Verall, "A Bayesian Reliability Growth Model for Computer Software," Same source as Ref. 15, p. 70.
17. J. Jelinski and P.B. Moranda, "Applications of a Probability-Based Model to a Code Reading Experiment," Same source as Ref. 15, p. 78.
18. M. Shooman, "Operational Testing and Software Reliability Estimation During Program Developments," Same Source as Ref. 15, p. 51.
19. Bell System Technical Journal, Dec. 1970. Articles on testing of TSPS No. 1 and ESS No. 1 ADF.
20. Donald E. Knuth, "An Emperical Study of FORTRAN Programs," Computer Science Department Report No. CS-186, Stanford University, 1970.
21. E. Yourdon, "Design of On-Line Computer Systems," Prentice-Hall, Inc., Englewood Cliffs, N.J., 1972, p. 515.
22. G. Weinberg, "The Psychology of Computer Programming," Van Nostrand Reinhold Co., New York, N.Y. 1971.
23. O. Dahl, E. Dijkstra, C. Hoarse, "Structured Programming," Academic Press, London, 1972.
24. H. Sackman, "Computers, System Science, and Evolving Society," John Wiley and Sons, New York, N.Y. 1967.

25. R. Rustin, "Debugging Techniques in Large Systems," Prentice-Hall, Inc., Englewood Cliffs, N.J., 1971.
26. W. Freiberger, "Statistical Computer Performance Evaluation," Academic Press, New York, N.Y., 1972.
27. H. Sackman, "Man-Computer Problem Solving: Experimental Evaluation of Time-Sharing and Batch Processing," Auerbach Publishers, Philadelphia, Pa., 1970.
28. NATO Conferences: (1) "Software Engineering Techniques," (Eds. Burton, J.N. and B. Randell), 1969; (2) "Software Engineering," (Eds. Naur, P. and B. Randell) Garmisch, Germany, Oct. 1968.
29. W. Cowell, "Proceedings of the Software Certification Workshop," Aug. 1972, Argonne National Laboratory, Argonne, Ill.
30. M. Hyman, IBM Federal Systems Division, Morris Plains, N.J., Private Communication.

TABLE 1-1
Distribution of Effort by Programming Phase for Various Projects

	Analysis and Design	Coding and Auditing	Checkout and Test
SAGE	39%	14%	47%
NTDS	30	20	50
GEMINI	36	17	47
SATURN V	32	24	44
OS/360	33	17	50
TRW Survey	46	20	34

TABLE 3-3

System	Mean Time Between Crashes
1. MIT MULTICS Time Shared System (Under Continuous Development)	3-9 hours
2. Honeywell MULTICS (Stable Version of Multics)	Very much greater
3. A data acquisition program (see Ref. 7)	30
4. A commercial time sharing company running CALL 360 on a 360/50	50
5. A commercial time sharing company running a business information system on a mini-computer with many attached disk files.	500

TABLE 4-1

Computation of Model Constants from the Data of Ref. 8

Program	Size	E_T/I_T	ρ_o
Supervisory A	210 K	6.14×10^{-3}	0.875×10^{-3}
Supervisory B	240	7.97	0.996
Supervisory C	230	7.48	1.25
Application A	240	13.20	2.20
Application B	240	7.70	1.54
Application C	240	7.00	1.00
Application D	240	12.90	0.995
Average		8.92	1.26

TABLE 4-2

Computation of Model Constants from the Data of Ref. 11

Program	Size	E_T/I_T	ρ_o
MA	4.03 K	25.4×10^{-3}	2.54×10^{-3}
MB	1.32	13.7	1.37
MC	5.45	17.1	1.71
MD	1.67	15.6	1.56
ME	2.05	34.6	3.46
MF	2.51	14.7	1.47
MT	2.10	12.4	1.24
MG	0.70	22.9	2.29
MH	3.79	13.2	1.32
MX	3.41	23.4	2.34
Average		19.3	1.93

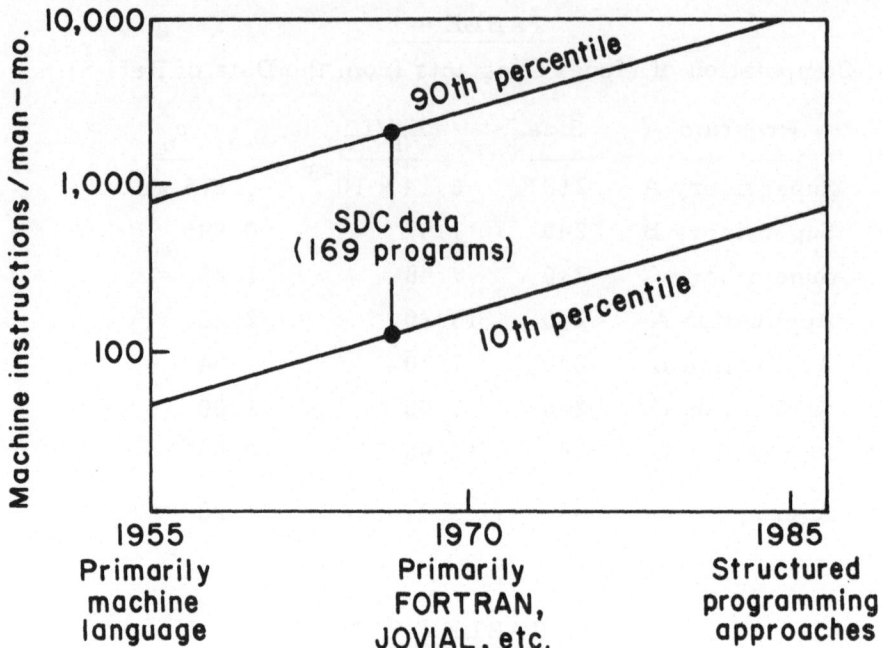

Fig. 1-1. Programming productivity per man month. See Ref. 6.

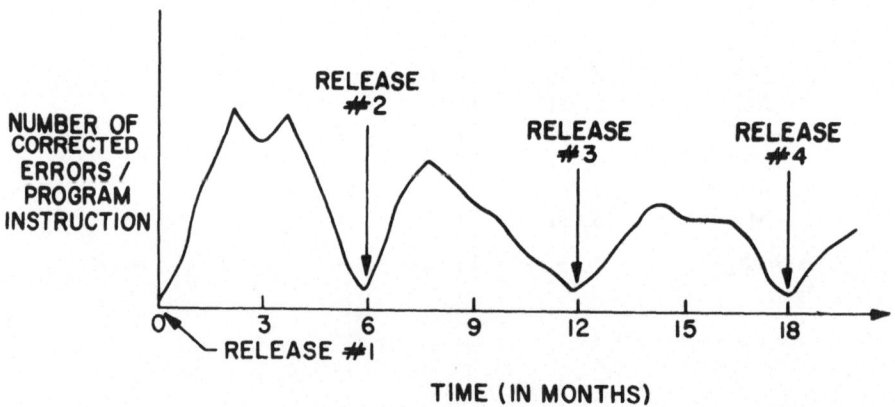

Fig. 4-1. Typical behavior of a compiler program.

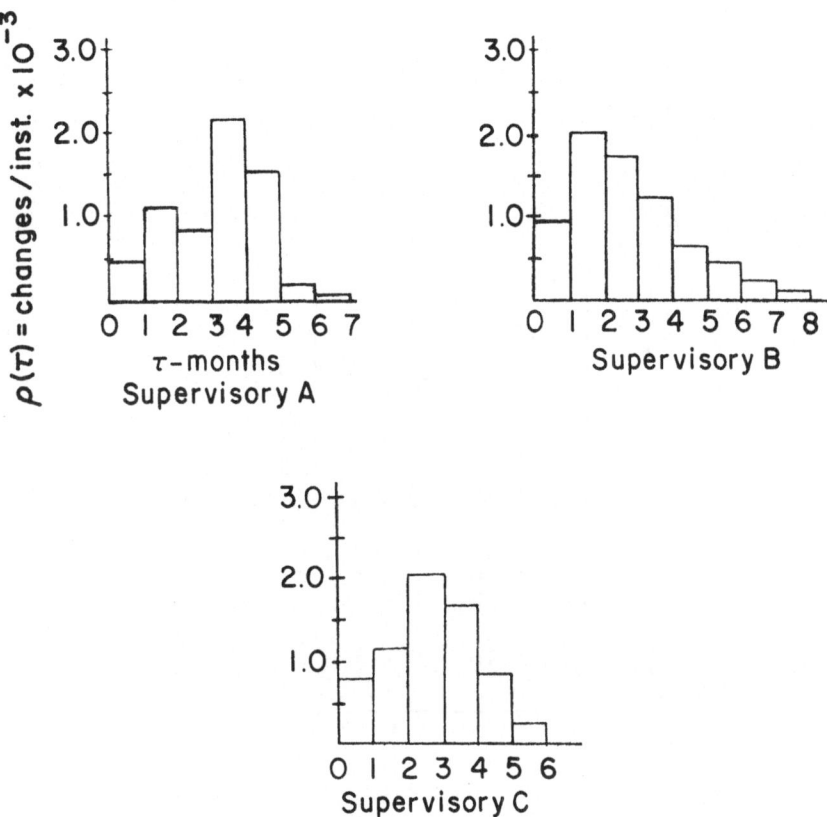

Fig. 4-2. Normalized error rate versus debugging time for three supervisory programs.

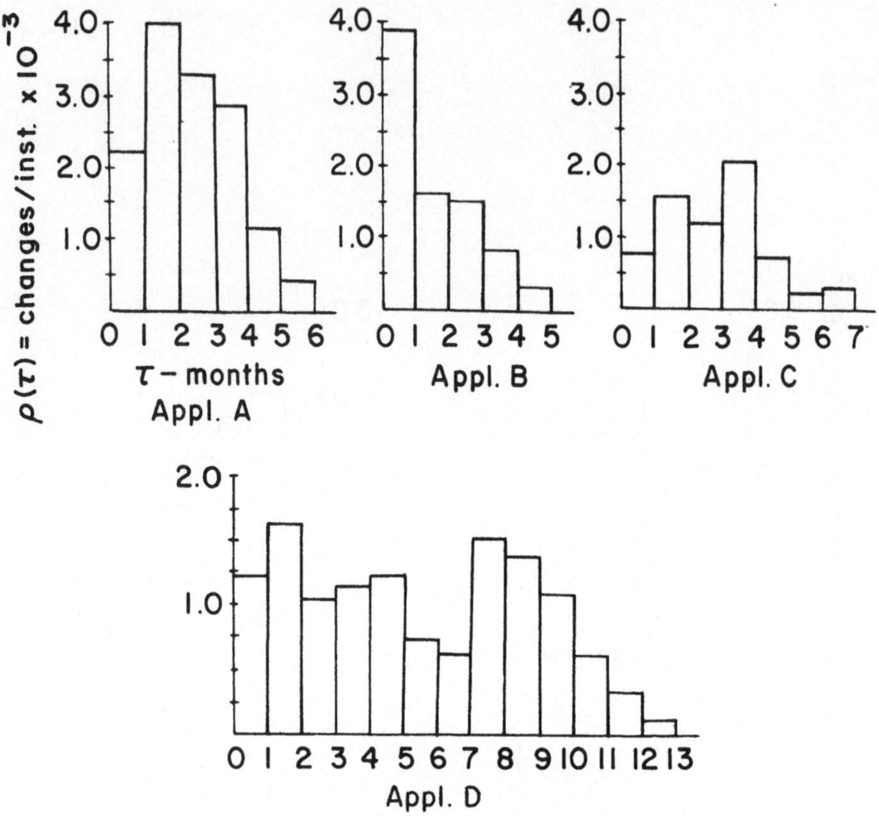

Fig. 4-3. Normalized error rate versus debugging time for four applications programs.

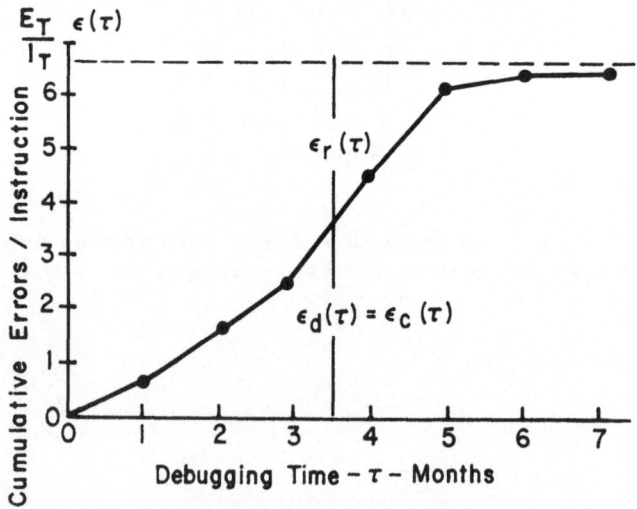

Fig. 4-4. Cumulative error curve for supervisory system A given in Fig. 4-2.

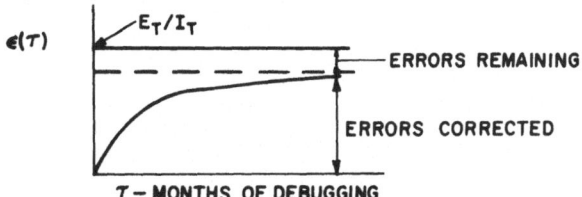

(a) APPROACHING EQUILIBRIUM, HORIZONTAL ASYMPTOTE, NO GENERATION OF NEW ERRORS.

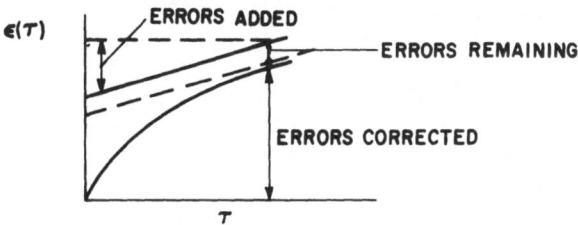

(b) APPROACHING EQUILIBRIUM, GENERATION RATE OF NEW ERRORS EQUALS ERROR REMOVAL RATE.

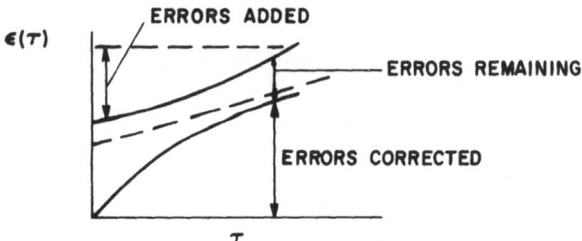

(c) DIVERGING PROCESS, GENERATION RATE OF NEW ERRORS EXCEEDS ERROR REMOVAL RATE.

Fig. 4-5. Cumulative errors debugged versus months of debugging.

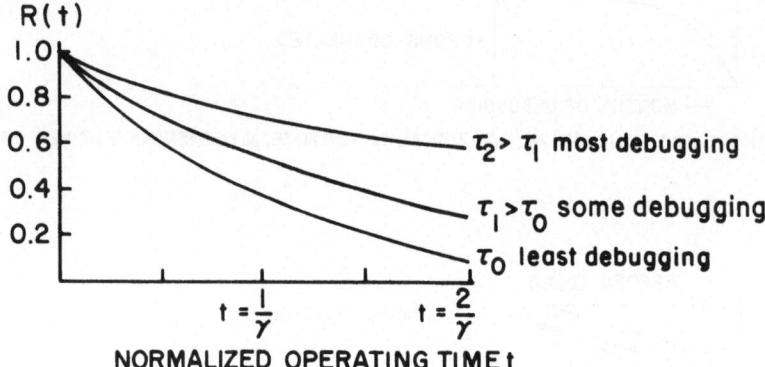

Fig. 5-1. Variation of reliability function R(t) with debugging time τ.

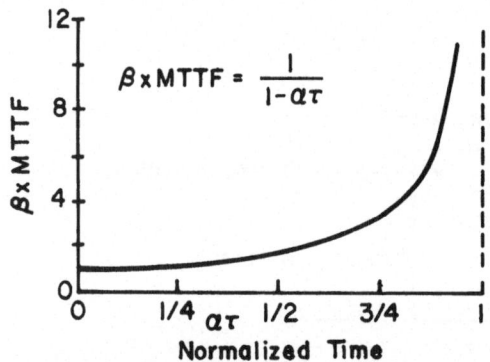

Fig. 5-2. Comparison of MTTF vs. debug time for the constant error debug rate model.

APPLICATION OF RELIABILITY THEORY TO COMPUTATION OF SYSTEM
OPERATION SECURITY IN A POWER INDUSTRY

S. L. Surana and A. Brameller

Department of Electrical Engineering and Electronics
University of Manchester
Institute of Science and Technology
Manchester, England

1. INTRODUCTION

One of the main objects in the operation of electrical power systems is to provide power to its consumers at a low cost with some degree of security. Breach of security or unacceptable operating conditions in a system may arise due to insufficient generating capacity, loss of system stability, lower than acceptable bus voltages, uncertanties in the forecasted loads, loss of transmission lines, etc. This paper deals exclusively with the breach of security occurring due to insufficient generating capacity caused by unforeseen forced outage of the units.

In the past few years there has been an increasing emphasis to use probability theory for evaluation of reliability and with few exceptions, most of the work is concentrated on long-term reliability calculations for system planning.

An operational engineer who is involved in day to day operation of power system is normally interested in short-term reliability assessment. In order to detect any potential danger area in the system and to take some suitable control action he must continuously know the level of system security into near future. At any future time, if it is felt that the security level has fallen below the desired level, a preventive control action may be initiated at some appropriate time. This control action may involve starting of additional generators or shedding of interruptable loads etc. On the other hand if the security level is considered higher than the desired level, some of the less efficient generators may be switched off to achieve lower operating cost. Thus probabilistic approach can provide consistent quantitative criteria for achieving a desired level of security.

Prediction of short-term reliability requires computationally efficient and accurate models. As in any statistical and probability calculations, the development of these models is based on certain assumptions about the distributions of various kinds of failures and repairs which affect reliability. It must be realized that these reliability models approach reality only to the extent that the actual distributions approach the assumed ones. In the past the models which have been developed are either computationally inefficient or assume repair times as exponentially distributed which is not true always. In this paper an attempt has been made to develop new models which are computationally efficient and accurate enough for all practical applications.

2. METHODS OF COMPUTATION OF OPERATION SECURITY

There are two different probabilistic approaches for computation of operation security. The first one is called the P-J-M approach[1] and was developed by a group of authors associated with the Pennsylvania-New Jersey-Maryland Interconnection. The second or the Security Function approach[2] was suggested by Patton. In this approach the system risk over a future time is evaluated by computing security function which is defined by Equation (1) below.

$$S(t) = \sum_{i=1}^{n} P_i(t) \cdot Q_i(t) \qquad \ldots (1)$$

where,

$P_i(t)$ = probability that the system is in state i at time t
$Q_i(t)$ = conditional probability that state i causes breach of system security at time t
n = number of possible system states

Since the aim is to evaluate operation security which is a function of future time, it is therefore necessary to use time dependent expressions in Equation (1) for system state probabilities. Both these approaches of evaluating operation security require the computation of future system state probabilities. At the zero time i.e. at the time of predicting operation security, the operational engineer would be aware of the various states which exist in the system i.e. which generators are in fully available state and which are in forced out state and hence the initial probabilities for these states can be assigned easily. They will be either unity or zero depending upon whether these states exist or not. Knowing the initial probabilities it is possible to determine those of individual generators at any future time using appropriate models developed in Section 3. Since it has been assumed that each generating unit fails and is repaired independently of other generating units, it is possible to apply the multiplication rule of probability theory to determine system state probabilities Pi(t) involving any number of generators.

The second factor Qi(t) in Equation (1) takes into account uncertainties in load forecasting and has a range 0 to 1. This can be included easily by assuming[3] the load forecasting deviation as normally distributed about the predicted load. For further details see reference 4.

3. COMPUTATION OF STATE PROBABILITIES

In the past, most of the models which have been proposed for computation of state probabilities were based on the assumption that the state residence times of the various system components are exponentially distributed. This means that the transition rates from one system state to another is constant. The recent analyses[5] of outage data have indicated that distributions other than the distribution of times spent in the operating state are non-exponential. Because of this, the simple technique of Markov process is no longer applicable directly. In a recent paper[6] Patton proposed an approximate solution for time dependent state probabilities. In yet another paper[7] Marco applied the theory of Semi-Markov processes for the same purpose. Both these techniques are computationally inefficient, particularly the latter one. In the calculation of security function described by Equation (1), summation has to be carried out over a large number of system states. It is, therefore, important that the probabilistic models of the generating units be computationally efficient, if they are to be of any practical use.

Due to non-exponential down time distribution the stochastic process becomes non-Markovian. Cox and Miller[8] have described several techniques to deal with non-Markov stochastic processes. One of the techniques called the "device of stages" or "method of phases" is relatively less complex and yields models which are computationally faster. In a very recent paper[9] Billinton and Singh applied this technique to the problem of transformer configurations.

3.1 Device of Stages or Method of Phases

This method consists of simulating non-exponential distribution by a suitably chosen set of exponential distributions. Let X be a non-negative continuous random variable which represents the time spent in a particular state and has k stages or phases. The division into these imaginary stages is a mathematical device and need have no significance. The lengths of time ($Y_1, Y_2 \ldots Y_k$) to complete the different stages are mutually independent random variables and are exponentially distributed with parameters ($\mu_1, \mu_2 \ldots \mu_k$) respectively. In the S-plane, probability density functions (p.d.f.) of these random variables are given by:

$$L\{f(y_i);s\} = \frac{\mu_i}{(s + \mu_i)}, \qquad i = 1,2\ldots k \qquad (2)$$

There are two fundamental ways in which these stages or phases can be simulated.

3.2 Simulation by Series Arrangement

In the series arrangement, the stages are traversed in turn i.e. when the first stage is completed, after time Y_1, the second stage is begun, and so on, the repair being completed at the end of the kth stage. The total repair time X, then has the distribution of the sum of k mutually independent random variables ($Y_1, Y_2 \ldots Y_k$). The probability density function of total repair time X can be obtained by using Convolution theorem:

$$L\{f(x);s\} = L\{f(y_1);s\} \cdot L\{f(y_2);s\} \ldots L\{f(y_k);s\} \qquad (3)$$

On substituting Equation (2) into Equation (3) gives

$$L\{f(x);s\} = \prod_{i=1}^{k} \frac{\mu_i}{(s+\mu_i)} \qquad (4)$$

The p.d.f. can be obtained by expressing Equation (4) in partial fraction form as:

$$L\{f(x);s\} \equiv \sum_{i=1}^{k} \frac{A_i \mu_i}{(s + \mu_i)}$$

where,

$$A_i = \prod_{j \neq i} \frac{\mu_j}{(\mu_j - \mu_i)}$$

The p.d.f. is therefore,

$$f(x) = \sum A_i \mu_i e^{-\mu_i x} \tag{5}$$

It can be shown[10] easily that the mean and the coefficient of variation of the distribution, the p.d.f. of which is given by Equation (5), are respectively,

$$\text{Mean} = E(X) = \sum_{i=1}^{k} \frac{1}{\mu_i} \tag{6}$$

$$\text{and coefficient of variation } r = \frac{\left[\sum_{i=1}^{k} \left(\frac{1}{\mu_i}\right)^2\right]^{\frac{1}{2}}}{\left[\sum_{i=1}^{k} \frac{1}{\mu_i}\right]} \tag{7}$$

When all the random variables are identically distributed i.e. when $(\mu_1 = \mu_2 = \ldots = \mu_k = k\mu)$, the Equations (5), (6) and (7) reduce to Equations (8), (9) and (10) respectively.

$$f(x) = \frac{1}{\lfloor k-1} (k\mu)^k x^{k-1} e^{-k\mu x} \tag{8}$$

$$E(X) = \frac{1}{\mu} \tag{9}$$

$$r = \frac{1}{\sqrt{k}} \tag{10}$$

Thus, by arranging stages in series, it is possible to simulate a distribution having any value of mean and coefficient of variation between 1 and $\frac{1}{\sqrt{k}}$.

3.3 Simulation by Parallel Arrangement

Since simulating exponential stages in series produces coefficient of variation less than unity, it should then be possible to obtain coefficient of variation greater than unity by simulating exponential stages in parallel[11]. Let there be only two independent stages connected in parallel. The lengths of time (Y_1, Y_2) to complete these stages are mutually independent random variables and are exponentially distributed with parameters $2\rho\mu$ and $2(1-\rho)\mu$ respectively, the constant ρ being restricted to range $(0<\rho\leq\frac{1}{2})$.

When a unit fails, it is assigned for repairs to one or the other stage at random, the choice going to 2ρμ branch, on an average, a fraction ρ of the time, and going to the 2(1-ρ)μ branch (1-ρ) of the times on an average. Let the repairs occur at the end of the single stage, whichever it may be. It is not difficult to see in this case the p.d.f. of the random variable X is given by:

$$f(x) = 2\rho^2 \mu \varepsilon^{-2\rho\mu x} + 2(1-\rho)^2 \mu \varepsilon^{-2(1-\rho)\mu x}, \quad (0<\rho\leq\tfrac{1}{2}) \qquad (11)$$

It can be shown easily that the mean and the coefficient of variation are respectively,

$$\text{Mean} = E(X) = \frac{1}{\mu} \qquad (12)$$

and coefficient of variation $r = [1 + \frac{(1-2\rho)^2}{2\rho(1-\rho)}]^{\frac{1}{2}}$ \qquad (13)

It is clear that in this case the coefficient of variation is greater than unity. Thus by arranging stages in parallel, it is possible to simulate a distribution having any value of mean and coefficient of variation greater than unity.

3.4 Models for Normally-Operating Generator

Consider a normally operating generator which is initially in the fully operating or the up state. It is assumed that the generating unit can occupy at any time (during the period considered) one of the following two mutually exclusive states.
1. generator operating and fully available or up state
2. generator on forced outage or down state

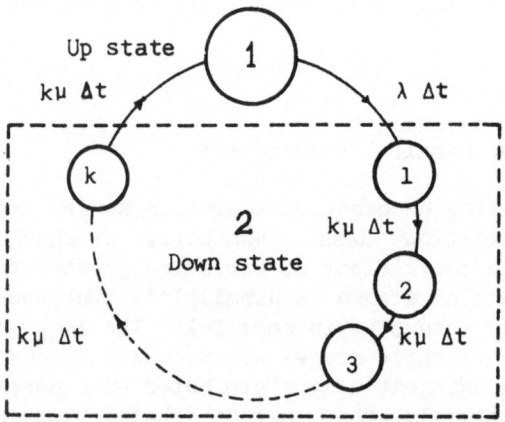

Fig. 1: Series stage-device repair model

The state space diagram of the model or process is shown in Fig. 1. The failure rate from state 1 to 2 is shown as constant λ. To take into account non-constant nature of repair rate, k virtual stages or phases in series each with repair rate $k\mu$ are introduced.

The time dependent state probability expressions may be obtained by considering all the possible events which can occur up to the time period t. This leads to multiple integral expressions in terms of failure and repair rates. From Fig. 1 the following state probability expressions may be written:

$P1(t)$ = P (no forced outage in time t)
    +P (one forced outage and one repair in time t)
    +P (two forced outages and two repairs in time t)
    +..........

and, $P2(t) = 1 - P1(t)$

Since the time period t for which state probabilities are required is much less than the average residence time in state 1, it is reasonable to assume that no more than one forced outage and repair cycle can occur during that period.

$\therefore$ $P1(t) \approx$ P (no forced outage in time t)
    +P (one forced outage and one repair in time t)
    $= 1 - P2(t)$

$$= 1 - \sum_{i=1}^{k} P2_i(t) \qquad (14)$$

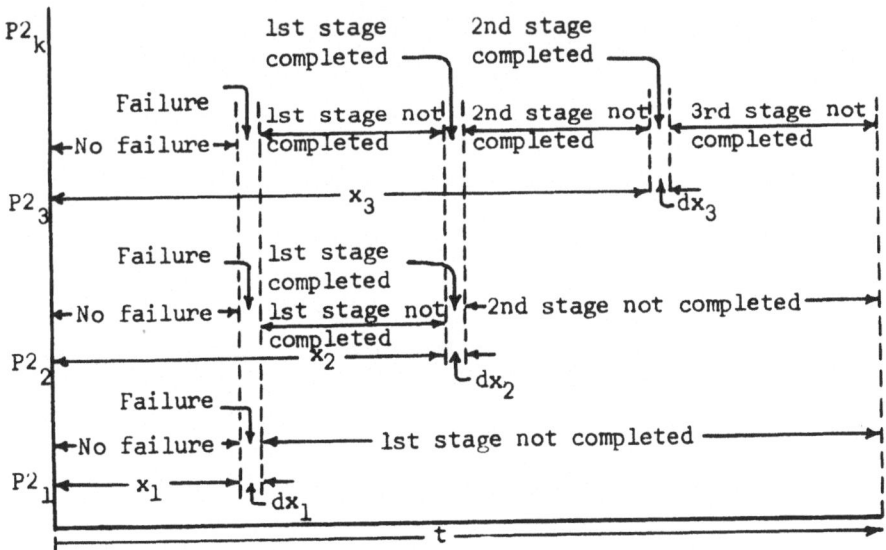

Fig. 2: Time interval diagram

The basic approach to calculate the probabilities $P2_i(t)$ is most easily visualized by reference to Fig. 2, which shows the various events which determine probabilities of stages 1,2......k. In order to derive expression for $P2_1(t)$, assume that the failure occurs at some arbitrary time $t=x_1$ and first stage repair is not completed in time $(t-x_1)$. The density function is then the probability of survival up to time x_1 multiplied by the probability of failure in dx_1 and probability of first stage repair not completed in time $(t-x_1)$

$$\therefore P2_1(t) = \int_{x_1=0}^{t} \begin{array}{l}\text{(probability of survival to } x_1\text{).} \\ \text{(probability of failure in } dx_1\text{).} \\ \text{(probability of first stage repair} \\ \text{not being completed in } t-x_1\text{)}\end{array}$$

$$= \int_{x_1=0}^{t} (\varepsilon^{-\lambda x_1}) \cdot (\lambda dx_1) \cdot \{\varepsilon^{-k\mu(t-x_1)}\}$$

$$= \lambda \varepsilon^{-k\mu t} \int_{x_1=0}^{t} \varepsilon^{(-\lambda+k\mu)x_1} dx_1$$

Similarly,

$$\therefore P2_2(t) = \lambda(k\mu)\varepsilon^{-k\mu t} \int_{x_2=0}^{t} \int_{x_1=0}^{x_2} \varepsilon^{(-\lambda+k\mu)x_1} dx_1\, dx_2$$

$$\vdots$$

$$P2_k(t) = \lambda(k\mu)^{k-1} \varepsilon^{-k\mu t} \int_{x_k=0}^{t} \int \int \cdots \int_{x_1=0}^{x_2} \varepsilon^{(-\lambda+k\mu)x_1}$$

$$dx_1\, dx_2 \ldots dx_k$$

Let $(-\lambda+k\mu) = a_1$ and evaluating the multiple integrals for $P2_1(t)$, $P2_2(t)\ldots P2_k(t)$ and substituting into Equation (14) we obtain

$$P1(t) = 1 - \frac{(1-b_1^k)}{b_1^k} \varepsilon^{-\lambda t} + \frac{1}{b_1^k} \varepsilon^{-k\mu t} \sum_{n=0}^{k-1} (1-b_1^{k-n}) \frac{(a_1 t)^n}{\lfloor n}$$

(15)

Hence,

$$P2(t) = \frac{(1-b_1^k)}{b_1^k} \varepsilon^{-\lambda t} - \frac{1}{b_1^k} \varepsilon^{-k\mu t} \sum_{n=0}^{k-1} (1-b_1^{k-n}) \frac{(a_1 t)^n}{\lfloor n}$$ (16)

where, $b_1 = \frac{a_1}{k\mu}$

It was pointed out in Sec. 3.2 that k stages are required to represent a distribution having coefficient of variation $1/\sqrt{k}$. With suitable combination of stages with complex transition probabilities, it is possible to represent the same distribution with appreciably fewer stages. This approach of using complex transition rates was first suggested by Cox[12]. The complex transition probabilities introduce trigonometric terms in the p.d.f. which reduces the number of stages. Since the p.d.f. of total repair time, which is convolution of the p.d.f. s of the times spent in the separate stages, is real, the complex transition rates must occur in pairs of complex conjugate. Although these transition rates are complex, the process can still be treated as Markovian and a set of differential equations may be written in the usual way. It can be shown that although in the solution of probabilities associated with the fictitious stages may be complex, the probabilities associated with real states of the system will be real.

One of the simplest two state models with complex transition rates of a generating unit is shown in Fig. 3. For this model it has been assumed that repairs take place in only two stages and transition rates from stage 2_1 to 2_2 and from 2_2 to 1 are μ_1 and μ_2 respectively, which are complex conjugate of each other.

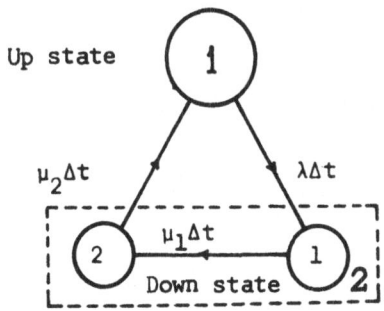

Fig. 3: Complex repair rates model

Let $\mu_1 = a-jb$ and $\mu_2 = a+jb$ where, a and b are both positive constants.

A set of differential equations relating the state probabilities may be written in much the same way as for real transition rates and is given below:

$$\frac{dP1(t)}{dt} = -\lambda P1(t) + \mu_2 P2_2(t)$$

$$\frac{dP2_1(t)}{dt} = \lambda P1(t) - \mu_1 P2_1(t)$$

$$\frac{dP2_2(t)}{dt} = \mu_1 P2_1(t) - \mu_2 P2_2(t)$$

On solving the above set of equations, the following expression for P1(t) is obtained

$$P1(t) = \frac{a^2+b^2}{A^2+B^2} + \frac{2a\lambda}{A^2+B^2} (\cos At + \frac{a^2-b^2-a\lambda}{2a\lambda} \sin At)\epsilon^{-Bt} \qquad (17)$$

where,

$$A = \sqrt{b^2 + a\lambda - \tfrac{1}{4}\lambda^2} \quad \text{and} \quad B = a + \frac{\lambda}{2}$$

and P2(t) can be obtained by subtracting P1(t) from unity.

Sometimes, repair data indicate coefficient of variation greater than unity. In this case the stages have to be connected in parallel instead of series. As explained in Sec. 3.3, simulation of only two stages in parallel with transition rates $2\rho\mu$ and $2(1-\rho)\mu$ would be sufficient. These two stages receive the failed unit on an average of a fraction ρ and $(1-\rho)$ of the times respectively. The modified model is shown in Fig. 4. The expression for P2(t) can be derived and is given below:

$$P2(t) = \frac{\lambda\rho}{(2\rho\mu-\lambda)} \{\epsilon^{-\lambda t} - \epsilon^{-2\rho\mu t}\} + \frac{\lambda(1-\rho)}{2(1-\rho)\mu-\lambda} \{\epsilon^{-\lambda t} - \epsilon^{-2(1-\rho)\mu t}\}$$

and P1(t) can be obtained by subtracting P2(t) from unity.

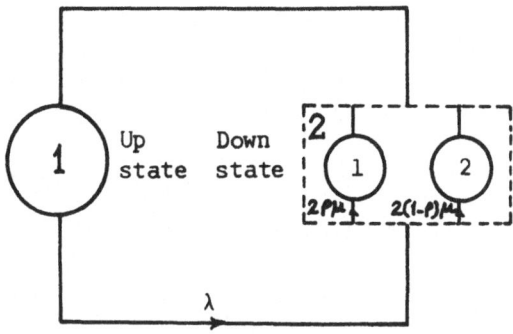

Fig. 4: Parallel stage-device repair model

3.5 Models for Stand-by Generators

Models for stand-by generators, in general, are more complex compared to normally operating generators. This is because their operation is influenced by several factors such as their dependence on normally operating generators, start up failures, change in failure rate due to change in state, start up time etc. In order to obtain a reasonably simple model it is assumed[3] that their operation depends upon only one normally operating generator at a time and their running failures and repairs are not considered. It is further assumed that the time required to bring them into operating state is known.

Consider a normally operating generator A backed up by a single stand-by generator B with constant, non-zero, start up time T_s. Let both the generators A and B be represented by only two states. The state-space diagram for the model is shown in Fig. 5. In the figure a letter corresponding to a generator with a bar at the top, at the bottom and without it, indicates that the generator is in down state, up state and available but not operating state, respectively.

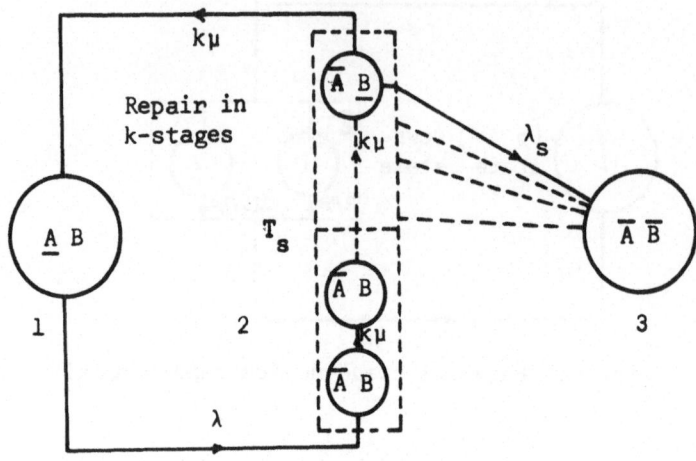

Fig. 5: Stand-by generator model

Assuming the normally operating generator to be initially in the operating state and the stand-by generator operates only when the normally operating generator is in failed state, the following state probability expressions may be written.

P [Gen. A in 1 and B up at t] = 0   (19)
P [Gen. A in 1 and B down at t] = P1(t)   (20)

where, P1(t) is given by Equation (15)

P [Gen. A in 2 and B up at t] = 0 for $t < T_s$ (21)
P [Gen. A in 2 and B up at t]; $t \geq T_s$

$$= (1-p_s) [P2(t) - \epsilon^{-\lambda(t-T_s)} P2(T_s)]$$ (22)

(for derivation see Appendix)

where, P2(t) is given by Equation (16) ; $P2(T_s)$ is the value of P2(t) at $t = T_s$ and p_s is the start up failure probability of the stand-by generator

and finally,

P [Gen. A in 2 and B down at t]

= P2(t) − P [Gen. A in 2 and B up at t]   (23)

Now consider the case of a normally operating generator backed up by several stand-by generators having the same finite start-up time. Let the operating policy be such that one stand-by generator should be started in priority order on failing the normally operating generator. It can be shown that for i^{th} stand-by generator to be up Equation (22) changes to:

P [Gen. A down and stand-by gen. i up at t for $t \geq T_s$]

$$= \prod_{j=1, i-1} P_{sj}(1-P_{si}) [P2(t) - \varepsilon^{-\lambda(t-T_s)} P2(T_s)] \qquad (24)$$

and the other Equations (19), (20), (21) and (23) remain unchanged.

4. COMPUTATION ON A SAMPLE SYSTEM

The sample power system considered for computation of security function consists of nine normally operating generators and two identical quick starting stand-by generators. The start-up time of these generators have been assumed to be zero. Consideration of conventional type stand-by generators having finite non-zero start-up time does not present any problem. They can be included simply by substituting the appropriate value of T_s in the equations developed for stand-by generators in Sec. 3.5. The parameters of the generating units are given in Table 1.

Table 1

Generating Unit Parameters

Unit No.	Full Load Capacity Output (MW)	Failure Rate λ(1/Yr)	Repair Rate μ(1/Yr)	No. of Stages k
1	550.0	4.0	219.0	8
2	520.0	4.0	638.0	8
3	443.0	2.6	638.0	10
4	320.0	2.6	638.0	12
5	148.0	2.5	585.0	16
6	118.0	2.5	585.0	18
7	280.0	2.6	638.0	14
8	100.0	1.2	151.0	18
9	80.0	1.2	151.0	20

Capacity of each stand-by generator = 80.0 MW
Capacity level L_1 = 100.0 MW
Capacity level L_2 = 200.0 MW

All the generating units have been assumed to be two state devices i.e. they are either fully available or totally forced out. The operating policy has been assumed to be such that no stand-by generator is started if the loss of total generating capacity is less than a certain level L_1. If the total forced out capacity is equal to or greater than L_1 but less than L_2 ($L_2 > L_1$) only one stand-by generator is started. Finally, when the total forced out capacity is equal to or greater than L_2 both the stand-by generators are started.

The latest hourly load forecast is assumed to be known and is presented in Table 2.

Table 2

Predicted Hourly Loads

Time (hr)	Predicted Load (MW)	Time (Hr)	Predicted Load (MW)
1	2000.0	6	1870.0
2	1980.0	7	1820.0
3	1940.0	8	1700.0
4	1900.0	9	1510.0
5	1840.0	10	1410.0

Since the load values have been assumed to be a prediction of future hourly loads, the load at any hour must be considered a random variable. For calculation of security function for the system described, it has been assumed that the uncertainty associated with the load forecast at each hourly load is same and equal to 1%. The security function computations for the system using the models developed are done up to a time period into the future equal to ten hours with and without considering the effect of stand-by generators. It has been observed that for the system conditions defined the inclusion of stand-by generators reduce the security function to nearly zero. Fig. 6 shows the plot of security function for the system vs the time into the future. To study the effect of uncertainty in load forecast on security function, the computations are repeated by assuming uncertainty in the load forecast to be 2%. While computing security function it has been assumed that for the entire time period considered, none of the nine normally operating generators is deliberately taken out of operation and no more than two generators can be on forced outage at the same time.

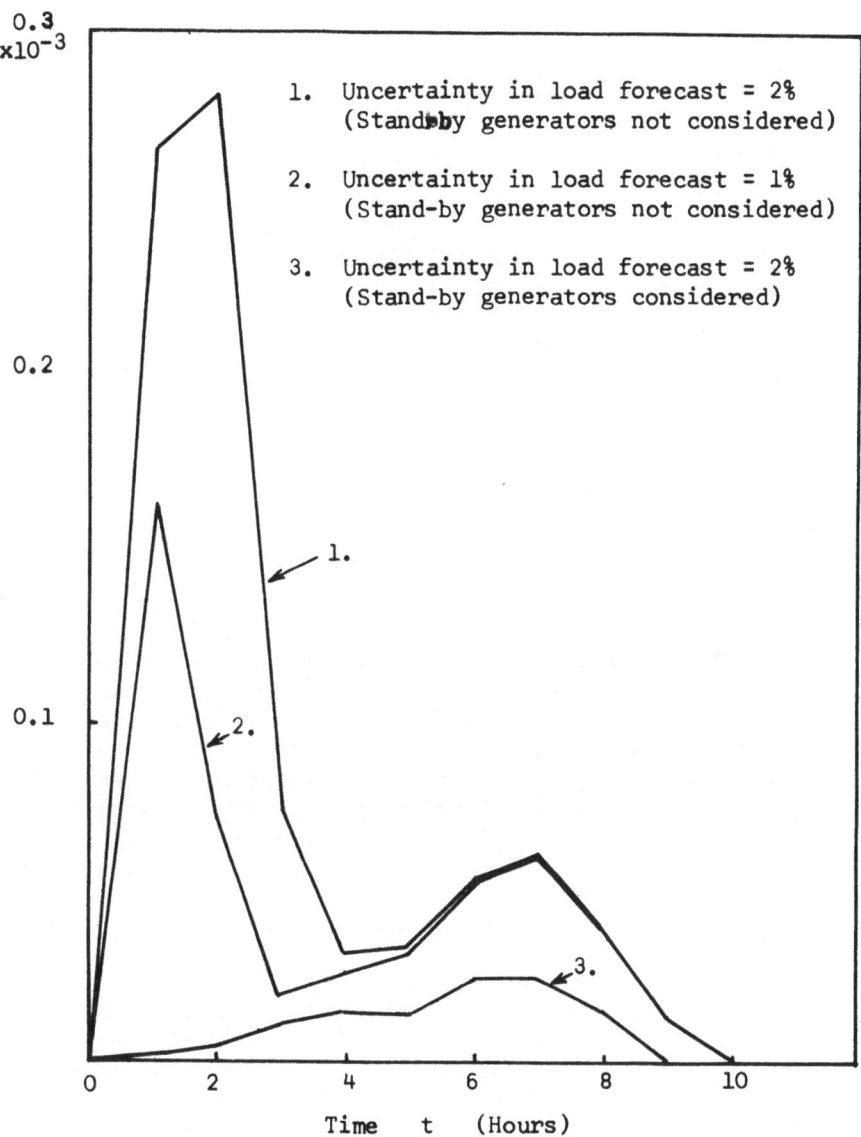

Fig. 6: Plot of Security Function for sample system

The security function computations are executed on a CDC7600 computer. The computation times for calculating security function for a time period equal to 24 hours into the future for systems having different number of generating units are obtained and plotted in Fig. 7.

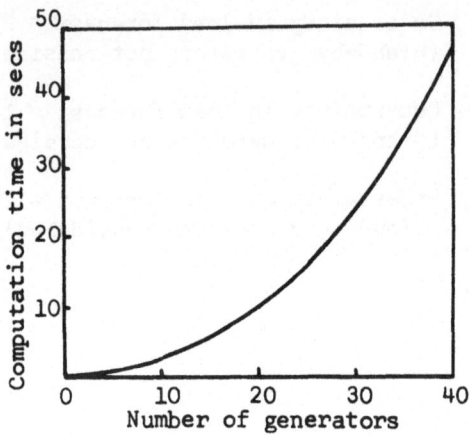

Fig. 7: Computation time for S.F. calculations

5. CONCLUSIONS

In this paper new probability models for generators to calculate time dependent state probabilities are developed. An important aspect of these models is that they allow the inclusion of non-exponential nature of repair time distributions. In developing these models, the stage device technique has been used which is capable of generating most practical distributions in a simple and routine manner by simulating exponential stages in series or parallel. It is also shown that by using complex transition rates it is possible to reduce the number of stages to be connected in series for generating a given repair time distribution.

Studies of several systems containing a different number of generating units proves that the models developed are computationally efficient. These can be applied successfully to most practical systems for evaluating level of system security into near future, thereby enabling operational engineer to take appropriate control action to keep the risk level below a desired value.

The accuracy of reliability analysis depends very much on the statistical information available irrespective of how powerful or sophisticated reliability models may be employed. The collection of trustworthy data and obtaining meaningful information from them, such as failure and repair rates, involves considerable problems.

It is often quite impossible to get information on system component failure and repair rates. Sometimes the information available is so little that it is not possible to estimate the required parameters to a desired accuracy. Even if sufficient data is available, its validity can be questionable because servicing personnel sometimes have a strong motive tending to bias the information. If however, correct data is available the developed models can be applied systematically and consistently to predict reliability of power system generation without resorting to techniques based on what is called "intelligent guesswork".

6. REFERENCES

1. Surana, S.L.:"Reliability of Power System Generation" Ph.D. Thesis, 1973, University of Manchester Institute of Science and Technology, Manchester.

2. Anstine, L.T., Burke, R.E., Casey, J.E., Holgate, R., John, R.S., and Stewart, H.G.:"Application of Probability Methods to the Determination of Spinning Reserve Requirements for the Pennsylvania - New Jersey - Maryland Interconnection", I.E.E.E. Transactions (Power Apparatus and Systems), Vol. 82, 1963, pp.720-735.

3. Patton, A.D.:"Short Term Reliability Calculation", I.E.E.E. Transactions (Power Apparatus and Systems), Vol. 89, 1970, pp.509-514.

4. Biggerstaff, B.E., and Jackson, T.M.:"The Markov Process as a Means of Determining Generating Unit State Probabilities for Use in the Spinning Reserve Application", I.E.E.E. Transactions (Power Apparatus and Systems), Vol. 88, 1969, pp.423-430.

5. Billinton, R.:"Power System Reliability Evaluation", Gordon and Breach, New York, 1970.

6. Patton, A.D.:"A Probability Method for Bulk Power System Security Assessment, III - Models for Stand-by Generators and Field Data Collection and Analysis", I.E.E.E. Transactions (Power Apparatus and Systems), Vol. 91, 1972, pp.2486-2491.

7. Patton, A.D.:"A Probability Method for Bulk Power System Security Assessment, II - Development of Probability Models for Normally-Operating Components", I.E.E.E. Transactions (Power Apparatus and Systems), Vol. 91, 1972, pp.2480-2485.

8. Di Marco, A.:"A Semi-Markov Model of a Three-State Generating Unit", I.E.E.E. Transactions (Power Apparatus and Systems), Vol. 91, 1972, pp.2154-2160.

9. Cox, D.R. and Miller, H.D.: The Theory of Stochastic Processes, Methuen and Co. Ltd., London, 1965.

10. Singh, C., and Billinton, R.:"Reliability Modelling in Systems With Non-Exponential Down Time Distributions", I.E.E.E. Transactions (Power Apparatus and Systems), Vol. 92, 1973, pp.790-800.

11. Cox, D.R., and Smith, W.L.:"Queues", Methuen and Co. Ltd., London, 1961.

12. Morse, P.M.:"Queues, Inventories and Maintenance", John Wiley, New York, 1967.

13. Cox, D.R.:"A Use of Complex Probabilities in the Theory of Stochastic Processes", Proc. Camb. Phil. Soc., Vol. 51, 1955, pp.313-319.

SHORTCUTS IN HUMAN RELIABILITY ANALYSIS*

Alan D. Swain

Systems Reliability Division, Sandia Laboratories,
Albuquerque, New Mexico 87115

ABSTRACT

Several techniques for making qualitative or quantitative esttimates of the likelihood that human tasks will be performed incorrectly are discussed. These techniques involve the use of judgments by subject-matter experts, ranging from the analyst himself to performers of tasks. The techniques are based on the human reliability model and methodology used by ergonomists at Sandia Laboratories.

INTRODUCTION

Human reliability technology is used to estimate the quantitative effects of human performance on the reliability of a man-machine system. The primary problem in using this technology has always been a dearth of usable human performance data. By usable is meant human performance data (e.g., defect rates) used in conventional reliability technology. Investigators in human reliability technology have spent much--some would say too much (Blanchard, 1972)--time and effort in building models but not nearly enough in developing usable human performance data.

In any event, the lack of human performance data has led to the development of shortcuts in human reliability analysis. Two types of shortcuts can serve as the starting point for a series of shortcut measuring methods which vary considerably in the amount of quantification. At one end of the quantification scale, a

---
*This work was supported by the United States Atomic Energy Commission. A more complete version, including a sample reliability analysis taken from an actual study, is available as Sandia Laboratories' report SLA-73-5530.

qualitative profile is developed in which tasks are characterized solely in terms of high, medium, or low error-likeliness (or scales with 5 or more points) and the other parameters listed below. At the other end of the quantitative scale, the various task parameters are expressed in probabilities or other ratio numbers, along with their associated probability density functions.

The human reliability analysis described is one in which the primary interest is estimating, qualitatively or quantitatively, four parameters:

1. <u>Task Reliability</u> - the likelihood that a task will be completed successfully within some allotted period of time.

2. <u>Error Correction</u> - the likelihood that incorrect task performance will be detected and corrected in time to avoid undesirable consequences of interest.

3. <u>Task Effects</u> - the likelihood that incorrect and uncorrected task performance will result in undesirable consequences to a system.

4. <u>Importance of Effects</u> - the degree of importance of the undesirable effects to a system in terms of cost or other criteria.

The human reliability model on which the shortcut methods are based was developed at Sandia Laboratories and called THERP - Technique for Human Error Rate Prediction. The model uses conventional reliability technology with modifications and additions appropriate to the greater variability and unpredictability of human performance as compared to equipment performance. The output of the model normally consists of quantitative estimates of the four foregoing parameters, based on the dependencies among human performances and among these and equipment performance, other system events, and various outside influences. The basic tool of the model (and of the shortcut methods) is probability tree diagramming. Branching in the tree is used to show the different events as well as different conditions or influences on these events. Thus, the values assigned to the events depicted by the tree limbs in the graphic representation of the model are conditional probabilities.

THE HUMAN RELIABILITY MODEL

The human reliability model used at Sandia Laboratories was outlined by Rook (1962) and expanded in subsequent Sandia reports (Swain, 1963, 1964, 1967a, 1967b, 1969, and 1971; Rook, 1964; Rigby, 1967; Rigby and Edelman, 1968; and Rigby and Swain, 1968) and in a book by Swain (1972). Following is a definition of Sandia's human reliability model:

> THERP (Technique for Human Error Rate Prediction) is a method to predict human error rates and to evaluate the degradation to a man-machine system likely to be caused by human errors in association with equipment functioning, operational procedures and practices, and other systems and human characteristics which influence system behavior.

The model can accept rank-order estimates of error-likeliness as expedient substitutes for estimates of error rates (Rook, 1964).

The steps in THERP are similar to the steps in conventional reliability analysis if human activities are substituted for equipment outputs:

1. Define system failure(s).

   These are the events for which the influence of human errors are to be estimated.

2. List and analyze the related human operations.

   This step is generally known as task analysis. Chapter VII in Swain (1972) describes the use of task analysis in THERP.

3. Estimate (predict) related error rates (or substitute estimates of error-likeliness).

4. Determine estimated effects of human errors on the system failure events of interest.

   This effort often involves integration of the human reliability analysis with a system reliability analysis.

5. Recommend changes to system and calculate new system failure rates.

   This statement reveals the iterative nature of the five steps and also the use of human reliability analysis as a tool in human engineering (ergonomics) design efforts.

THE PROBABILITY TREE

The probability tree is the major tool for application of the human reliability model. The tree serves three major purposes: (1) It is a convenient way of showing what leads to what, and its use decreases the probability of the analyst overlooking some important human behavior or other system event; (2) its use greatly simplifies mathematical computations and decreases the probability of mathematical errors by the analyst; and (3) it enables the analyst to estimate conditional probabilities readily and avoids the rather complicated equations normally associated with such calculations.

To illustrate THERP and to show how the probability tree is used, two simple, hypothetical examples from a production setting follow.

Example 1

Assume that a production worker is putting finishing touches on an electronic assembly. Consider that his task on a production line is made up of two subtasks, A and B:

Subtask A. Connect two cables which can be reversed.

Subtask B. Plug in two tubes which can be reversed.

Other errors are possible, but for the purpose of simplicity it is assumed that the only errors possible are the two reversal errors listed above.

This production line task can be diagrammed as shown in Figure 1, assuming no other subtasks. The tree diagram shows that the worker can perform subtask A correctly (as denoted by the small letter a) or incorrectly (as denoted by the capital letter A). (In Sandia's human reliability work, small English letters are used for correct behavior and the correct behaviors normally are branched to the left. Capital English letters are used for incorrect behavior and are normally branched to the right. English letters may represent events as well as probabilities. Greek letters are often used for other events and their probabilities, as example 2 illustrates).

Having performed subtask A correctly or incorrectly, the worker can then perform subtask B correctly or incorrectly. The definitions and construction of this probability tree indicate that the worker's error rate for subtask B is the same whether he has performed subtask A correctly or incorrectly. This assumption of independence of tasks--that the performance of one task correctly or incorrectly has no bearing on subsequent performance of a second task--is an example of the simple multiplicative model.

In this example, the only route for task success is the product of a and b; that is, both must be done correctly. This task success probability equation is stated as $S = ab$, and the failure equation is $F = 1 - ab$ or $F = aB + Ab + AB$. Of course, F indicates only the worker's failure and does not necessarily mean that a

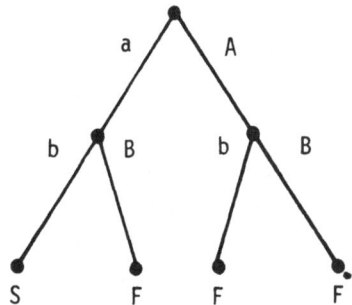

a = probability of successful performance of subtask A
A = probability of unsuccessful performance of subtask A
b = probability of successful performance of subtask B
B = probability of unsuccessful performance of subtask B
S = probability of task success = ab
F = probability of task failure = 1 - ab = aB + Ab + AB

Figure 1. Probability Tree Diagram of a Hypothetical Production Task

defective product would be turned out. An inspector's visual inspection or the employment of test equipment might catch the error (parameter 2, listed earlier). Therefore, in a real example, other limbs and branches would be added to the tree diagram to represent the entire production process schematically.

The simple multiplicative model as described above might be valid for that particular example or, at least, its use might not introduce an unacceptable amount of error into the analysis in question. However, unlike much of the reliability work done with equipment, the use of the simple multiplicative model would often lead to faulty conclusions when used in human reliability work. So it is necessary to employ a model which enables the analyst to take into account those cases when interaction among various human actions does have an important effect on the outcome of the analysis.

Example 2

Figure 2 shows a situation where there are two workers, alpha (α) and beta (β). One is a female production worker and the other a male production worker. In keeping with the U.S. antidiscriminatory laws, let us assume that there really is no difference between these two workers. They turn out the same quality of product; one produces, on the average, no more defects than the other, as indicated by an unbiased analysis of their outputs.

However, let us assume that the inspector is a women's libber anathema--a male chauvinist pig (MCP). His probability of detecting defects differs as to whether he is inspecting the product of the male worker or that of the female worker. That is, his records ascribe more defects to the female worker. The symbols used in the tree diagram show how this behavioral interaction is indicated. If the inspector did not have a built-in bias or if he did not know whose product he was inspecting, the probability tree could be simplified to the form of Figure 1, where a would equal the probability of a good product being produced by either worker and b would indicate the probability of correct detection by the inspector.

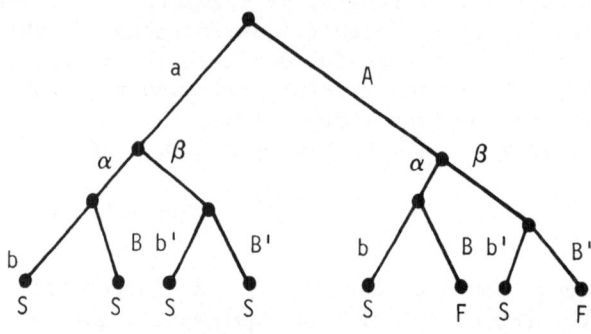

a = probability of production task done correctly by either worker
A = probability of production task done incorrectly by either worker
α = probability that production task is done by female worker
β = probability that production task is done by male worker
b = probability of inspection task done correctly given α
B = probability of inspection task done incorrectly given α
b' = probability of inspection task done correctly given β
B' = probability of inspection task done incorrectly given β

S = probability that defective assembly did <u>not</u> get by inspector
F = probability that defective assembly did get by inspector

S = sum of all paths leading to S, or a + A(αb + βb')
F = sum of all paths leading to F, or A(αB + βB')

Figure 2. Probability Tree Diagram of a Hypothetical Production Situation

It can be noted that in the example shown in Figure 2 failure F is the joint probability that a bad unit is made and gets by the inspector. It is considered that the inspector's rejection of a good unit (paths aαB and aαB') is not a failure, because his rejected good units would be reinspected and, if really nondefective, would be accepted for shipment to the customer. On the other hand, these paths could have been defined as some other type of failure. What is a failure is that which the analyst (or management) defines as a failure.

In both of these examples (Figures 1 and 2), quantitative estimates of S and F can be calculated by substituting the appropriate probability estimates for each of the symbols and performing the simple arithmetic indicated in the two figures.

The occasion of interaction, illustrated by Figure 2, defines "conditional probabilities"; that is, the probability of one event depends on what happened during preceding events. The mathematical rules for handling conditional probabilities can result in rather complex equations. In human reliability work at Sandia, this complexity (with its attendant high probability of error by the analyst) is avoided. In the case of the MCP inspector, the separate branches for α and β indicate that the inspector really has two different tasks; therefore, they are diagrammed as separate tasks. By doing so, one has, in effect, accounted for the interaction effect in the tree diagramming and has avoided the complicated probability equations which deal directly with interaction.

The above examples were, of course, made deliberately simple. More realistic examples are included in documents referenced earlier, where the probability trees have considerably more limbs. However, even in real-life examples, pictorial complication is reduced by using hierarchies of trees where a limb in one probability tree may represent another, entire tree.

The limbs in the above two tree diagrams represent human events only. In a man-machine system there are many other kinds of events. The advantage of using conventional reliability technology is that probability estimates of human events can be melded into probability equations which provide estimates of the probabilities of other events. All one needs to do is be sure of what leads to what, and the probability tree aids the analyst in this task.

Thus, in a real reliability study, tree diagrams might contain limbs representing the probabilities of operator errors, inspector errors, test equipment defects, prime equipment defects, different ambient temperatures, and so on. Any one tree diagram, or some set of tree diagrams, would be used to estimate the probability of achieving some undesired system consequences.

SINGLE VALUES RATHER THAN DISTRIBUTIONS

In the two examples described above, only one estimate of error probability or probability of occurrence for some other event was indicated. Yet most events have distributions of values if data are collected over a period of time. For example, the MCP inspector in example 2 may be more kindly disposed toward the female worker on some days and may not be so strict in his surveillance of what she produces. Also, the performance of the worker herself may vary from day to day. Nevertheless, in most applications of human reliability studies, it is sufficient to use an estimate of the average error rate or average probability of some event and disregard the distribution of this probability. This shortcut approach uses what some reliability specialists have called the point-estimate model of reliability.

When more than one point on a distribution must be used to obtain sufficiently accurate estimation, the points can be treated as different events and diagrammed as different limbs on a tree. For example, if a task is to be performed under more than one stress level, different branches of the tree would be used to show the different error rates under each stress level as well as the probability of occurrence of each stress level.

In Sandia's human reliability studies, the time to complete a task has usually been treated by using one or more single values to characterize the time distribution. Normally, the maximum time allowable to complete a task can be readily defined. In such a case, an estimate is made of the probability of task success within the allowable time frame. In other cases, a series of tasks must be completed successfully within some period of time, and an estimate is made of the probability that all of the tasks will be successfully completed within that time. Rarely is it necessary to place limits; for example, the probability is 90 percent that the tasks will be completed within a stated time. The usual influence of time pressures on the types of tasks included in Sandia's human reliability studies is to increase the probability that an error will be made in some procedure and that this error will not be caught.

Another shortcut used in human reliability technology is that the behavior of a given human being ordinarily cannot be predicted except insofar as that particular person is representative of the group of individuals for whom the predictions are being made. Stated in another way, an estimate of human reliability for a task is the estimate of the reliability with which that task will be performed by a person drawn at random from the population of performers in question.

Thus, when an analyst estimates the reliability with which a typist will type a letter, the estimate will apply to the "average" typist. A particular typist may be better or worse than the average, but this fact is usually not of interest in human reliability studies. If the error rate of a <u>particular</u> individual must be known, data can be collected on that one individual.

DERIVING ESTIMATES OF ERROR-LIKELIHOOD

The above discussion suggests that for many practical applications, certainly those using human reliability analysis as an ergonomics tool, single values of error-likelihood will sufficiently characterize the distributions of errors. What kind of data should be used for these single values? Most analysts would feel more comfortable with error rates collected in actual job situations or obtained from experiments where the work situation of interest was simulated as closely as possible.

For certain uses, however, more qualitative indices of human behavior, as well as other system events, will often be adequate and can be employed as an economic measure or when error rates cannot be obtained. For example, in the early planning stages in some U.S. military systems, 5-point scales of parameters like the four listed earlier were used to rate various proposed design features. The 5-point scale might be as follows: 1 - unacceptable, 2 - poor, 3 - average, 4 - good, 5 - excellent. A profile made up of the scale values for each parameter would provide a rough quantitative index of qualitative worth of each proposed design feature.

Ergonomists at Sandia quickly discovered that design planners were not satisfied with qualitative statements about error-likeliness. Nor were they satisfied with the statement, "Further research is needed," as an excuse for not providing quantitative data. The system planners wanted error rate estimates that would fit into system reliability equations, and they wanted these estimates in days rather than weeks or months. Hence, the development of the human reliability model known as THERP.

This model was originally developed to use estimates of error rates as the estimated probabilities of human behaviors. Since the time and cost of conducting controlled experiments to obtain such estimates were not often afforded the users of THERP, various approximations (shortcuts) had to be developed.

The remaining sections describe some shortcuts used in deriving estimates of error rates. These shortcuts involve various levels of approximation of the true error rates. The first (highest) level of approximation occurs when there is so-called "hard data" consisting of error rates collected under the best of conditions.

Approximation is involved because there is always some possibility of error in generalizing data obtained from one study to some different application, no matter how similar the circumstances. Nevertheless, the best (i.e., most trustworthy) data for error rates of tasks would be from the same tasks performed in the "same" setting. Such data are difficult to obtain because records of human performance (even defect rates from production jobs) seldom provide the necessary information to derive error rates. This dearth of human performance data still exists despite numerous appeals (dating from the 6th Annual Meeting of the Human Factors Society in November 1962 (Swain et al., 1963)) for the U.S. government to sponsor a program to build a data bank of error rate and other human performance data.

Some worthwhile attempts to develop data banks have been made. Although the AIR Data Store (Munger, Smith, and Payne, 1962) is probably the best-known data bank, it has serious limitations (Swain, 1967b) and is no longer used in Sandia human reliability studies. Most of the data banks include derived reliability figures for very small (molecular) elements of behavior. For example, in the AIR Data Store an average of 10 to 15 molecular elements make up a typical step in a task. In Sandia's human reliability studies reliability estimates are made for more molar lumps of behavior (e.g., a step or even a task consisting of several steps).

Without the "hard data" described above the analyst can turn to the available psychological literature. Naturally the risks are greater in using data from these sources. Considerable expertise is needed to apply error rates reported in the literature. Studies conducted in research laboratories in university psychology departments, for example, seldom provide data on error rates that are directly usable in practical reliability work (Payne and Altman, 1962). However, such studies may provide the basis for the analyst to estimate correction factors to apply to error rate data from other sources.

Error rate data from experiments conducted in real or simulated operational settings are to be preferred to data obtained under more artificial conditions. Next to data obtained from actual job operations, this kind of experimental data is most valuable.

USE OF SUBJECT-MATTER EXPERTS TO OBTAIN DERIVED DATA

The use of collective judgment by subject-matter experts is one way to make up for the paucity of empirical error-rate data. Subject-matter experts are persons who are especially knowledgeable about the tasks for which error rates are to be derived. Thus, if one wanted to derive error-rate estimates for new tasks in an assembly plant, the experts would be persons who perform or had performed similar tasks in an assembly plant. The ergonomist himself can be considered a subject-matter expert if he has done a thorough task analysis of the tasks in question.

It is not suggested that job performers be asked to make estimates of actual error rates for tasks. Such estimates would tend to vary widely and would generally underestimate the true error rates. Some subject-matter experts would refuse to make such estimates. However, persons who will not guess at the probability of a particular event will readily rank-order several events in terms of increasing difficulty, hazard, or other dimension and will do this ranking with great confidence. Furthermore, several studies have demonstrated sufficiently high agreement among subject-matter experts to make these judgments for ratings useful. These studies include aviators' rankings of inflight emergencies in terms of the relative extent to which internal stress would be produced (Rigby and Edelman, 1968), naval personnel rankings of the error-likeliness of shipboard operation and maintenance tasks (Blanchard et al., 1966, and Smith et al., 1969a), and engineers', technicians', and mechanics' estimates of the relative human reliability in preventive and corrective maintenance tasks on TITAN II missiles (Irwin et al., 1964).

DERIVING ORDINAL SCALES

The above studies indicate that it would be feasible to have a number of people with detailed knowledge of various tasks in an industrial setting rank-order tasks in terms of error-likeliness. Rook (1964) has developed an approximate model for THERP which uses rank-ordering of tasks in terms of error-likeliness and rank-ordering of system consequences of the errors (i.e., the first three of the four parameters listed earlier). Normalized ranks of error-likeliness are used in place of error rates. In place of the probability of detection and correction of an error, its normalized rank order on these dimensions is used. Finally, instead of the probability that an uncorrected error will result in some undesirable consequence, its normalized rank order in comparison with other uncorrected errors is used. The arithmetic is the same when using the normalized rank values as when using the ratio data provided by probabilities of human errors and other events.

The use of normalized rank values for these three reliability parameters gives an approximate ranking of error importance and thus enables the analyst to estimate which potential errors are the most important. Although some rigor is lost because rank-order data rather than rate data are used, the use of the Rook rank-order analogue to THERP provides sufficient information for many applications; e.g., in deciding which of several design concepts should be developed.

When using Rook's rank-order model of THERP, within each dimension (e.g., likelihood of error) each of the events to be ranked is concisely described on a card. The simplest approach is to have each subject-matter expert merely sort cards which describe the events to be ranked within each dimension into some arbitrary number of categories ranging from least probable to most probable. The more categories used, the more precise the approach will be, at the cost of requiring more judgments from each subject-matter expert. The limiting case, of course is when the number of categories equals the number of tasks. The judges' rankings within each dimension would then be averaged or otherwise combined to derive an ordinal scale for each dimension which reflected the combined judgments of all the subject-matter experts.

A more precise ordinal scale can be derived by use of the paired-comparison method of psychological scaling. In this method, each task is compared with every other task in terms of the dimensions being evaluated; e.g., the three parameters described earlier. If the number of events is small (say, 20 or less), the event description cards are presented in pairs to each subject-matter expert. Each judge is requested to indicate which of the two events is the more probable. For 20 events, the number of complete comparisons required would be $20(20 - 1) = 380$. Such a cumbersom task would probably reduce a judge's motivation, therefy mitigating his willingness to provide reliable comparisions. Therefore, a partial comparison would be used to reduce the number of paired-comparisons by half; in the above case, to 190. The formula, $n(n - 1)/2$, is used with the assumption that comparing event A with event B will give the same result as comparing event B with event A; i.e., that the events are commutative.

If the number of paired-comparisons using the $n(n - 1)/2$ formula is too cumbersome, various approximations to the paired-comparison method can be used (Guilford, 1954, and Torgerson, 1958). Rigby and Edelman (1968) describe the use of a pairing matrix which reduced 992 comparisons to 290 pairs, or to 145 if one is not interested in order effects.

DERIVING INTERVAL SCALES

The advantage of the paired-comparison method is that the ordinal judgments of subject-matter experts can be combined to derive an interval scale for each dimension of interest. In fact, if the paired-comparison method is used, very little extra effort is required to derive interval scales. Guilford, Torgerson, and Edwards (1957) show various methods and techniques of psychological scaling. Rigby and Edelman, Smith et al. (1969a and 1969b), and especially Blanchard et al. (1966) present details on the use of psychological scaling for use in human reliability work.

In Figure 3 the top line represents an ordinal scale such as could be derived from simple rank-order data. The letters A to M indicate the ranks of the events, but the distances between each of the ranked events is not known. All that is known is that A is less error-likely than B, B is less error-likely than C, and so on. The middle line represents an interval scale computed from paired-comparision data. The computations are described in the above references. Now one has a measure of the relative distance between pairs of tasks. However, there is still no estimate of the absolute error, or error rate, for each task. For many design purposes this level of approximation is sufficient.

DERIVING RATIO SCALES

Occasions when it will be necessary to use estimated error rates rather than rank-order or interval data include supporting system reliability studies and making design decisions involving tradeoffs between different design concepts each with different projected costs and development times. By using the interval scale of error-likeliness described above, it is possible to derive a ratio scale; i.e., a scale with a true zero point where the scale values are additive. This derivation is done by calibrating the interval scale with known error rate figures for some of the tasks in question. Generally, it is possible to obtain error rate figures for at least some of the tasks by using data sources described earlier.

The rescaling technique can involve generation of an equation of the form $Y = f(X,b)$, where Y is a value from the actual error-rate scale, X is a value from the interval, and b is some mathematical constant. Alternatively, the rescaling can be done graphically. By either method (see Figure 3), the estimated interval data is moved to the established ratio scale consisting of the known error rates for the few tasks for which error rate data are available. (In the figure, the known error rates are indicated on the bottom scale.) In this transformation the interval distances between the various tasks must remain relatively the same when the interval scale is moved to the ratio scale. Otherwise, either the derived interval scale values are in error or the error rates used as calibrators are inappropriate, and the transformation should not be used.

This approach, which uses rank-ordering of events along some dimensions to derive an interval scale and then converts the interval scale to a ratio scale with the use of calibrators, is not

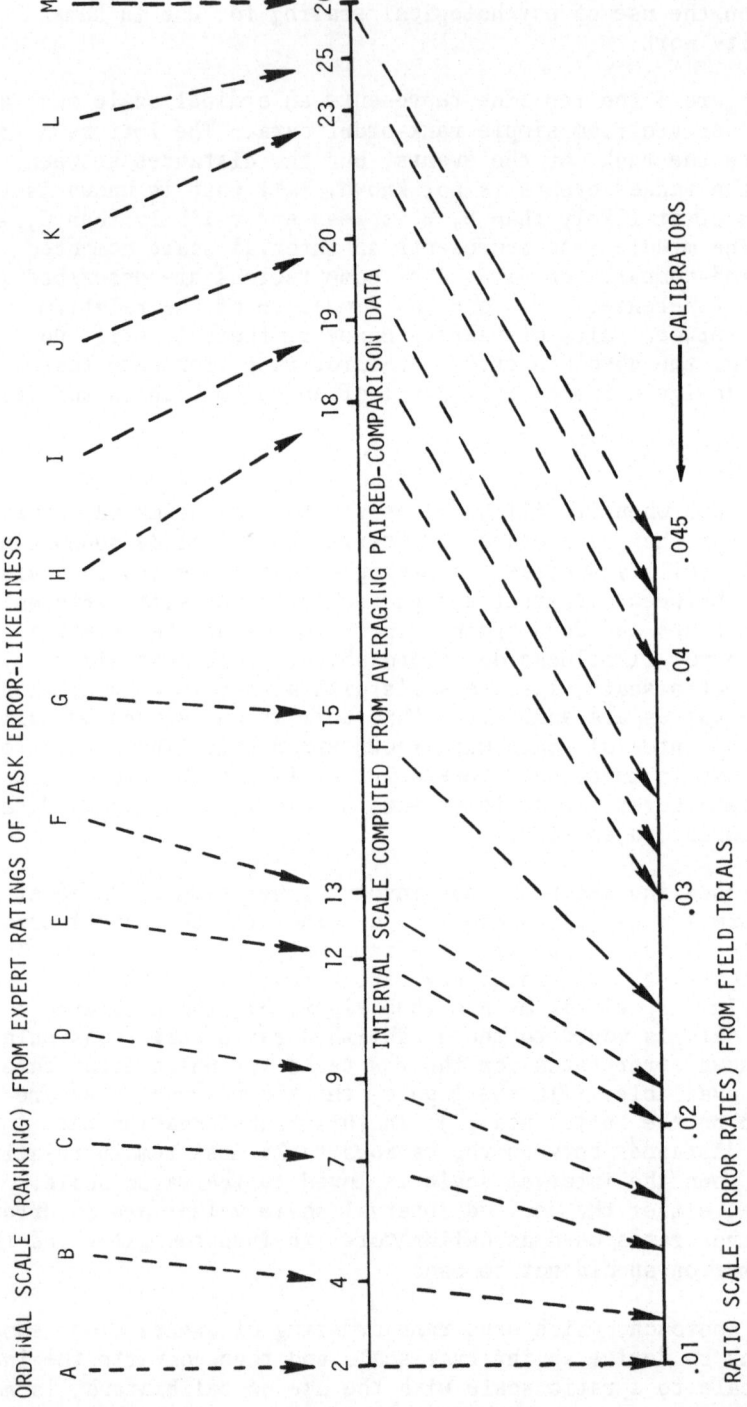

Figure 3. Error Rates Derived From Expert Judgments of Error-Likeliness

restricted to tasks performed by humans. It can also be applied to system and equipment reliability studies and may provide more precision and validity than some of the studies which require reliability specialists to make direct estimates of the absolute rates of equipment defects and other system events.

In some cases, as when immediate design action is required, the collective judgment by subject-matter experts may not be available and the ergonomist will have to rely on his own judgment. In such a case, he can rank-order the tasks in terms of error-likeliness by using Rook's rank-order analogue, described earlier. In early phases in the development of a man-machine system, this level of approximation will often suffice.

Another shortcut which is especially useful in using human realiability analysis as a design tool is known as worst-case analysis. In one study at Sandia Laboratories, for example, the possible events in a production line leading to a highly undesirable system consequence were identified and each human task in this chain assigned an inflated error rate estimate of .10. In addition, various possible "end runs" were also given an inflated error rate estimate of .10. The following is an example of an end run: A tray of components somehow mistakenly got delivered from station 9 to station 11, skipping station 10, and this error was not caught by the operator at station 11. The .10 inflated error rate meant that 1 in 10 times, this error would be made and not be detected at station 11. This worst-case analysis of the entire production process, including what we call an end-run analysis, showed that even with the inflated error rates, the resultant probability of the undesirable consequence in question was several orders of magnitude less than the tolerable probability of this consequence. Therefore, it was possible to recommend the use of a particular component as being consistent with the tolerable probability of the system failure in question. Had the worst-case analysis resulted in an unacceptable probability of system failure, the analyst would then have had the option of performing another human reliability analysis using more accurate estimates of the human error rates for the tasks in question.

One may sum up by saying that the derivation of estimates of error-likelihood range from the ergonomist's best guess, based on a thorough study of the factors involved, to so-called hard data, based on actual measured error rates for the tasks involved. Whichever approach is used, efforts should be made to make it as systematic and objective as possible.

CONCLUDING COMMENTS

Shortcut methods for human reliability studies involve various levels of approximation to real reliability data. In some of these approaches the approximation is gross. Nevertheless, when the primary purpose of a human reliability analysis is to assist in system planning, design, and evaluation before actual performance data can be collected under controlled conditions, such approximations have been found to be both acceptable and useful by system planners and designers.

Moreover, the usual lack of data on error rates for tasks and task conditions of interest forces one to derive estimates of error-likelihood from any source available. This report advocates the use of subject-matter experts as an approach which draws upon (1) tried and proven psychological scaling techniques to develop interval scales of error-likeliness or other indices of human performance and (2) calibration of these scales with some actual performance data so as to derive ratio scales of estimated error probabilities or other quantified measures of human performance.

REFERENCES

Blanchard, R. E., Survey of Navy Needs for Human Reliability Models and Data, Report No. 102-1, Naval Underwater Systems Center, New London Laboratory, New London, Connecticut, Dec. 1972

Blanchard, R. E., Mitchell, M. B., and Smith, R. L., Likelihood-of-Accomplishment Scale for a Sample of Man-Machine Activities, Dunlap and Associates, Inc., Santa Monica, Calif., June 1966

Edwards, A. L., Techniques of Attitude Scale Construction, Appleton-Century-Crofts, Inc., New York, 1957

Guilford, J. P., Psychometric Methods, McGraw Hill Book Co. Inc., New York, 1954

Irwin, I. A., Levitz, J. J., and Freed, A. M., "Human Reliability in the Performance of Maintenance," Proceedings, Symposium on Quantification of Human Performance, Electronic Industries Assn. and Univ. of New Mexico, Albuquerque, New Mexico, August 1964, 143-198

Munger, S. J., Smith, R. W., and Payne, D., An Index of Electronic Equipment Operability: Data Store, AIR-C43-1/62-RP(1), American Institute for Research, Pittsburgh, Penn., Jan. 1962

Payne, D. and Altman, J. W., An Index of Electronic Equipment Operability: Report of Development, AIR-C43-1/62-FR, American Institute for Research, Pittsburgh, Penn., Jan 1962

Rigby, L. V., "The Sandia Human Error Rate Bank (SHERB)," R. E. Blanchard and D. H. Harris (eds), Man-Machine Effectiveness Analysis, Los Angeles Chapter, Human Factors Society, Los Angeles, June 1967, 5-1 to 5-13

Rigby, L. V. and Edelman, D. A., "A Predictive Scale of Aircraft Emergencies," Human Factors, 1968, 10, 475-482

Rigby, L. V. and Swain, A. D., "Effects of Assembly Error on Product Acceptability and Reliability," Proceedings of the 7th Annual Reliability and Maintainability Conference, American Society of Mechanical Engineers, New York, July 1968, 3-12 to 3-19

Rook, L. W., Reduction of Human Error in Industrial Production, SCTM-92-62(14), Sandia Laboratories, Albuquerque, New Mexico, June 1962

Rook, L. W., "Evaluation of System Performance for Rank-Order Data," Human Factors, 1964, 6, 533-536

Smith, R. L., Westland, R. A., and Blanchard, R. E., Technique for Establishing Personnel Performance Standards (TEPPS), Technical Manual, Report PTB-70-5, Vol. I, Personnel Research Division, Bureau of Naval Personnel, Wash. D.C., Dec. 1969

Smith, R. L., Westland, R. A., and Blanchard, R. E., Technique for Establishing Personnel Performance Standards (TEPPS), Procedural Guide, Report PTB-70-5, Vol. II, Personnel Research Division, Bureau of Naval Personnel, Wash. D.C., Dec. 1969

Swain, A. D., A Method for Performing a Human Factors Reliability Analysis, Monograph SCR-685, Sandia Laboratories, Albuquerque, New Mexico, Aug. 1963

Swain, A. D., THERP, SCR-64-1338, Sandia Laboratories, Albuquerque, New Mexico, Aug. 1964

Swain, A. D., "Field Calibrated Simulation," Proceedings of the Symposium on Human Performance Quantification in Systems Effectiveness, Naval Material Command and the National Academy of Engineering, Wash. D.C., Jan. 1967, IV-A-1 to IV-A-21

Swain, A. D., "Some Limitations in Using the Simple Multiplicative Model in Behavior Quantification," W. B. Askren (ed), Symposium on Reliability of Human Performance in Work, AMRL-TR-67-88, Aerospace Medical Research Labs, U.S. Air Force, Wright-Patterson AFB, Ohio, May 1967, 17-31

Swain, A. D., *Human Reliability Assessment in Nuclear Reactor Plants*, SCR-69-1236, Sandia Laboratories, Albuquerque, New Mexico, April 1969

Swain, A. D., "Development of a Human Error Rate Data Bank," J. P. Jenkins (ed), *Proceedings of U.S. Navy Human Reliability Workshop 22-23 July 1970*, Naval Ship Systems Command, Office of Naval Research, and Naval Air Development Center, Dept. of the Navy, Wash. D.C., Feb. 1971, 113-148

Swain, A. D., *Design Techniques for Improving Human Performance in Production*, Industrial and Commercial Techniques Ltd, 30 Fleet St., London EC4Y 1AD, 1972

Swain, A. D., Altman, J. W., and Rook, L. W., *Human Error Quantification, A Symposium*, SCR-610, Sandia Laboratories, Albuquerque, New Mexico, April 1963

Torgerson, W. S., *Theory and Methods of Scaling*, John Wiley & Sons, New York, 1958

RECENT DEVELOPMENTS IN THE PREDICTION OF HUMAN OPERATOR PERFORMANCE

D. R. Towill

Professor of Engineering Production, UWIST, Cardiff, UK

1 Some Difficulties in Predicting Future Performance

In seeking to predict human operator performance the problem is complex because,

(a) on a day-to-day basis, there are random fluctuations in performance[1]

(b) on a long term basis, there is a trend towards increased performance, so that any model used must be capable of accounting for the transient nature of performance level[2]

(c) deterministic effects, such as false ceilings on performance[3], and periodic variations[4], observable in the performance data, and which are often thought to be due to assignable (though at that point in time unknown) causes are evident

(d) considerable variations between operators in the performance of the same task, even for comparable experience, are to be found[5].

(e) environmental, psychological, and physiological conditions in the practical situation can rarely be matched in laboratory trials, so that extrapolation of laboratory results is not always a realistic guide to human operator performance. For the real world, there is consequently a dearth of usable data, which is a general feature of human operator reliability studies[6]

(f) although an intuitive theory of human operator performance improvement exists[7], it is extremely tentative, and is inadequate as a general predictor[8]

2 Purpose of the Model

In the light of the above difficulties, it is suggested that the prediction problem be studied by recognising the limitations of any 'a priori' forecast, and updating the forecasts immediately experimental evidence becomes available. Specifically, it is argued that the measurements give valuable clues to the rate at which performance is likely to improve given the current conditions. If the future performance forecast is unsatisfactory, then we must take action to improve the conditions in some beneficial manner. Any model used for such prediction cannot give the reason for good or bad performance, it can only be used to indicate the need for action, or to establish the consequences of the action. The model must therefore complement, not replace, the human factors expert. At the same time, to help the human factors expert, the model must predict, not simply compress historical data.

Human operator performance improvement may be expressed in a number of ways. For the purpose of this paper, performance is defined by a performance index which increases as a function of experience, eventually approaching asymptotically some limiting value. Typical indices fulfilling this requirement are 'time on target' during vehicular control[8], probability of continuing with an aircraft landing in poor visibility[9], and the quantity of electrical items inspected during a shift[5]. The index is applicable to individuals, and to groups of people, as, for example, in running printing presses[10], start-up of steel mills[11], and packaging of manufactured goods[12]. The model used here for human operator future performance prediction relates the performance index to cumulative time spent performing the task.

3 Time Constant Model

As initially defined in reference (4) the observed performance index, $Y(t)$, and the model prediction, $Y_M(t)$ are related by the equation

$$Y(t) = Y_M(t) + N(t) \quad\quad\quad (1)$$

where $N(t)$ takes account of all variations from the model, one source of which may be modelling errors. The time constant model is defined by

$$Y_M(t) = Y_c + Y_f (1-e^{-t/\tau}) \quad\quad\quad (2)$$

Y_c is the initial performance, (Y_c+Y_f) is the final performance, which is approached asymptotically, and τ is the learning time constant. To indicate the improvement in performance possible with experience, it is sufficient to say that it is common for Y_f to be greater than Y_c. At time $t=\tau$, the performance predicted by the model is $Y_c + 0.63 Y_f$, and at time $t=3\tau$, predicted performance is $Y_c + 0.95 Y_f$. τ varies considerably from task to task, and may be days, weeks, or months.

4 What the Model Tells Us

The time constant model defined by equation (2) adequately describes many tasks studied by the author and his colleagues. Provided the parameters may be updated as further observations become available, so that management may act to correct unsatisfactory situations, the parameters Y_c, Y_f, and τ may be used to indicate many factors[13] including,

(a) relocation of unsatisfactory operators
(b) need for retraining
(c) manpower scheduling
(d) revised delivery dates
(e) revised costing
(f) effectiveness of training schemes

A knowledge of $N(t)$, the model residuals, can also be used to advantage. $N(t)$ may be random, sinusoidal, or indicate a false ceiling[14] by virtue of a plateau effect. As a general guide, $\sigma_{N(t)}$ has been established as of the order of 4% Y_f for a well controlled inspection task, and 10% Y_f for a poorly controlled task. There is some correlation between successive residuals even in the well controlled cases, suggesting that $N(t)$ cannot be completely modelled by the assumption of white noise[14]. $N(t)$ controls the range between which future performance is expected to lie. There is some evidence to suggest that good training schemes reduce scatter and hence the standard deviation of $N(t)$ in addition to resulting in a higher average value of the performance index during training. A formula for plateau detection has been derived and applied to an industrial case study[15], which also showed that corrective action resulted in a subsequent performance recovery also adequately fitted by the time constant model. Sinusoidal scatter does pose problems for certain ranges of frequency, but may be alleviated by suitable filtering of the parameter estimates[4].

5 Parameter Estimation Techniques

An advantage of describing the performance improvement phenomenon by the time constant model has been the immediate recognition of

the parameter estimation problem as the identification of a dynamic
system with first order lag and gain terms from the early part of
the step response when the observations are masked by considerable
scatter. Three methods already developed successfully are

    (a) least squares error fit[4]
    (b) Kalman filter formulation[15]
    (c) impulse moment updating[16]

Of these, the Kalman filter and the impulse moment updating method
require far less computational effort than the least squares error
fit. The impulse moment updating method works well even with
scanty, noisy, data, but requires two smoothing constants to be
chosen. On the one hand this is a disadvantage since the smoothing constants fundamentally control the trade-off between rapid
estimation and noise rejection, but on the other hand if the simple
computations are repeated for several combinations of smoothing
constants, then the expected range of future performance is also
estimated[8]. To date, the Kalman filter and impulse moment updating methods rely on a suitable estimate of Y_c being available,
as did the initial least squares error algorithm, which has recently been updated as a three parameter curve fit procedure.

6 Conclusions

The paper has reviewed a new tool available to the human factors
engineer for forecasting human operator future performance.
Since a range of forecasts can be made, the probability of meeting
a particular performance level at some future point in time can be
established, although there is obviously scope for further work in
this area as the model meets with wider acceptance and further
case studies become available.

7 Acknowledgements

Colleagues F W Bevis, B Hitchings, and Dr H Sriyananda have greatly
contributed to understanding of human operator performance modelling. Their references make excellent reading material for further
study of the subject.

8 References

1. **N A Dudley**   PhD Thesis, University of Birmingham, 1955

2. **J R de Jong**   Ergonomics, Vol 1, No: 1, 1957, p 151

3. **J H Glover**   Int Jnl Prod Res, Vol 4, No: 4, 1965, p 279

4. **F W Bevis, C Finniear, D R Towill**   Int Jnl Prod Res, Vol 8, 1970, p 293

5. F W Bevis   MSc Thesis, UWIST, 1970

6. A D Swain   Paper presented at NATO Advanced Study Institute on Systems Reliability, University of Liverpool, July 1973

7. E R F W Crossman   Ergonomics, Vol 12, 1959, p 153

8. F V Taylor   American Psychologist, Vol 15, 1960, p 643

9. D R Towill   Paper presented at Vth IFAC Symposium on Automatic Control in Space, Genova, Italy, 4-8 June 1973

10. F K Levy   Management Science, Vol 11, p 136, 1965

11. N Baloff   IEEE Trans EM17, November 1970, p 132

12. T Kadota   JIE, Vol XIX, No: 8, 1968, p 407

13. D R Towill and F W Bevis   To be published Int Jnl Prod Res, July 1973

14. B Hitchings   MEng Thesis, UWIST, 1972

15. H Sriyananda   PhD Thesis, UWIST, 1972

16. D R Towill   IEEE Trans. EM-20, No: 2, May 1973, p 44

SOME DESIGN ASPECTS OF ELECTRICAL POWER DISTRIBUTION
SYSTEM RELIABILITY

Dr C.R.Wakeman   Dr M.A.Laughton

Midlands Electricity   Queen Mary College
Board, U.K.   University of London

1. INTRODUCTION

The performance of any engineering system is characterised by a number of parameters and the design objective may be broadly defined as the production of some optimum combination of these parameters. In the economic design of a system, to any incremental improvement in one of these parameters there can be estimated a cost of implementation and, at least in principle, an economic "benefit" which accrues from the improvement can also be estimated. The major design objective then becomes one of maximising the difference between the total benefits which accrue from the operation of the system and the total costs of operating the system. The design process may be subject to certain external constraints, such as those of safety, which limit the maximization process.

Some of the performance parameters of an electrical power system, for example its energy efficiency, are well defined and it is a relatively straightforward matter to estimate costs and values of benefits which arise from system modifications to improve those parameters. Other parameters such as reliability are less well defined and their economic effects, particularly the resulting benefits, are much more difficult to quantify. There are other parameters, maintainability and flexibility for example, which have received very scant mathematical attention and which thus remain completely within the planner's subjective judgement. This paper is concerned with the economic process involved in the choice of the optimum values of the reliability parameters for an electrical power system.

The general principles of the interplay between capital investment and plant capacity margins are well appreciated[1]. Though the failures of supply caused by equipment breakdown do not have the same simultaneous impact on the overall system as insufficient generation, transmission or distribution plant capacity their economic effects are precisely similar. The smaller the expedient taken against the contingency of equipment failure, the smaller will be the investment and hence the smaller the cost to the consumer.

Clearly, there is some level of reliability below which any consumer will regard his supply as economically or socially intolerable. A perfectly reliable supply is attainable only by infinite investment and, hence, infinite cost to the consumer. The optimum standard of reliability lies somewhere between these two levels.

Even in situations where a quantitive valuation of supply failure costs is untenable, a reliability appraisal can be of considerable importance. It does not follow, for example, that the most costly of alternative schemes is necessarily the most reliable. There are many costly means of reducing the reliability of a power system. There is also the situation in which it is required to determine the section of a system (transmission or distribution, for example,) in which further investment would produce greatest rewards in improved reliability.

2. THE VALUE OF RELIABILITY

Attaching a value to electrical distribution system reliability presents many great problems. The main problems, however, stem from the diversity of consumers which any distribution system may supply. It is, in principle, possible to undertake an economic appraisal of each consumer but, because of the large numbers of consumers involved, the cost of performing such studies would far outweigh the advantages which would result from optimisation of the system reliability. Thus, if reliability-value-analyses are to be of use in practice they must be applicable to general classes of consumers. Only consumers with especial merits, for example by virtue of exceptionally high load or exceptional reliance on supply continuity, would merit an individual economic study.

A second difficulty which prevents a simple and accurate valuation of reliability is that of relating costs to the many intangible effects which result from loss of supply. For the domestic class of consumer almost all the losses which result from supply failure might be termed intangible. Only to actual damage which results from supply failure, for example to food in the process of being cooked or under deep-freeze, can an objective assessment be made, whilst the major effects of supply loss are domestic inconvenience to which no direct financial loss may be attributed. For consumers in the production industries loss of supply has certain direct financial implications on the loss of production (these implications are discussed in detail later) but there are still the intangible factors, such as customer dissatisfaction resulting from late delivery, to which no single valuation may be applied.

In common with other branches of reliability analysis the accuracy required of reliability valuation is not high. Indeed, the complexity of the socio-economic processes involved precludes an accurate valuation. When comparing alternative design schemes, however, it will often be found that the alternative schemes have widely differing reliabilities and thus they have widely differing expected failure costs. It is possible, for example, to provide only integer numbers of alternative supplies. The provision of a further alternative supply will usually increase the reliability by many factors and the decision that it is economically feasible to provide the further alternative will be valid for a wide range of reliability values. The example of a design study, given in section 4, illustrates the large steps in reliability that arise with alternative schemes.

The situation may also arise in which a consumer, and in particular an industrial consumer, finds his reliability of supply unsatisfactory and is prepared to provide the investment necessary for the provision of a better supply. Frequently, however, an investment of this type will improve the reliability of supply to consumers other than the consumer who financed the improvement. Some allowance should be made, in estimating the payment required, to account for the benefits to these other consumers and a social-benefit costing may be used for this purpose.

2.1 Social benefit calculation vs supply revenue calculation.

It is necessary to decide the basis on which a reliability valuation is to be made. One form of calculation would be a simple commercial calculation performed solely from the view-point of the supplier's economics.

Given the reliability parameters, system loads, unit cost of electricity to the distributor and unit cost to the consumer it is a simple matter for the supplier to estimate the expected loss of revenue from a particular design scheme. Even the supplier's economics, however, are appreciably more complicated than this simple calculation would indicate. A low supply reliability will undoubtedly influence the decision of the consumer in the choice of energy supply he will employ in future expansions. (A large proportion of energy demand is transferable in this manner, for example, space heating, furnaces, ovens, many welding applications etc.). Very little is known about the relationship between supply unreliability and loss of growth and, at present, it would not be possible to introduce this factor into a valuation of reliability.

An alternative form of reliability valuation would be by means of a social-benefit type calculation. Social-benefit calculations have been suggested for use in similar public service industries: for estimating the social benefit of constructing an underground railway[2] and for investment studies in public water supply.[3] In this context, the basic principle of a social-benefit costing is that the costs and revenues which result to all members of the community, and not just to the investor, are considered in the calculation.

In the case of supply system reliability the reasons for preferring a social cost-benefit to a single commercial cost-benefit study may be summarised as follows:-

(i) at present price levels a single commercial costing would justify, from the reliability point of view, only the most rudimentary of systems.

(ii) investment in improved reliability is frequently large and usually provides an improved supply to more than one consumer.

(iii) the major benefits from a reliability investment accrue directly to the consumer and not to the supplier who makes the investment.

On superficial examination the first condition might appear to be a somewhat a priori condition since it is the object of a costing study to obtain the most economically suitable type of system and if the study indicates a very rudimentary type of system then this indication should be accepted. It might be argued that a more reliable system would be justified if the price of electricity supplied by the system were increased. This would not, however, mean that the additional investment is justified. The price of electricity is determined not to any great extent by a market structure but by the costs involved and the rate of return required on the capital employed. One, not inconsiderable, factor which determines the costs and the amount of capital employed is reliability. Hence, a general price level can only be obtained once a general reliability level has been allocated.

The method of social costing to determine optimum reliability levels may be defined as the maximization of the net social benefit from an investment where the net benefit has been defined (2) as "the predicted quantified addition to national wealth of the investment". The demand for electrical energy is closely correlated with national product and hence, if growth is a managerial objective in electricity supply, a social costing of reliability may not be incompatible with the long-term commercial objectives of the supply authority. Indeed if a consumer behaves rationally with regard to his future expansions a social benefit study may not be incompatible with a commercial costing study which includes the effect of future sales loss from the unreliability of the system.

2.2 Commercial calculation

There are rather special situations where a commercial type calculation is the appropriate form of calculation for the evaluation of reliability. These cases are normally restricted to distribution systems which are owned and operated by the consumer. Systems of this type are frequently found in large industrial organisations where the system under study is entirely internal to the organisation. The designer of these "industrial systems" is in a greatly advantageous position compared with the designer of a public supply system since he is more likely to have a reasonably full knowledge of the economics of the process which are dependent upon the supply.

One published example of a commercial type-valuation of reliability has been given [4] for a large distribution system which is internal to an oil-refinery. The value of reliability is estimated from financial losses which result from a failure and is obtained from three components:-

(i) A fixed cost for all failures except those of very short duration (i.e. of duration long enough to have a significant effect upon production). This factor covers costs incurred by damage to plant and damage to the product being processed.

(ii) A cost proportional to the length of the outage. This factor covers the costs incurred by loss of production during the outage.

(iii) A second fixed cost which is proportional to the start-up time of the plant. This factor covers the loss of production during the re-starting of the process after restoration of the supply.

Clearly, this type of valuation is much more precise than would be expected from the costing of a general class of consumer by a public supply authority. Even a costing of this form, however, does not include valuation of those "intangibles" such as customer dissatisfaction resulting from late delivery of the manufacture.

2.3 Valuation by simple input-output analysis

One method for the valuation of reliability of supplies to industrial consumers has been suggested by Sheppard.[5] The method employs the premise that when supply to an industrial consumer fails the employees are rendered idle and the actual financial loss to the consumer is the sum of the wages paid to these non-productive workers. Calculations on this basis lead to valuation of reliability of the order of 60 times the price of the energy not supplied.

This method, however, appreciably underestimates the true financial loss to the consumer since two considerable factors are ignored. The cost of a manufactured article (i.e. the financial return to the producer) may be considered composed of four basic constituents namely:-

(i) The cost of input materials (including energy costs)
(ii) The cost of labour
(iii) The fixed costs which include capital and depreciation charges on buildings and plant, rates, etc.
(iv) The net profits of the producer (i.e. after allowance for capital and depreciation charges).

Failure of supply causes the producer to lose not only the expenditure on labour but also the costs involved in idle equipment and the value of the profits which might be expected to result from operation of the production activity. Thus the potential earnings need to be included in the calculation.

The value of the reliability of supply to industrial consumers can be simply calculated, as the "value added" by the industry per KWhr consumed, by considering the input-output statistics of the industries. Table 1, based upon the census of production [6] (1958), shows the value of reliability of supply obtained using this simple principle for various classes of industry. For comparison the values of reliability as predicted by using only the wages paid are also given in the Table. In common with the method of Sheppard the most drastic assumption which needs to be made in the valuation of reliability by this method is that during the period of supply failure all production is irretrievably lost. The validity of this assumption will depend upon two main factors:-

(i) The degree to which the process is dependent upon the supply.

(ii) The ability of the process to recover lost production after supply restoration.

Clearly the degree of dependence on supply will vary greatly from industry to industry and indeed from firm to firm within an industry. At one extreme most "production line" firms (e.g. vehicle and textile production) will be brought to a complete standstill by a supply failure whilst many construction projects will continue with little or no disturbance. (in 1958 few temporary electricity supplies were provided on construction sites and this accounts for the very high value added per unit

Standard Industrial Classification		Electrical Energy Consumption (GWh.) **	Purchases of Materials & Fuel (£M) *	Wages Paid (£M) *	Total Sales (£M) *	Value Added per kWh Consumed (£/kWh)	Wages Paid per kWh Consumed (£/kWh)
XVII	Gas, water, etc., (excl. electricity)	1450	250	111	473	0.154	0.077
101,102	Coal Mining	4170	196	529	820	0.149	0.127
103,109	Other mining and quarrying	486	39	37	112	0.171	0.076
XIII	Bricks, pottery, glass, cement, etc.	2270	257	175	592	0.147	0.077
311-3	Iron and Steel	4780	1129	301	1757	0.131	0.063
321,322	Non-ferrous metals	1030	417	81	560	0.139	0.079
IV	Chemicals & Allied industries	6890	1510	291	2310	0.115	0.042
VII	Shipbuilding & marine engineering	607	217	173	496	0.460	0.285
VI	Engineering, etc.	3790	1650	1058	3470	0.480	0.280
VIII	Vehicles	2240	1338	533	2233	0.400	0.238
IX	Miscellaneous metal industries	1790	708	255	1183	0.265	0.142
X, XI, XII	Textile leather & clothing	2610	1648	609	2771	0.430	0.233
III	Food, drink & tobacco	1970	2936	367	4260	0.675	0.185
XV	Paper, printing & publishing	1260	566	337	1256	0.548	0.268
XVII	Construction	176	1051	882	2780	10.200	4.450
XVI	Other industries	2810	303	133	543	0.085	0.047
	TOTAL (ALL INDUSTRIES)	33500	14800	6144	26830	0.360	0.180

Table 1. Valuation of Reliability for Industrial Consumers

Sources:- ** Electricity Council Annual Report 1963/64
* Census of Production - Summary Tables 1958

424

consumed shown in Table 1. The construction industry
today, however, is much more dependent upon continuity
of supply).

The ability of the consumer to recover lost production will depend to a large extent upon the amount of
'slack' in the production process. A firm whose production facilities are fully employed will find it difficult to recoup any lost production whilst a firm which
operates only one shift can recover the loss by overtime
working at the main additional cost of the overtime wages.
Indeed an inefficient firm may be capable of recovering
lost production without recourse to overtime. For the
efficient firm which is highly dependent upon supply,
the valuation of reliability given by this method will
be most applicable.

A further important factor which must be borne in
mind if these reliability values are to be used for design
purposes is that of the time of day and day of the week
on which the failure occurs. Many industrial organisations work only one shift per day and the effect of a
supply failure outside the hours of work will have little
direct effect upon production. Supply failure will affect,
for example, the space heating within a factory but if
the duration of the failure is not longer than one or
two hours this is unlikely to present intolerable working conditions during the following shift.

There are of course many cases where supply failure
may have more drastic effects than merely loss of production. The majority of these problems may be summarised
as:-

(i) Damage to plant or product.
(ii) Very long delay in restarting the process.

In nylon spinning, for example, even a very short supply
interruption can cause irreparable damage to a valuable
quantity of product. Also, in the petrochemical industry,
short failures can require a very long start-up time with
a consequently high loss of production.

In situations where the above conditions apply the
simple input-output valuation will give misleading
results but in the majority of cases will probably
provide a useful indication.

2.4 Other Possible Approaches

The methods given in sections 2.2 and 2.3 are applicable primarily to the valuation of supply to industrial consumers. The input-output method could be applied to certain commercial consumers such as stores and cinemas but here the nuisance element in supply failure tends to predominate and for other commercial consumers, such as offices and banks, it is difficult to define any input and output statistics. None of the methods suggested so far are applicable to domestic consumers.

Some other possible methods of reliability valuation are given below:-

(i) <u>Survey of Consumer Contingencies</u>. One possible method of valuation is to determine the value which the consumer himself places on supply failure by examining the costs he is willing to incur in order to mitigate the effects of supply failure. Industrial and commercial consumers install private emergency generation equipment or batteries for use in the event of a supply failure. On a less conscious basis, domestic consumers frequently have available sources of heat other than from electricity and store candles for emergency lighting. A survey of the cost of these emergency services could give an indication of the value which the consumer himself places on reliability. This, however, would probably be of little use as a basis for design. Usually the consumer makes his provisions with little or no quantitative knowledge of the reliability of his supply. The same provisions and in similar circumstances will probably be found over a wide range of supply reliabilities.

(ii) <u>Market Survey Experiments</u>. Market survey techniques have been used for assessing the value to the community of many commodities. It is possible that these techniques may be of assistance in deciding a reliability value, particularly to domestic consumers.

(iii) <u>Value of Leisure</u>. The central problem in assessing the value of domestic consumer reliability hinges on the assessment of the monetary value of comfort and leisure. This problem is basic to most social-welfare calculations and as yet has received no satisfactory general solution.

Those members of the community who are provided with the opportunity of working voluntary overtime, in a sense establish their own value of leisure time. Overtime will be marked only whilst the incremental increase in income is greater than the incremental value of the leisure time which has to be foregone. However, even if this marginal value of leisure could be readily calculated, the following two drastic assumptions would need to be made before this value could be used as a basis for reliability valuation:-

(a) That the enjoyment of all leisure time in a household is completely ruined by a supply failure.
(b) That all members of the household (not just the wage earners) have known marginal values for leisure.

These assumptions are of questionable validity.

(iv) <u>Committee action</u>. Because of the many factors which affect the reliability costs and since there are many indications which may be of assistance in evaluating these costs some form of committee decision has obvious advantages over the use of any single technique.

In Sweden the Committee system has already been used[13] to assign a mean national figure for reliability value and at present is evaluating the values of supply to various classes of consumer. The mean value which has been agreed upon corresponds to 7.5p per kW of load lost plus 15p for every kW-hour of unsupplied energy.

The interesting point has been made [8] that if these Swedish figures are applied to the United States north-eastern blackout in 1965 an economic loss of $80 million is obtained. The Federal Power Commission Report on the blackout states that "estimated economic losses run as high as $100 million". The north-eastern failure covered a wide diversity of consumers and since the consumptions of electrical energy per capita in Sweden and the U.S.A. are comparable (equating some measure of the dependence of a community on electrical supply with its consumption) the close agreement between the two estimates is not altogether surprising.

3. A METHOD FOR COMPARING RELIABILITY COSTS WITH INVESTMENT

The costs incurred through the unreliability of a system are a stream of randomly valued costs which are themselves randomly distributed in time throughout the working life of the system. The distribution of each cost is dependent upon the distribution of system restoration times and on the reliability cost function. The distribution of these costs in time is determined by the distribution of the times between system failures.

In the economic design of a system, on a reliability basis, it is necessary to compare the investment costs incurred at the time of construction of the system with the unreliability costs which result from that system over its working life. For such a comparison it is necessary to make allowance for the "time preference" of money and the method of "Present Value"[9] is employed. Thus, a cost R, say occurring at a time t after the construction of the system is considered to be equivalent to a cost of:-

$$\frac{R}{(1+i)^t}$$

known as the present value, occurring at the time of inception of the system. The quantity i is known as the discount rate. The present value of each cost in the stream may be calculated in this manner and the sum of these present values provides a single cost occurring at the time of construction of the system which is equivalent to all the costs incurred by the system throughout its working life.

The System lifetime to be used for design studies depends upon the particular type of system and 20-40 years covers most power system studies. However, because of the discounting factor, costs which are incurred in the later part of the system life contribute less to the total present value than costs incurred in the earlier part and thus the choice of system lifetime is not critical. For the purpose of reliability studies the stream of costs may be taken in perpetuity with the introduction of little error.

In this case where the costs and times are random variates the mathematical expectation of the total present value of the costs will be taken as the basis for the comparison of alternative schemes. The following analysis is based upon the times between the occurrance of successive costs being exponentially distributed and the values of the costs are taken as distributed according to some general form.

3.1 Basis of the Analysis

Consider the stream of n costs: $R_1, R_2, \ldots, R_n$ and let the time of occurrence of R_1 be t_1 and the time between the occurrence of R_1 and R_2 be t_2, etc., Thus, the n th cost will occur at time T_n where:

$$T_n = t_1 + t_2 + \ldots + t_n$$

and the present value, P_n, of the n th cost is given by:-

$$P_n = \frac{R_n}{(1+i)^{T_n}} \qquad \ldots (1)$$

The total present value, P, of a perpetual stream of costs is given by:-

$$P = \sum_{n=1}^{\infty} \frac{R_n}{(1+i)^{T_n}} \qquad \ldots (2)$$

Since R_n and T_n are random variables the present value, P, will also be a random variable, and from the definition of expectation it is known[8] that $E(X \pm Y) = E(X) \pm E(Y)$, Equation (2) leads to:-

$$E(P) = \sum_{n=1}^{\infty} E\left[\frac{R_n}{(1+i)^{T_n}}\right] \qquad \ldots (3)$$

There is assumed to be no dependence between the time of occurrence of a cost and the magnitude of that cost. Thus, equation (3) may be written:

$$E(P) = E(R) \sum_{n=1}^{\infty} E\left[(1+i)^{-T_n}\right] \qquad \ldots (4)$$

By expanding, in series form, the expression $(1+i)^{-x}$ and comparing the terms of the expansion with those of the expansion of e^{-ix} it is seen that for small ix the discounting factor may, to a good approximation, be taken as e^{-ix}. When i = 10%/annum and x = 10 years the difference between the two terms is of the order of 4%. Using this approximation the expression for the present value of the costs reduces to the analytically more amenable form:-

$$E(P) = E(R) \sum_{n=1}^{\infty} E\left[e^{-iT_n}\right] \qquad \ldots (5)$$

From the definition of the expectation of a function[8]:-

$$E\left[e^{-iT_n}\right] = \int_0^{\infty} e^{-iT_n} P_n(T_n) dT_n$$

where $P_n(T_n)$ is the probability density function of T_n.

This integral is equivalent to the Laplace Transform of $P_n(T_n)$ with the Laplace Operator, s, replaced by the discounting rate. Thus, the above integral may be written as:-

$$E\left[e^{-iT_n}\right] = L\left[P_n(T_n)\right]_{s=i} \qquad \ldots\ldots (6)$$

The time T_n to the occurrence of the n th cost is given by the sum of n independent random variables each with probability density function $f(t) = \lambda e^{-\lambda t}$ where $1/\lambda$ is the mean time between the occurrence of successive costs. The probability density function $P_n(T_n)$ is obtained from [10] from the n-fold convolution of $f(t)$ and may thus be expressed in Laplace Transform terms as:-

$$P_n^*(s) = \left\{f^*(s)\right\}^n \qquad \ldots\ldots (7)$$

Substituting for $f^*(s)$ in equation (7) and substituting (7) in (6) gives:-

$$E\left[e^{-iT_n}\right] = \left\{\frac{\lambda}{\lambda + i}\right\}^n \qquad \ldots\ldots (8)$$

Denoting the mean cost by R, the expectation of the present value of the costs is obtained from equation (5) as:-

$$E(P) = R \sum_{n=1}^{\infty} \left\{\frac{\lambda}{\lambda + i}\right\}^n = \frac{R\lambda}{i} \qquad \ldots\ldots (9)$$

Thus the expected present value of costs may be simply obtained from the mean cost and the mean time to system failure. This present value cost is independent of the form of the distribution of the costs but expression (9) has been developed on the basis of exponentially distributed times between the occurrence of costs.

Simulation tests have shown[11] that the assumptions made regarding system lifetime and use of the exponential discounting function are for all practical purpose realistic.

4. AN EXAMPLE OF RELIABILITY DESIGN

A simple example of the reliability design of an 11 kV distribution system has been performed the results of which are given in this section. The example is intended to indicate a simple design study, by examination of alternative schemes, which may be performed entirely by hand calculation.

4.1 The Hypothetical load distribution

A simple hypothetical load distribution is taken as the basis of the study. The load is assumed to have a peak value of 120 MVA and to be uniformly distributed over a circular region of area 24 sq. miles. A bulk supply point is assumed at the centre of the region and power is distributed at 11 kV to blocks of 500 kVA where it is transformed and distributed to individual consumers at medium (415 V/240V) voltage. The assumed shapes of the blocks of 500 kVA of load are indicated in Fig. 1. The 11 kV/415V sub-stations are taken to lie approximately at the centroids of the load blocks. It is assumed that the 11kV cables are laid in straight lines between sub-stations.

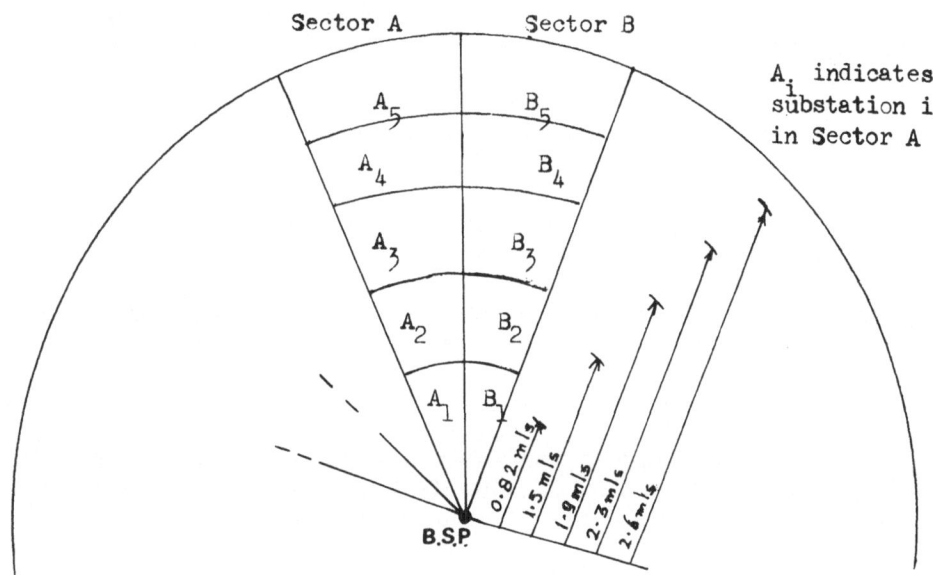

Fig.1 Diagram of Hypothetical Load Distribution

4.2 Simple radial system without alternative MV supplies.

The most elementary scheme which may be used for the supply of the five blocks of loads in each sector of the load model is the simple radial network shown in Fig. 2.

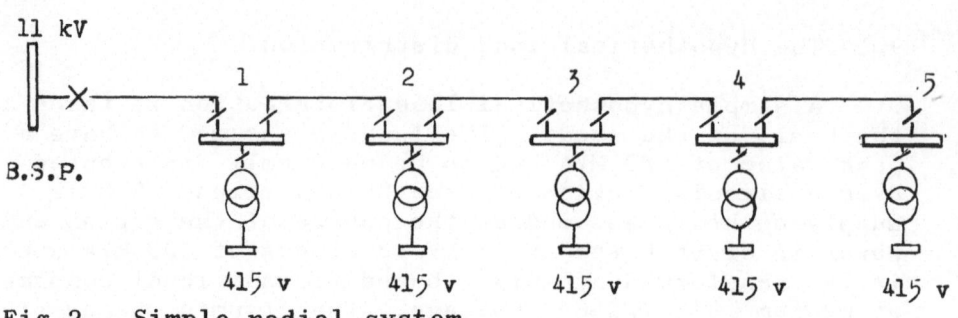

Fig 2.   Simple radial system

All switches on this system, except the main feeder oil circuit breaker, are considered to be non-automatic and require manual operation for either opening or closing. Failure of any 11 kV component causes the main breaker to open and the entire radial is out of service until manual switching takes place. After switching, the consumers on the side of the fault remote from the supply point require the completion of repairs before the restoration of their supplies.

The assumed failure rates and mean repair times of the components are shown in Table 2. The switch failure rates lead to an annum failure rate of equipment at substations 1 to 4 of 0.0047/substation and for substation 5 (which has one less switch) of 0.0038/substation. A mean 11 kV switching time of 1 hour is assumed.

For the supply failures at all the supply points a total present value of the MW-hrs lost is calculated by using the expression (9) and assuming unit cost for every MW-hr of unsupplied energy. The total present value of all the MW-hrs lost by the system is the criterion used as the basis for comparing alternative design schemes.

A reliability analysis of this simple radial system was performed and Table 3 shows the mean supply failure rate and mean restoration time for each substation on the radial. Table 4 shows the contribution of each of the components in the system to the supply parameters and to the total present value of the MW-hrs lost.

The main salient feature of this type of system which is evident from Table 4 is the proportion of MVA-hrs loss caused by cable failure. Only 1% of the present value of MW-hrs lost is attributable to switchgear, and transformer failures.

Equipment	Failure rate	Mean repair time*
Cables	0.1/mile/yr	36 hr
Oil immersed isolator or switch	0.0009/yr	4.7 hr
Transformer	0.002/yr	10 hr
Busbars	Assumed zero	

Table 2. Component Parameters[5]

* These times are taken as the mean time between the commencement of repair (i.e. after completion of switching) to the restoration of the component to service.

Sub-station no	Supply failure rate (/yr)	Mean restoration time of supply (hrs)
1	0.286	11.1
2	0.286	19.6
3	0.286	25.0
4	0.286	29.6
5	0.286	33.9

Table 3. Supply Reliability Parameters for Simple Radial System.

The operational inflexibility of this system is emphasised by the long restoration times shown in Table 3.

A minor modification to this type of system which might be considered is to replace the oil-switch connected in the transformer "T-off" with either an oil-circuit breaker or a switch fuse. An obvious effect of this modification is to remove the necessity for manual switching to restore the other supplies on the radial when a transfomer failure occurs. Unfortunately, both switch fuses and oil circuit breakers have lower reliabilities than oil immersed isolators. Sheppard[5] quotes a typical overall failure rate for a switch fuse of 0.008/unit/annum and 0.003/unit/annum for an oil

Equipment	Failure Rate (/yr)	Average MVAh lost/fault*			Present Value of MVAh lost**
		Restored by switching	Restored by Repair	Total	
Cable 0-1	0.082	0	54	54	55.4
" 1-2	0.069	0.3	43.2	43.5	37.6
" 2-3	0.044	0.6	32.4	33	18.2
" 3-4	0.037	0.9	21.6	22.5	10.4
" 4-5	0.031	1.2	10.8	12	4.7
S/G at sub-s 1	0.0027	0	7.1	7.1	0.2
" 2	0.0027	0.3	5.6	5.9	0.2
" 3	0.0027	0.6	4.2	4.8	0.2
" 4	0.0027	0.9	2.8	3.7	0.1
" 5	0.0018	1.2	1.4	2.6	0.1
Transformer 1	0.002	1.2	3	4.2	0.1
" 2	0.002	1.2	3	4.2	0.1
" 3	0.002	1.2	3	4.2	0.1
" 4	0.002	1.2	3	4.2	0.1
" 5	0.002	1.2	3	4.2	0.1
				Total =	128

Table 4. Contributions of Component Failures for the Simple Radial System.

* Assuming an average load of 60% rated value (i.e. average of 300 MVA in each block).
** Using a discounting rate of 8%/annum.

circuit breaker. Though a precise analysis of switchgear reliability should include a breakdown of this failure rate into components to cover failure of the switchgear to operate when required, operation of the switchgear when not required and static failure of the switchgear when operation is not involved; these figures may be used as a useful illustration. Table 5 shows the supply reliability, on the low voltage side of the transformer, at each of the 11 kV/415 V substations for a radial system employing switch fuses. A similar result is given in Table 6 for a system employing oil circuit breakers. Comparing Tables 3 and 5 it is seen that the effect of introducing switch fuses is to increase the system failure rates and to somewhat reduce the restoration times. The total present value of the losses resulting from the radial with switch fuses is approximately 129 MVA-hrs and thus according to this measure the system reliability is slightly reduced by investing in switch fuses. A similar, though less pronounced, result is obtained for the system employing oil circuit breakers. The present value of losses for the system with oil circuit breakers is approximately 128 MVA-hrs.

Sub-station no	Supply failure rate (/yr)	Mean restoration time of supply (hrs)
1	0.313	10.4
2	0.313	18.1
3	0.313	23.2
4	0.313	27.5
5	0.313	31.0

Table 5. Reliability of Radial System Employing Switchfuses.

Sub-station no	Supply failure rate (/yr)	Mean restoration time of supply (hrs)
1.	0.288	11.1
2	0.288	19.5
3	0.288	24.9
4	0.288	29.5
5	0.288	33.6

Table 6. Reliability of Radial System Employing OCB's

4.3 Simple Radial with Alternative MV-supplies.

The main disadvantage of the system of Fig 3 is that in the event of a cable or switchgear failure it is necessary for the repair to be completed before supplies to loads situated on the side of the fault remote from the bulk supply can be restored. This disadvantage is minimised by the provision of an alternative source of MV-supply to the loads from each of the substations.

Fig 4 illustrates the assumed mode of MV-interconnection for blocks of load in the hypothetical load distribution. The blocks of load are normally supplied independently from one substation. Thus, for example, substation A_5 supplies only the shaded region shown. In the event of the loss of supply to an 11 kV substation the block of load is considered to be split into two equal portions. Each of these portions is then manually linked to the nearest MV-network fed from a different radial. For example, if supply is lost to substation B3 the portion of the load X normally fed from B_3 will be fed from A_3 and the portion Y will be fed from C_3.

Since the MV linking operations will involve the linking of several MV feeders for every block of load which needs to be restored, the mean time for the restoration of MV supplies will clearly be a function of the magnitude of the load to be restored. For the purpose of this example the mean restoration time of MV-supplies is assumed to be 1 hour per 500 kVA of capacity to be

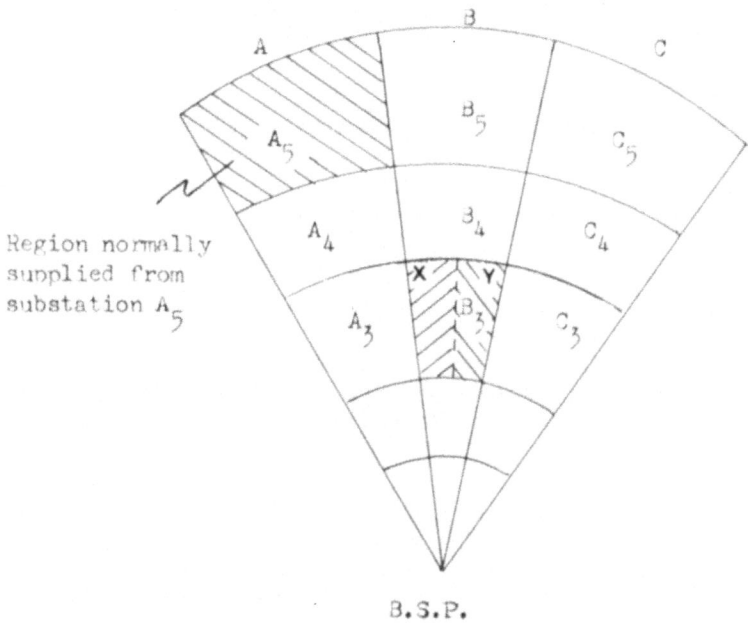

Fig. 4 Representation of MV-Interconnection Scheme

restored. Linking is considered not to begin until 11 kV switching operations on the faulted 11 kV radial or associated Transformer have been completed.

If a radial has substations which for some reason are unavailable then it will be impossible to provide alternative MV supply to an adjacent radial if a fault occurs on the latter. In this case, either repair will need to be completed on a faulted radial or the alternative source of MV supply will have to be made available before the restoration of all supplies. An alternative MV-source of supply may be unavailable for two reasons:-

(i) Because of a fault on its 11 kV feeder or transformer

(ii) Because of maintenance on its 11 kV feeder or transformer

A further source of supply failure is the failure of the alternative MV-source when it is in use as the source of supply.

A maintenance frequency of 0.3/annum for each of

the 11 kV 415V substations with a mean duration of 5 hours is assumed. Any unreliability in the MV system is not considered.

Table 7 shows the analysis of system switching times for the failure of components on the radial and Table 8 shows the failure rates and mean restoration times for each of the substations on the radial. Table 8 is obtained from the switching times alone and the rates and restorations times, seen in Table 9 to be negligible, for which both supply via the radial and supply via the MV interconnection are unavailable are ignored.

The additional costs incurred in providing this type of system over the simple radial of section 4.2 arise from:-

(i) Increasing the load-carrying capacity of the 11 kV radials to enable them to carry the additional load when an adjacent radial fails. Since the blocks of MV-load from each 11 kV substation are divided into two halves and fed from separate radials the capacities of the 11 kV cables need to be increased by 50%.

(ii) Increasing the capacity of each 11 kV/415V transformer and associated MV cabling. For the reasons outlined in (i) above, transformer capacity needs to be increased by 50%. The total MV cable capacity will need to be increased, however, by 100% since, usually, separate MV cables will be used to feed the two halves of each block of load.

(iii) The need to provide short additional cabling and link boxes to facilitate MV-interconnection between adjacent load blocks. The cost of these additional cables and link boxes is shared between the two radials which are thus interconnected.

By comparing tables 4 and 7 it is seen that this minor modification to the simple radial system reduces the present value of the MW-hrs lost to approximately 11% of the value for the simple radial. A discussion of the reliability savings and the investments is given in section 4.6.

439

Equipment	Failure Rate (/yr)	Loads Restored by 11kV switching		Loads Restored by MV Linking		MVAh Lost Per Fault	Present Value of MVAh Lost
		S/S No.	Time hr	S/S No	Time Hr		
Cable 0-1	0.082	-	1	1,2,3,4,5	1,2,3,4,5	6	6.15
1-2	0.069	1	1	2,3,4,5	1,2,3,4	4.5	3.9
2-3	0.044	1,2	1	3,4,5	1,2,3	3.3	1.8
3-4	0.037	1,2,3	1	4,5	1,2	2.4	1.1
4-5	0.031	1,2,3,4	1	5	1	1.8	0.7
S/G at .1	0.0027	-	1	1,2,3,4,5	1,2,3,4,5	6	0.2
2	0.0027	1	1	2,3,4,5	1,2,3,4	4.5	0.15
3	0.0027	1,2	1	3,4,5	1,2,3	3.3	0.11
4	0.0027	1,2,3	1	4,5	1,2	2.4	0.08
5	0.0018	1,2,3,4	1	5	1	1.8	0.04
X-former 1	0.002	2,3,4,5	1	1	1	1.8	0.05
2	0.002	1,3,4,5	1	2	1	1.8	0.05
3	0.002	1,2,4,5	1	3	1	1.8	0.05
4	0.002	1,2,3,5	1	4	1	1.8	0.05
5	0.002	1,2,3,4	1	5	1	1.8	0.05
					TOTAL=		14.5

Table 7 Analysis of Switching Times for Radial System with MV Interconnection.

Substation No	Supply Failure Rate (/yr)	Mean Restoration Time of Supply (hrs)
1	0.286	1.26
2	0.286	1.85
3	0.285	2.55
4	0.286	3.44
5	0.286	4.18

Table 8. Supply Reliability Parameters for Radial System with MV Interconnection.

Substation No	Rate of Occurrence (/yr)	Mean Duration (hrs)
1	0.0004	4.7
2	0.001	5.8
3	0.002	6.2
4	0.003	6.3
5	0.004	6.4

Table 9. Simultaneous Failure of both 11kV and MV Alternative Supplies for the Radial System with MV Alternative.

4.4 Open-ring without alternative MV supply.

The main disadvantage of the configuration of the system of section 4.3 is the relatively long linking times required under certain fault conditions. For faults on a radial which occurs close to the point of supply a large amount of load needs to be transferred by manual linking and, as shown in Table 4, loads at the remote end of the radial have long restoration times.

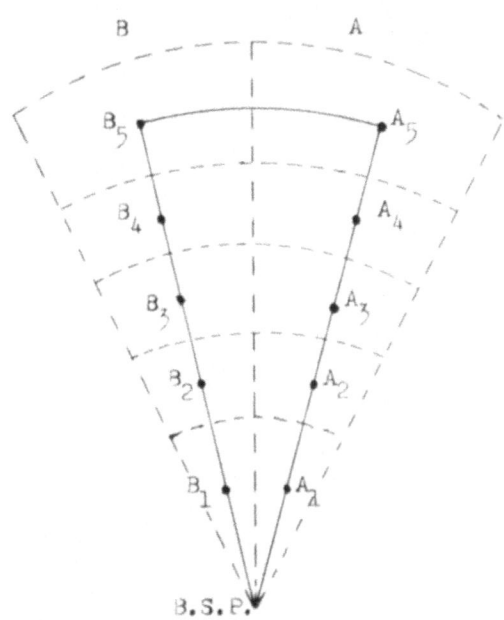

Fig 5. Open-Loop system supplying two sectors of load.

This disadvantage is to a large extent mitigated by the use of an open loop system. Two adjacent sectors of load supplied from one loop system are illustrated in Fig 5. The two radials are provided with a means of 11 kV interconnection via a cable joining substation A_5 and B_5. For reasons of lower fault level and simplicity of protection, however, the two radials are operated independently under normal conditions by maintaining one of the switches on the interconnector at either A_5 or B_5 in the open position. In the event of a fault on one of the radials the faulty section is manually isolated, loads on the supply side of the fault are fed, as usual, directly from the bulk supply point and loads on the side of the fault remote from the supply point are fed, via the interconnector, from the adjoining sector. For cable faults, all loads can be restored entirely by 11 kV switching. For switchgear or transformer faults, however, completion of repair is necessary before the load at the faulted sub-station can be restored.

In estimating the reliability performance of this system the failure of the interconnecting cable needs to be included. The costs associated with interconnector failures will be considered equally shared between the two radials since, at any one time, the interconnector is electrically connected to only one of the radials but at all times performs the same service of interconnection to both radials.

The mean 11 kV switching time will be increased slightly by the introduction of the interconnector but this is not included in the calculation. The reliability of the interconnection facility is obtained from:-

Failure rate of interconnector (length 0.34 mls) = 0.034/annum

Mean time for which interconnection is available =

$$\frac{0.034 \times 36 + 0.286 \times 33.9}{0.32} = 34 \text{ hrs}$$

Table 11 shows the analysis of the system switching times and the system switchgear and transformer repair times necessary for the restoration of all supplies. In Table 10 is shown the results of an analysis of the rates and times for which both directions of supply around the loop to a particular substation are unavailable This type of system does not permit maintenance of switchgear and transformers without interruption of supplies and since in computing Table 10 maintenance outages have been ignored the parameters for joint failure are somewhat artificial. Preventive maintenance may have impact upon the system in one of the following ways:-

(i) Preventive maintenance is not performed and there may be an increase in component failure rates.

(ii) Prearranged outage of consumers is undertaken to permit maintenance. These outages must have a cost attached to them but this cost will be somewhat less than the cost of unarranged outages.

Taking no account of maintenance implications the total present value of the MW-hrs lost by each sector is 6.49.

The additional costs of providing this open-loop system as opposed to the simple radial system are incurred by:-

(i) Providing the additional cable between Substations A_5 and B_5.

SUB-STATION No.	UNAVAILABILITY RATE FROM BSP (/yr)		UNAVAILABILITY RATE VIA INTERCONNECTOR		UNAVAILABILITY TIME FROM BSP (hrs)		UNAVAILABILITY TIME VIA INTERCONNECTOR		JOINT FAILURE		
	FORCED*	MAIN-TEN-ANCE	FORCED*	MAIN-TEN-ANCE	FORCED*	MAIN-TEN-ANCE	FORCED*	MAIN-TEN-ANCE	RATE (/yr) $\times 10^{-4}$	TIME (hr)	P.V. OF MWh LOST
1	0.082	0	0.504	0	36	–	34.4	–	3.3	17.6	0.02
2	0.154	0	0.43	0	35.4	–	34.3	–	5	17.4	0.03
3	0.20	0	0.385	0	35.2	–	34.4	–	6	17.4	0.04
4	0.24	0	0.345	0	35	–	34.5	–	6.6	17.4	0.04
5	0.27	0	0.311	0	34.8	–	34.6	–	7	17.4	0.04

Table 10. Simultaneous Failure of Both Directions of Supply for Substations on a Ring.

EQUIPMENT	FAILURE RATE(/yr)	LOADS RESTORED DIRECTLY FROM B.S.P.	LOADS RESTORED VIA INTERCONNECTOR	MANUAL SWITCHING TIME(hr)	REPAIR TIME BEFORE RESTORATION (hr.)	MWh LOST PER FAULT	PRESENT VALUE OF MWh LOST
Cable 0-1	0.082	-	1,2,3,4,5	1.0	0	1.5	1.54
1-2	0.069	1	2,3,4,5	1.0	0	1.5	1.29
2-3	0.044	1,2	3,4,5	1.0	0	1.5	0.83
3-4	0.037	1,2,3	4,5	1.0	0	1.5	0.69
4-5	0.031	1,2,3,4	5	1.0	0	1.5	0.58
S/G at sub-s 1	0.0027	-	2,3,4,5	1.0	4.7 for 1	2.9	0.1
2	0.0027	1	3,4,5	1.0	4.7 for 2	2.9	0.1
3	0.0027	1,2	4,5	1.0	4.7 for 3	2.9	0.1
4	0.0027	1,2,3	5	1.0	4.7 for 4	2.9	0.1
5	0.0027	1,2,3,4	-	1.0	4.7 for 5	2.9	0.1
Transformer 1	0.002	2,3,4,5	-	1.0	10 for 1	4.5	0.11
2	0.002	1,3,4,5	-	1.0	10 for 2	4.5	0.11
3	0.002	1,2,4,5	-	1.0	10 for 3	4.5	0.11
4	0.002	1,2,3,5	-	1.0	10 for 4	4.5	0.11
5	0.002	1,2,3,4	-	1.0	10 for 5	4.5	0.11
Interconnector	0.034	1,2,3,4,5	-	$1.0 \times \frac{1}{2}$	0	0.8	0.34

TOTAL = 6.32

Table 11 Analysis of switching and repair times for the open loop system

(ii) Increasing the load carrying capacity of each radial to permit it to carry the full load of the other radial in addition to its own load.

4.5 Open-ring with alternative MV Supplies.

The main disadvantage of the elementary open-ring system of section 4.4 lies in its operational inflexibility. As noted previously it is not possible to perform maintenance on switchgear or transformers without interrupting the supplies to some consumers. Further, it is necessary to complete repair on faulted switchgear or transformers before all supplies can be restored. These disadvantages may be overcome by the provision of MV interconnection between 11 kV substations. The scheme of interconnection is assumed to be the same as that described in section 4.3.

The Table 12 shows an analysis of the 11 kV switching times and the MV linking times involved in the restoration of supplies. In Table 13 is shown the results of an analysis of failure of both sources of 11 kV supply to each of the substations. Finally in Table 14 is shown the parameters for failure of all sources of supply, both HV and MV, to each portion of load. It is seen that those portions of load whose alternative MV supply is derived from a separate ring to its 11kV supply are much more reliable than those portions whose MV is derived from the same ring.

Summing the results shown in Tables 12 and 14 gives a total present value of MV-hrs lost by each sector of 5.86

It is seen from Table 14 that the chance of permanently losing all sources of supply is very small and when the MV alternative is derived from a completely separate ring the loss of all supplies is quite negligible.

4.6 Comparison of system costs.

The systems analysed in Sections 4.2 to 4.5 have varying supply reliabilities and cost varying amounts to implement. An improvement in reliability will only be justified if the investment cost incurred in providing that investment is exceeded by the value of the improved

Table 12 Analysis of switching times for open-ring system with MV interconnections

EQUIPMENT		FAILURE RATE (/yr)	LOADS RESTORED FROM B.S.P.	LOADS RESTORED VIA INTERCONN-ECTOR	11kV SWITCHING TIME (hrs)	MV LINKING TIME (hrs)	MWh LOST PER FAULT	PRESENT VALUE OF MWh LOST
Cable	0-1	0.082	–	1,2,3,4,5	1.0	0	1.5	1.54
"	1-2	0.069	1	2,3,4,5	1.0	0	1.5	1.29
"	2-3	0.044	1,2	3,4,5	1.0	0	1.5	0.83
"	3-4	0.037	1,2,3	4,5	1.0	0	1.5	0.69
"	4-5	0.031	1,2,3,4	5	1.0	0	1.5	0.58
S/G at sub-s	1	0.0027	–	2,3,4,5	1.0	1 hr for 1	1.8	0.06
"	2	0.0027	1	3,4,5	1.0	1 hr for 2	1.8	0.06
"	3	0.0027	1,2	4,5	1.0	1 hr for 3	1.8	0.06
"	4	0.0027	1,2,3	5	1.0	1 hr for 4	1.8	0.06
"	5	0.0027	1,2,3,4	–	1.0	1 hr for 5	1.8	0.06
Transformer	1	0.002	2,3,4,5	–	1.0	1 hr for 1	1.8	0.04(5)
"	2	0.002	1,3,4,5	–	1.0	1 hr for 2	1.8	0.04
"	3	0.002	1,2,4,5	–	1.0	1 hr for 3	1.8	0.04
"	4	0.002	1,2,3,5	–	1.0	1 hr for 4	1.8	0.04
"	5	0.002	1,2,3,4	–	1.0	1 hr for 5	1.8	0.04
Interconnector		0.034	1,2,3,4,5	–	$1.0 \times \tfrac{1}{2}$	0	0.8	0.34
							TOTAL =	5.80

447

SUB-Station No.	UNAVAILABILITY RATE FROM BSP (/yr)*		UNAVAILABILITY RATE VIA INTERCONNECTOR*		UNAVAILABILITY TIME FROM BSP (hrs)*		UNAVAILABILITY TIME VIA INTERCONNECTOR*		JOINT FAILURE		
	FORCED	MAIN-TEN-ANCE	FORCED	MAIN-TEN-ANCE	FORCED	MAIN-TEN-ANCE	FORCED	MAIN-TEN-ANCE	RATE (/yr) $\times 10^{-4}$	MEAN TIME (hr)	PRESENT VALUE OF MWh LOST
1	0.082	0	0.504	2.7	36	5	34.4	5	4.6	13.8	**
2	0.154	0.3	0.43	2.4	35.4	5	34.3	5	8.1	12.9	**
3	0.20	0.6	0.38	2.1	35.2	5	34.4	5	9.5	12.9	**
4	0.24	0.9	0.35	1.8	35.0	5	34.5	5	11	12.3	**
5	0.27	1.2	0.31	1.5	34.8	5	34.6	5	11	12.0	**

Table 13. Failure of both 11kV Supplies to Substations on an Open-Ring with MV Interconnections.

* These quantities are for permanent outages, i.e. after completion of switching operations.

** Supply may be available via the MV interconnection.

SUB-STATION NO.	MV ALTERNATIVE DERIVED FROM THE SAME RING			MV ALTERNATIVE DERIVED FROM AN ADJACENT RING		
	FAILURE RATE (/yr)	MEAN DURATION (hrs)	PRESENT VALUE OF MWh LOST	FAILURE RATE (/yr)	MEAN DURATION (hrs)	PRESENT VALUE OF MWh LOST
1	0.6×10^{-4}	18	1.9×10^{-3}	5×10^{-10}	8.3	0.8×10^{-8}
2	2.5×10^{-4}	15.1	7.1×10^{-3}	12×10^{-10}	8.1	2×10^{-8}
3	4.7×10^{-4}	14.0	1.2×10^{-2}	17×10^{-10}	8.0	3×10^{-8}
4	7.3×10^{-4}	13.4	1.8×10^{-2}	20×10^{-10}	8.0	3×10^{-8}
5	10×10^{-4}	12.9	2.4×10^{-2}	21×10^{-10}	8.0	3×10^{-8}
		TOTAL	6.3×10^{-2}		TOTAL	1×10^{-7}

Table 14. Failure of all Supplies both MV and 11kV for Substations on an Open Ring with MV Interconnections.

reliability. Using the present value concept as the basis for the reliability assessment the values of each kW-hr not supplied will be determined at which various of the system improvements may be considered justified.

Most of the additional costs incurred in providing progressively more reliable systems, for example additional or larger 11 kV cables, greater transformer capacity etc., can be estimated by a simple commercial costing. The assessment of the cost of providing MV interconnection between blocks of load, however, presents difficulties since in the load model no assumptions have been made concerning the layout of the MV distributors. In practice, the MV cabling costs will depend to a much greater extent than the 11 kV cabling on the street layout and the geographical configurations of the consumption points.

The costs incurred in providing MV interconnection arise mainly from uprating the MV cables and small additional costs in providing short cable lengths and linking boxes to connect with adjacent blocks of MV network. It will be assumed that excavating, jointing, reinstatement costs, etc., are independent of the size of the distributor. It has been stated(12) that the cost of providing an MV supply to a new house may be broken down approximately as:-

~~~
Cable      £20
Labour     £18
Jointing   £14
Sundries   £1

Total      £53
~~~

It will be assumed that the introduction of an MV alternative causes the equivalent cost per house of MV network to be increased by a further 150%. Thus the cost per house of providing MV cable where an alternative is not provided may be broken down approximately as:-

~~~
Cable      £8
Labour     £18
Jointing   £14
Sundries   £1

Total      £41
~~~

Now, taking the after diversity maximum demand for a typical consumer as $3\frac{1}{2}$ kVA the number of consumers in a block of load fed from one 11kV/415V substation would be approximately 140. The additional cost incurred in providing the alternative MV supply is thus of the order of £1,700 for each block of 500 kVA of load.

The cost of the 11kV/415V transformers is assumed to be linearly related to the transformer capacity as:-

        500 kVA   £600

        750 kVA   £800

       1000 kVA   £1000

Using these figures as a basis, comparisons of the various types of system are readily made as follows:-

(i) <u>Comparison of simple radial and radial employing Transformer Breakers</u>. In section 4.2 it was shown that introduction of switch fuese or oil circuit breakers to protect the 11 kV/415V transformers marginally increases the total present value of MW-hrs lost. Thus, from a reliability point of view this system modification is not justifiable. It is possible that the rapid action of an oil circuit breaker or the current-limiting effect of a switch fuse will significantly reduce the extent of the damage to a faulted transformer but since repair of a faulted transformer usually involves replacement of a complete coil it is doubtful whether this is a justification for using the more costly switchgear.

(ii) <u>Comparison of a simple radial and radial with MV alternative</u>. Since in the system with MV alternative the 11 kV cables have to carry only 50% additional load following the loss of an adjacent feeder it can be seen that an 0.15 in^2 aluminium cable is sufficient for both the simple radial and the radial with MV alternative. The additional costs arise solely from larger transformers and larger MV distributors and for each sector of load total approximately £9,500. The reduction in present value of KW-hrs which results from this investment is 114,600 and thus this additional investment will be justifiable, in terms of improvied reliability, if the value of reliability is greater than 8.3p per kW hr. of unsupplied energy.

(iii) <u>Comparison of simple radial and elementary open-ring</u>. The additional costs incurred in providing better reliability by the formation of an open ring from two adjacent radials lie solely in the 11 kV cabling. The two sections of 11 kV cable on each of the radials nearest to the point of supply need to have their sizes increased from 0.15 in aluminium to 0.3 in^2 aluminium. A length of 0.343 miles of 0.15 in^2 cable also needs to be added to form the ring from the two radials. The cost of laying this additional cable is taken as £3.3/yard[12]. Assuming that excavating, laying, jointing, reinstatement, costs etc., are independent of cable size the cost of increasing cable size from 0.15 in$_2$ aluminium to 0.3 in^2 may be taken as £0.68/yard [12]. Two additional switches are also required at the ends of the interconnector but the cost of these is assumed to be negligible compared with the additional cable costs.

The total additional investment cost amounts to approximately £5,800 and there is a reduction of 122,600 kW-hrs. in the present value of unsupplied energy. Thus this additional investment would be justifiable for reliability valuations of greater than 4.8p/kW-hr.

(iv) <u>Comparison of Elementary open-ring with alternative MV</u>. The additional costs of providing the MV alternative to supplies fed from the open-ring are the same as those for providing the MV alternative in (ii) above, and total £9,500. There is a reduction of 630 kW-hrs in the present value of unsupplied energy and thus the additional investment would be justifiable for reliability values in excess of approximately £15/kW-hr.

5. CONCLUSIONS

The attempts to value reliability, discussed in Section 2, tend broadly to indicate values within the range 15p to £1/kW-hr. of unsupplied energy. For this particular example all valuations within this range would lead to the decision that a more reliable system than the simple radial is justified. The open-loop system without MV alternative supplies and the more reliable radial system with alternative MV supplies are both justifiable on a reliability basis. It appears, however, that, on the basis of this reliability analysis above, the open-ring system with MV alternative cannot be justified.

The analysis given indicates how the reliability criterion may be introduced quantitatively into the design problem. General design policies cannot, however, be drawn from this simple example. Account has not been taken of the unreliability of the MV equipment and introduction of this factor will tend to promote those systems which include MV alternatives. The load density will clearly have a great effect upon the choice of the optimum design scheme. Load densities much greater than that used for this example will frequently be encountered and more reliable systems will frequently be justifiable since it will be cheaper to provide alternative supplies to the load points.

REFERENCES

1. Edwards, R., 'The Electricity Supply Industry - The Present Position and Aims for the Future', Conference of Electrical Development Association 1964.

2. Foster,C.D., Beesley, M.E.,'Estimating the Social Benefit of Constructing an Underground Railway in London', Journal of the Royal Statistical Society, Series A, 126 Pt.I, 1963.

3. Hirshleifer, J., DeHaven, J.C., Millman,J.U, 'Water Supply: Economics, Technology and Policy' (Book), University of Chicago Press, 1960.

4. Dickinson, W.H.,'Economic Evaluation of Industrial Power System Reliability', Trans. A.I.E.E., Pt II, Nov. 1957.

5. Sheppard,H.J., 'The Economics of Reliability of Supply Distribution (Great Britain)', Conf. on the Economics of Reliability of Supply, Sept 1967, I.E.E. Conference Publication No. 36.

6. Board of Trade, Census of Production 1958 - Summary Tables.

7. The Electricity Council, Annual Report and Accounts 1963/4.

8. Von Mises, R.,'Mathematical Theory of Probability and Statistics', (Book) Academic Press, 1964.

9. Merrett,A.J, Sykes,A.,'The Finance and Analysis of Capital Projects', Longmans, 1965.

10. Cox, D.R,'Renewal Theory', (Book) Metheun & Co, 1962.

11. Wakeman,C.J.'Reliability Analysis of Power System Networks', Ph.D Thesis, University of London,1969.

12. Barnes, C.C.,Hill,E.,Sutton,C.T.W.,'Cables and Accessories: Some Trends in Standardisation and Ratings", Proc.I.E.E., April 1969.

13. Jancke, G., 'The Swedish Transmission System', Conference on the Economics of Reliability of Supply, Sept 1967, I.E.E. Conference publication No. 34.

RELIABILITY, REDUNDANCY AND CASCADED PRIORITIES IN PHYSIOLOGICAL SYSTEMS

DR. ROBERT E. C. WEAVER
DEPARTMENT OF CHEMICAL ENGINEERING
TULANE UNIVERSITY
NEW ORLEANS, LOUISIANA 70118 USA

Reliability analysis maintains in many respects its early orientation toward maintenance of a minimum performance level at minimum cost and maximum time between failures. Failure in turn is considered to be the "absolute failure" to meet a unique performance criterion and is usually presumed to be characterized by statistically defined frequencies. System reliability probabilities are then constructed from these stochastic failure frequencies of the system components along with similar stochastic descriptions of the demands upon the system.

The resulting technology for reliability assessment provides a valuable methodology for choosing among different levels and patterns of component redundancy, for evaluating the incentive for higher grade components and for prescribing policies for replacement and repair. The requirements for many fixed-mission vehicles and processes are conveniently addressed in this framework to be sure, but the technology responds less well for the operations management of a number of systems of developing engineering interest, prominently those involving an interface with the physiology of living systems. This essay seeks to suggest focus for such reliability analysis effort upon particular areas of agricultural production management and upon diagnosis and prescription of therapy protocol for the human physiology.

These opportunities for physiological management can be succinctly characterized in the present context in terms of their operation within a scheme of cascaded priorities wherein policy changes are invoked "on-line" in response to the system state.

The elegant system which has evolved in the human physiology and the curiously related processes that are evoked in the scientific management of agricultural production show little simple standby redundancy that is not used for enhanced performance (when not called upon for basic survival). There is a bases here to inspire as well as to challenge the more generalized concepts in design and analysis for reliability. In both these life-system contexts, it is inappropriate to ride with any fixed design (however well evolution may have fitted the system with redundancy and versatility) and thus simply accept actuarial failures. Instead the very concept of failure is replaced by one of challenge which generally operates within the following guidelines:

1. Corrective action other than replacement is acceptable and is typically accomplished along alternate pathways for vital functions which are exercised during stress at some cost in overall performance.

2. There is a continual drive toward performance enhancement leading to both rehabilitation and to growth whenever there is spare capacity.

3. There are definite indices of system health--system state variables which can be and are monitored as a basis for control. These characteristics are interestingly enough those used in engineering circles to characterize "adaptive control", to-wit; the alteration of control through a diversity of control levels and pathways (1) which are called upon in response to a performance index (2) operating on measurements of the state of the system's health and reserve capacity (3).

The existence of this parallel unfortunately does not immediately open up as productive a bag of tricks as might be hoped for because evolution of any really attractive general adaptive control methodology has been slow indeed. With appropriate deference to such schemes as model-reference adaptive control, most such designs have foundered on their demand for technological capabilities of some consequence including prominently the following:

1. A reasonably good simulation treatment for the system is needed: Processes with the complexity to justify an interest in adaptive control are often non-linear and of high interactive dimension. Even the definition of resident degrees of freedom is no idle specification (as the distillation designer can attest). Further some measure of system compaction is usually called for to obtain a practical treatment.

2. Measurement technology and concomitant analysis capabilities have developed slowly along such varied lines as classical regression methodology (non-linear programming), Kalman/Wiener filtering and deconvolution schemes such as quasilinearization and it is difficuilty tor example to find a system identification textbook or treatise encompassing the breadth of the field that the applications engineer must concern himself with.

3. Even the objective function synthesis which provides the basis for recognizing heirarchic levels of control has not been addressed adequately, probably because not enough specific systems have been treated in explicit enough fashion to provide a basis for generalization.

Nonetheless all of these functions can be expected to be a part of most schemes for engineering the recruitment of system capability from lower priority functions during stress and the return after rehabilitation to reserve development and enhanced performance. Workers in adaptive control can profit much from the perspective of the reliability analyst and can in return bring to reliability assessment some of the methodology that is likely to be appropriate in assessing highly-tuned and self-regenerative systems. Several specific efforts at addressing the performance of living systems will now be sketched by way of specific illustration.

ANALYSIS OF CROP PRODUCTIVITY

The course of the International Biological Program reflects how the systematic analysis of agricultural operations is moving to a reasonable level of determinism in at least six major crop systems. The goal of the program is to arrive at some scientific basis for integrated crop and pest management. The impact of so many substantially stochastic inputs (the climatological vectors and pest import boundary conditions) makes this no simple simulation programming study; at the same time the inapplicability of simple redundancy or other common devices of the reliability engineer makes the problem intractable to a simple stochastic requirement/availability analysis.

As a starter, the basis for a surprisingly deterministic description of the plant physiology does appear to be at hand. The cotton production system with which our laboratories are currently involved builds upon the successful compilation of experimental evidence on growth and maturation of the cotton plant (in a USDA-"SIMCOT" simulation). Briefly, a plant map is constructed in the simulation in which a pattern of "physiological days" is experienced en route to successful fruiting.

The effects of accompanying abscission of cotton squares (buds) through drought or nutritional deficits and of pest damage are simulated along with the impact of various control measures. Viables models also exist for parallel treatment of soil mositure profiles, seedling emergence, effects of weed competition and various cultivation practices, the parallel growth of insect pest populations (given boundary conditions) and finally the logistics of harvesting and ginning. The reliability engineer would find familiar ground in the models' characterizations using density functions (number of plants vs. potential cotton yield per plant) but will find a fit challenge in accounting for the time evolution of these profiles in response to both the stochastic meteorological effects and of course the operational controls to be applied. In this particular case, the total number of plants represents total vagetative growth which is related to ultimate yield but not in a single valued relationship. This points up the need to recognize that the time course of these non-linear processes is itself an ingredient in the final state. But nonetheless a basis for making these projections involving the overlay of various distributions does probably exist in data from an increasingly sophisticated agriculatural research community.

ANALYSIS OF CIRCULATORY CHALLENGE

The results of circulatory stress arising out of hemorrhagic shock and severe muscular demand on a somewhat weakened heart in the human physiology is another critical problem for which reliability technology may be coming of age. The scenario runs something as follows: Following cardiac insufficiency, cerebral deficiencies in oxygen and perfusion pressure lead to vasoconstriction in the renal, hepatic and cutaneous circulations. The reduced renal flow in turn leads to fluid accumulation and the compensatory development of stronger cardiac function through the Starling reserve (wherein cardiac contractility responds up to a point to filling pressure). But the systemic pressure development in hypertension leads to some respiratory discomfort syndromes, periferal edema and ultimately renal damage as well as the obvious loss of reserve for subsequent challenges.

One is dealing here with stochastic inputs of various intensities which challenge a system able to manage initial insults with remarkable resilience, but for which there is long term consequence when the system cannot be nursed back to satisfactory normal performance levels. A particular clinical history will illustrate the elusive character of reliable interpretations: Recently a patient arrived at the hospital with advanced circulatory congestion which made respiration so difficult as to require him to sleep upright. A young intern prescribed a diuretic which in effect reduced the congestion through fluid removal. The patient felt much relief in his pulmonary function. When an "old hand" at the hospital chastened the young intern (who to be sure was following the textbook), the patient **rallied** to his defense attesting that he had not felt better in years. To prove it, he got out of bed and strode across the room. Even this exercise proved trying for his system and he collapsed in the face of a cardiac insufficiency. As troublesome as the congestion was, it was still the physiological vehicle for maintaining a requisite stimulus to the heart!

Interestingly enough, physiological simulation has proceeded to a point where the above history was consistently portrayed in the Tulane version of PHYSBE (a Windkessel model with control and pulmonary/renal detail). Parenthetically, this model has also suggested that left heart function alone can maintain the system reasonably adequately paralleling considerable clinical evidence that the trauma of open-chest surgery is not to be justified in seeking to correct right heart failure.

Students of the control of the circulation see the curious heirarchy of control mechanisms which collectively account for the system's dramatic ability to respond to severe stress. These have been portrayed in the aforementioned simulations essentially as follows:

Blood pressure must be maintained both to implement the capillary perfusion mechanism (representing a beautiful interplay between hydrostatic and osmotic driving forces) as well as simply to impel flow. The arterial pressure is therefore sensed in baroreceptors in the aorta and carotid sinus, these signals being integrated in a mid-brain circulatory center which moderates the tone of the parasympathetic and sympathetic branches of the autonomic nervous system. The parasympathetic innervation serves primarily as a brake on heart rate and is stimulated by hypertension. Hypertensive signals simultaneously have an inhibitory effect on the sympathetic system which provides stimulus to ventricular contractility, arterial tone and venous compliance through neurochemical mediators (the catecholamines) which are also secreted by the adrenal medulla to provide a parallel but longer term humoral action. Along with this pressure regulatory system, other mid-brain activity directs respiratory function based on stimuli from chemoreceptors responsive to oxygen partial pressure and the CO_2/pH levels in the plasma bicarbonate buffer system. The cerebral circulatory shunt (as well as respiratory rate) receives direction from this control center. Local vasocontrol is also exhibited at the arterioles of the major capillary beds in a manner which is responsive to pressure, oxygen levels and metabolite levels. Finally, the fluid balance in this contained compliant system is regulated through the kidney function which processes nearly a liter/minute (of the total circulation level of 6 liter/minute). Practically all of the contained glucose and aminoacids and nearly 99% of water and salt in a 100 ml/minute filtrate are reabsorbed in a delicate interplay of pressure and osmotic potentials with a biological pump wherein middle molecular weight metabolites are removed with little loss of either low molecular weight nutrients or high molecular weight proteins which are returned to the system at venous pressures. When unimpeached by vasoconstriction from a hypotensive stress, the kidney further exhibits the remarkable resilience of glomerulo-tubular balance wherein reabsorption rate increases virtually proportionately with filtration load and sodium recovery increases also with the corresponding filtrate concentration.

The point being made is the following: Circumstances frequently compel the use of prosthetic (dialysis) or pharmocological (diuretic administration) treatments for early term relief which can in turn invoke longer term imbalances in the naturally tuned mechanism. The clinician wants to respond to the patient's anxious query as to his chances with a reasonably reliable recognition of protective restraints that should be observed during recovery.

The relevant reliability analysis will recognize that one is facing a set of boundary conditions (the extracorporeal interface) which are stochastic, but that the physiology itself must be examined fairly deterministically. The amalgamation of boundary condition effects with system properties may be tenable for linear systems, but for a coupled non-linear system the confounded product will have little interpolative value. Since the number of modes of failure have strikingly different time constants and are interactive, they resist any but the grossest of actuarial statistics and it is literally fatal to simply accept these.

THE INTERFACE WITH RELIABILITY ENGINEERING

This essay has specifically sought to encourage more specific consideration of how current reliability technology can be used and then developed further to address both the farm manager's decisions on keying fertilizer and pesticide applications with climatological inputs and other external vectors and also the clinician's problems in assembling a therapy protocol that acknowledges at least the major risks in circulatory trauma.

A logical place to start is to recognize the technology which may be considered to be state-of-the-art in the reliability engineering field today. Almost anyone's list would acknowledge the following which draw substantially from Green and Bourne's significant text.

The reliability designer turns to increasing success probabilities most often through the parallelisms of redundancy. These can take many forms operating on majority-vote, on simple veto, or even averaging and system capacity can be either continually engaged with a reserve or simply stored in standby. The redundancy preferably has some versatility also in the interest of reliability. Time domain analysis for regenerable systems has been also recognized as desirable and a straightforward synthesis of mean failure rates and related downtime or availability factors has been formulated for use in series with the probabilities for the external forcings.

In assessing existing systems for reliability, the basis for overall system probability of success is constructed by comparing required performance (the environmental component) with achieved performance of the system. Straightforward formulations have been assembled for example for the prospect of bounding a given operational function (expressed as a frequency distribution depending on **system** state) between required upper and lower limits. This involves the capability of combining distributions into composite values. The impact of fault frequencies and repair times have also been incorporated into analysis of these substantially open loop problems.

Our brief review of the characteristics of physiological systems have hopefully indicated how one is faced with inherently closed loop systems with the following features which pose challenges for simple application of existing reliability technology.

1. The performance analysis of living systems must recognize that there is a cascade of priorities ranging from the survival level to rehabilitation and high performance development which

variably guide the control or resource allocation during stochastic stresses of occasional great intensity. The recruitment of "availability" then is highly correlated (via adaptive mechanisms) with "requirements" contrary to the independence usually accorded these functions in the aforementioned reliability assessment procedures.

2. The stochastic component for the illustrative life-systems presented is primarily tied to a credible characterization of meterorological factors (for crop studies) and of human stress factors (of the sort that surely are available from NASA flight studies) and a generally acceptable data base needs to be assembled in these areas apart from the physiological systems themselves.

3. Simulation bases for the physiology do seem to exist for interfacing the impact of the stochastic forcings in the closed-loop, nonlinear and interacting systems. Any knowledge that can be gained on multiple entry forcing into non-linear system dynamics will certainly be of considerable value here. Needless to say, this recognition of the value of any progress toward generalization in non-linear dynamics is not new in any way.

In all, the reliability analysis of these life-systems appears to overlay significantly with efforts toward adequate adaptive control design. This field in isolation has encountered much frustration due as often as not to inadequate attention to objective function definition. A shot-gun marriage of adaptive control research with the reliability discipline in the interest of treating living systems should prove a happy one.

DISTRIBUTION OF TIME-TO-DAMAGE METHOD IN RELIABILITY OF PIPES*

S. A. Wilson, Statistician

General Electric Company
Nuclear Energy Division
San Jose, California, U.S.A.

1. ABSTRACT

    The "Distribution of Time-to-Damage Method" for reliability
    estimation is described. The method relies on knowledge of
    a damage prediction equation, and the uncertainties in the
    variables appearing in the equation. The uncertainties may
    be subjective judgments as well as proven or known distribu-
    tions. For illustration purposes, the method has been used
    to predict the probability of growing a crack through the
    wall of a cylindrical pipe in a chosen time period, due to
    low-cycle fatigue. For this mechanism, the variables of
    materials, applied stress, and initial crack condition have
    been considered. Results are in terms of probability of
    leak due to low-cycle fatigue in a 40-year plant lifetime,
    at the one cross section having highest calculated stress in
    a reference piping system. Results include the relative con-
    tribution of the variance of each participating variable to
    the variance of cycles to leak, implying approximate relative
    contribution to the probability of leak. Analytical tech-
    niques include the structuring of subjective input frequency
    distributions for physical variables, propagation of errors,
    and Monte Carlo using importance sampling to find the ordi-
    nate of a tail of small probability, such as 10^{-2}, in the
    distribution of a function.

---

*This work was supported by the U.S. Atomic Energy Commission
under contract AT(04-3)-189, Project Agreement 37, Reactor Primary Coolant System Rupture Study.[1] Superscript numbers identify references following the text.

2. THE DISTRIBUTION OF TIME TO DAMAGE METHOD

The subject method was devised because of lack of adequate historical failure rate information. It relies on design uncertainties. It is outlined on Figure 1 and is described as follows:

a. Draw a probability diagram. One for piping is illustrated in Figure 2. The diagram shows a block for each stage of damage, such as crack initiation, leakage and severance. Lines connect the blocks to show the logic of how one stage of damage can lead to another. Three ideas relate to each line: (1) a damage mechanism, such as low cycle fatigue, (2) a time-to-damage prediction equation, such as for the number of cycles to grow a crack from initial depth through to leakage, and (3) the probability of this event in a chosen time, such as 40 years. Thus many lines connect each block, one for each damage mechanism.

b. Devise a damage prediction equation for each line, i.e., for each mechanism for each stage of damage. Figure 3 shows the basic damage prediction equation used for this illustration, for cycles to leak, N_w.

c. Choose a piping study section, and assign a nominal value and an uncertainty distribution to each variable

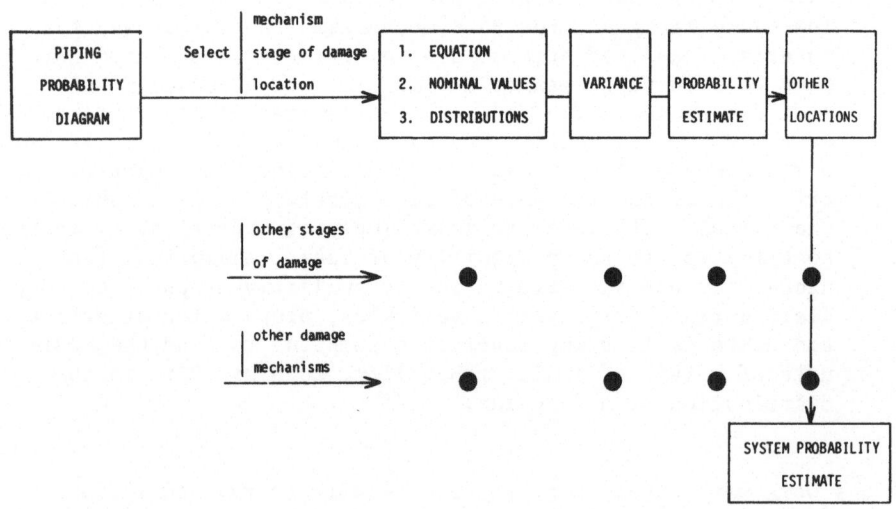

Figure 1. Distribution of Time to Damage Method

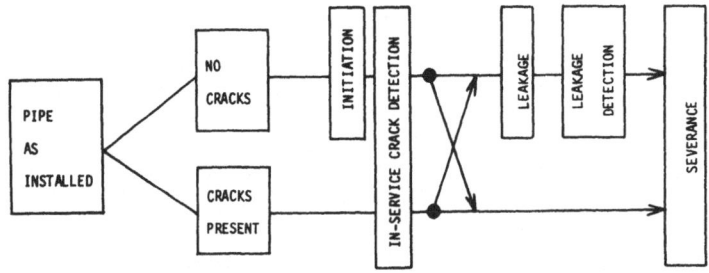

Figure 2. Piping Probability Diagram
For One Pipe Study Section and Damage Mechanism

in the equation. Nominal values used for this illustration, along with sources of nominal values and distributions, are shown in Table 1. Note that design values, results of experiments, and engineering judgment and experience were all drawn upon. Thus this method uses the best available information about how well the actual value of a variable is believed to agree with its nominal value.

d. Select only those variables making the major contributions to the variability of the equation, e.g., of N_W. This is done by a propagation of errors estimate of the variance of the equation shown in Figure 4.

e. Form the distribution of the equation from the nominal values and distributions of its variables. Or only estimate the fraction of the equation distribution which lies less than 40 years, as illustrated in Figure 5. This fraction is the desired probability and will be estimated by Monte Carlo using importance sampling.

Thus, in a systematic way, the "Distribution of Time-to-Damage" method calls for interrelations among stages of damage to be diagrammed, damage mechanisms to be identified, and design uncertainties to be carried ahead by a damage prediction equation to estimate the probability of moving from one stage of damage to the next.

There are additional benefits of this reliability estimation method. The time to damage equations form the basis for rational design equations. And the greatest contributors to lack of reliability, both in terms of damage mechanisms and variables within equations, can be identified for more exact definition.

Basic crack growth rate equation:

$$da/dN = A (\Delta K)^B \text{ inches/cycle}$$

Linearize for parameter estimation:

$$\log da/dN = \log A + B \log \Delta K$$

where da/dN = crack growth rate

ΔK = stress intensity factor range, a function of $\Delta s_{PM}\sqrt{a}$ where Δs_{PM} is stress and a is crack depth

log A and B = intercept and slope parameters

Mean-adjusted form to make A and B independent:

$\log da/dN = \log C + B [\log \Delta K - (\text{mean} \log \Delta K)]$ where log C is the intercept at mean log ΔK. For simplicity, only log A is referred to in the text as the intercept.

In original formulation:

$$da/dN = C (\Delta K/\overline{\Delta K})^B$$

where $\overline{\Delta K}$ is antilog mean log ΔK.

Invert and integrate for cycles to grow from initial crack depth, a_o, through wall thickness, w, to leakage:

$$N_w = \frac{(\overline{\Delta K})^B}{C} \int_{a_o}^{w} \left(\frac{1}{\Delta K}\right)^B da, \text{ cycles to leak}$$

N_w = cycles to leak

a_o = initial crack depth

w = wall thickness

Linear expression found to be nearly exact:

$$\log N_w = 27.4687 - 1.0084 (\Delta \log C)$$
$$- 4.5069 (\log \Delta s_{PM}) - 6.2615 \sqrt{a_o/w}$$
$$- 2.47 e^{-B} - 0.03 B,$$

where $\Delta \log C$ is the random change in intercept from the nominal intercept for the material and temperature shown in Figure 4 and Δs_{PM} is primary membrane stress range appearing in ΔK,

$\Delta s_{PM} = s_E \times s_E^* + $ hoop stress/2. Log is to base 10.

Figure 3. Low Cycle Fatigue Crack Growth Rate Equation

Table 1

NOMINAL VALUES USED, AND SOURCES OF NOMINAL VALUES AND DISTRIBUTIONS

Variable		Nominal Value Used	Source		
			Design Value	From Test Data	From assigned Uncertainty Distribution
Slope parameter, B	For SS 304 at R.T.	4.45		Nom. Distr.	
Parameter, A (Intercept is log A)		5.5×10^{-26}		Nom. Distr.	
Expansion stress, s_E, psi		27450	Nom.		Distr. (S_E^* or S)
Wall Thickness, w, in.		0.432	Nom.		Distr.
Diameter, d, in.		6.624	Nom.		Distr.
Pressure, P, psig		1000	Nom.		Distr.
Initial crack depth ratio, a_o/w		*			Nom. and Distr.
Mean crack occurrence rate, $\bar{p}$, cracks/in^2		0.1			Value assg.

*0.036 without, 0.0146 with ultrasonic testing, arising from the distributions shown in Figures 5d and 5e

Nom. = Shows source of nominal value used

Distr. = Shows source of uncertainty distribution used.

$$\sigma_{N_W}^2 \doteq \left(\frac{\partial N_w}{\partial A}\right)^2 \sigma_A^2 + \left(\frac{\partial N_w}{\partial B}\right)^2 \sigma_B^2 + \text{term for each variable}$$

for each variable independent.

Figure 4. Propagation of Errors
Approximate Variance of a Function, $\sigma_{N_w}^2$

The distribution of time to damage method is similar to methods known as "stress vs. strength,"[2] and "requirement vs. capability,"[3] in that areas under continuous probability distributions are used to estimate reliabilities.

The crack growth stage of low cycle fatigue was chosen for first application of this method because of its importance and the technical feasibility of proposing a cycles-to-leak equation.

The following sections describe in more detail steps c, d, and e above on assigning uncertainty distributions, finding the equation variance and making the Monte Carlo probability estimate.

3. NOMINAL VALUES AND DISTRIBUTIONS OF THE VARIABLES

The piping study section chosen was the one subject to highest design stress in a reference piping system. The surface area of the study section was 1 inch in length by 1/4 circumference, the latter chosen as the portion in tension from the expansion stress.

Table 1 defines the variables and shows the source of the design value and distribution of each. Figures 6 - 10 show the distributions of those variables found by the variance

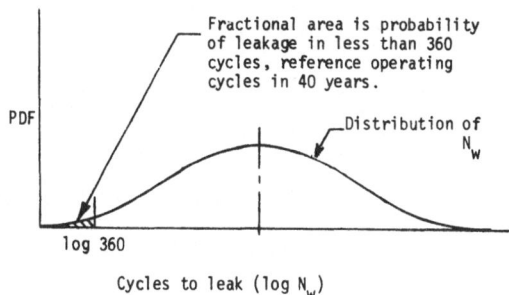

Figure 5. Probability of Leakage in 360-Cycle Service Life-Time as the Tail of a Distribution of Cycles to Leak

solution - described in the next section - to make the substantial contributions to the variance of cycles to leak, and thus to the probability of leakage. The following paragraphs describe each variable.

Parameters A and B: The low cycle fatigue testing from which the mean and variance of parameters A and B were estimated was conducted under another part of the program. The values estimated are for a particular material, temperature level, and loading pattern. The zero-to-tension loading pattern was chosen as most characteristic of piping loading between non-operating and operating conditions. The distribution used for parameter A is shown in Figure 6.

s, w, d and P: The assigned uncertainty distributions for s, w, d and P were for the ratio: actual value/design value. Thus they are coefficients on the design values, and have values in the vicinity of 1. A distribution was sketched for each cause of uncertainty which was believed important such as, for expansion stress: stress analysis model error, computation error, installation arrangement error, restricted flexibility error, etc. Also, for each cause, a conditioning probability was assigned to reflect the probability with which

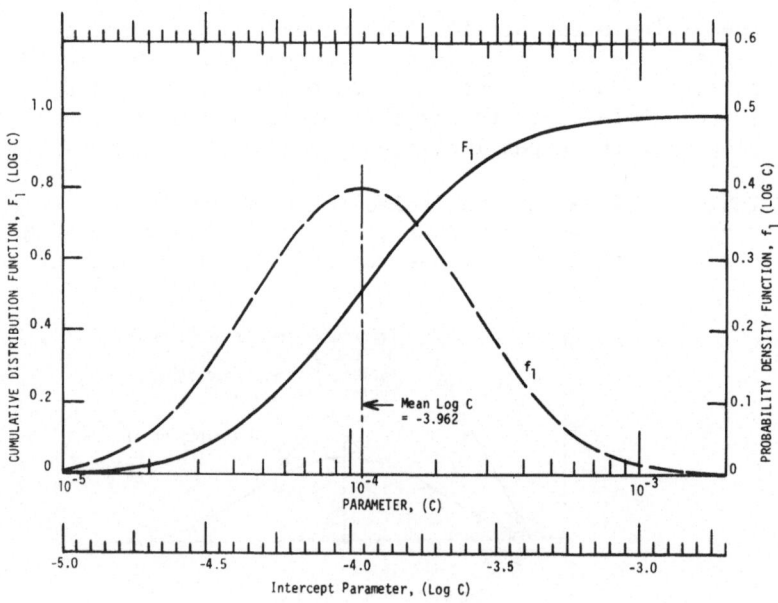

Figure 6. PDF and CDF of Intercept, Log C (= Log A + 21.30)

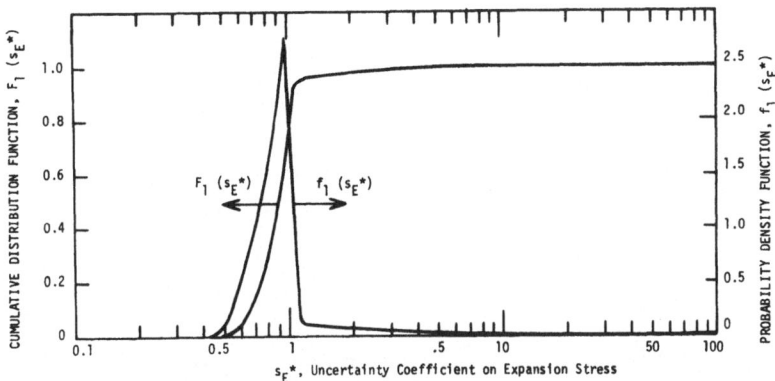

Figure 7. CDF and PDF of s_E^*

each cause was judged to occur in a piping loop, joint, etc. Conditioning probabilities were assigned for two cases: with and without checking and inspection throughout the design and fabrication cycle. The distribution used for s is shown in Figure 7.

a_0/w: The assigned uncertainty distributions for initial crack depth ratio, a_0/w, are on a scale of that ratio, as shown in Figure 8. Two depth ratio distributions were considered: with and without ultrasonic testing. In addition, crack frequency was assumed to be Poisson distributed, with an assigned mean crack occurrence rate of r per square inch. Since the deepest crack in a region of otherwise uniform properties will be the first to leak, we were interested in the distribution of maximum a_0/w in the chosen study surface, which can be derived from the Poisson and a_0/w distributions. The resulting distributions for maximum a_0/w are shown in Figures 9 and 10.

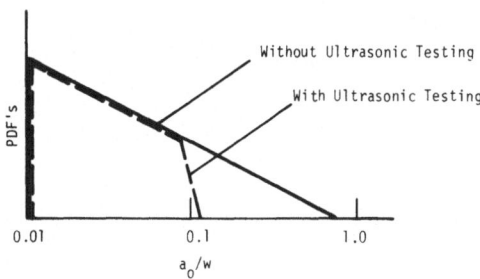

Figure 8. Judgmental Uncertainty Distributions for a_0/w, Initial Crack Depth Ratio

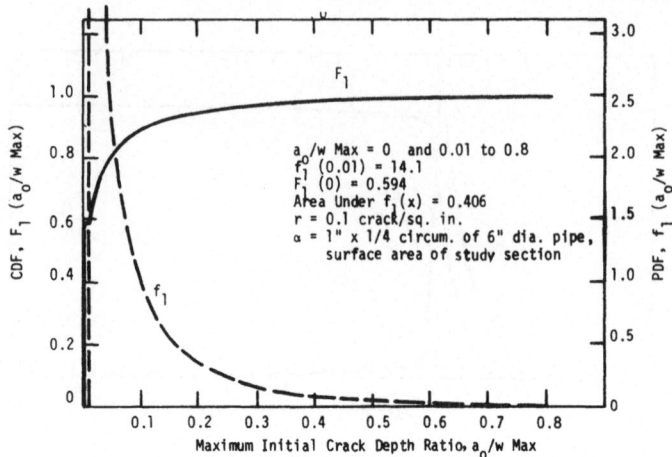

Figure 9. PDF and CDF of a_o/w Max Without U/S Test

For both the variance solution and the Monte Carlo work it is desirable to write the N_w expression so that parameters A and B are independent. This was accomplished by the mean-adjusted form shown in Figure 3 which requires having the mean of the log-ΔK's for all the cycles to grow the crack from a_0 to leakage at w. This mean depends on slope B, and on initial and final log ΔK values (which depend on a_0, w, stress and the ΔK formulation); it does not depend on intercept log A. Attention to the lack of independence between A and B in the standard formulation for N_w is important when studying the sensitivity of this function to these variables.

After recognizing the importance of mean log ΔK in the sampling procedure, it was found that variance in slope B made

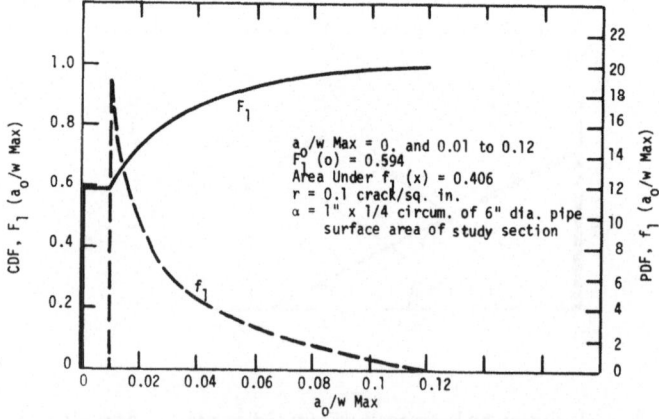

Figure 10. PDF and CDF of a_o/w Max With U/S Test

only a small contribution to the variance of N_w, and that a linear expression could be used for N_w as shown in Figure 3, which simplified the Monte Carlo work compared to using numerical integration. The formulation for ΔK appears in reference 4; it is a function of $\Delta s_{PM} \sqrt{a}$.

4. THE VARIANCE SOLUTION

In order to minimize computation time in the Monte Carlo work, it was desirable to retain only those variables whose distributions made large contributions to the variability of the equation. An approximate solution for the variance of the equation, often called the propagation of errors method,[5] was appropriate. The general formulation for the variance of an equation is shown in Figure 4; it can be seen that a separate term appears for each variable. The formulation shown is for the case when all variables are independent, as applies here for the one study section. The percentage contribution found for each variable for both a_o/t distribution cases is tabulated in Table 2.

It can be seen that for the distributions assigned, the major contributor for the case of no ultrasonic testing is initial crack depth ratio, a_o/w; lesser contributors are stress and intercept parameter log A; and contributions by uncertainties in slope parameter B, pipe diameter, wall thickness, and operating pressure are negligible. Accordingly, the Monte Carlo work used only the distributions of A, stress and a_o/w, with a_o/w identified for better definition. For the case of ultrasonic testing, the variance of log N_w is reduced to 31% in magnitude, and the contribution of the variance of a_o/w is reduced so as to nearly equal that of stress.

Table 2

CONTRIBUTIONS TO VARIANCE OF log N_w

	% WITHOUT ULTRASONIC TEST	WITH ULTRASONIC TEST
a_o/w, initial crack depth	78	42
log A, intercept	6	16
Δs_{PM}, expansion stress	15	41
	99	99
B, w, d, P	1	1
	100	100
MAGNITUDES, Var. log N_w	1.	0.31

5. THE PROBABILITY ESTIMATE

If we could assume that the distribution of N_w, cycles to leak, was clearly related to the normal distribution, we could use the mean and variance values found in the preceding section to find the probability that N_w lies less than 360, the operating cycles in 40 years. But because the distribution of N_w is analytically inaccessible, the distributions of the variables are not standard, and we must estimate small probabilities, less than 10^{-2}, assuming a distribution for N_w appeared unwise. It was decided to estimate the required probability by Monte Carlo with importance sampling, a variance-reduction method.[6,7]

In ordinary Monte Carlo for estimating a tail probability of a function, a trial consists of selecting a value at random from the distribution of each independent variable, solving the function using those values, assigning 1 to the outcome if it lies in the tail region, 0 if it does not, and finally summing and dividing by number of trials to obtain the fraction of trials in the tail, which is an estimate of the tail probability. Clearly, if this true probability is small, a large number of trials is required for a precise estimate. Importance sampling overcomes this problem by directing the random choices to the region of each variable which leads more often to function solutions lying in the tail of interest. Then, rather than 1 being assigned to each such outcome, a weight is assigned which is the reciprocal of the density of the weighting function used to focus the random choices. This process is illustrated in Figure 11. Fortunately, N_w is monotonic in A, stress and a_0/w, so the region to sample from for each of these variables to give low N_w values is known.

The difficulty in using importance sampling lies in choosing appropriate weighting functions. In general, a weighting function should be uniformly distributed over that region of each variable which leads to function values in the tail of interest, and distributed in a manner falling away at first steeply and then more gradually over variable values leading to function values increasingly distant from the tail of interest. For this problem an ad hoc procedure was adopted whereby beta distributions having one parameter equal to 1 were systematically modified so as to be uniformly distributed over values in the equal-probability tail of each variable which, when chosen together, would always give N_w less than 360; and distributed in a decreasing exponential-appearing manner over the balance of the distribution, this portion being steeper (and so reflecting the other beta parameter value) if, for a variable, choosing values in it increasingly

ESTIMATE Pr [X $\leq x_0$]

1. Choose 0-1 uniform random number, U*.
2. Using weighting function having density f_2 and cumulative F_2, transform U* to F1*.
3. Using the distribution of X, having density f_1 and cumulative F_1, transform F1* to x*.
4. If x* $\leq x_0$ assign weight W = $1/f_2$; if x* $> x_0$ assign W = 0.
5. Repeat for n trials.
6. Estimate Pr X $\leq x_0$ = $(\Sigma W)/n$.
7. f_2 was chosen so x*$\leq x_0$ usually results, and for other properties described in section 5.

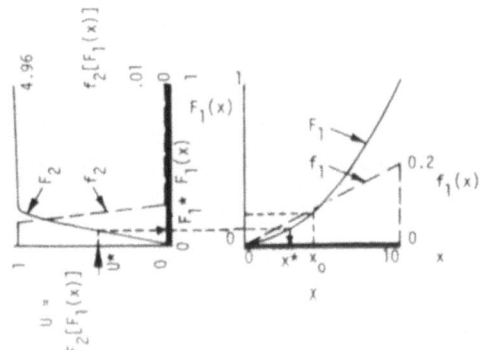

WEIGHTING FUNCTION   **DISTRIBUTION OF X**

Figure 11. Monte Carlo Using Importance Sampling

distant from the uniform portion would lead to N_w values rising more rapidly past 360.

A 3000-trial estimate of the probability that N_w is less than 360 cycles was made for the reference input conditions. The result, for the probability of leakage at the chosen 1 in. wide X 1/4 circumference study surface, in 40 years due to low cycle fatigue, was 0.0077, with 95% limits on the Monte Carlo portion alone at 0.0084 and 0.0112 assuming normality of estimate. Variance of estimate was reduced 82% by using importance sampling.

6. EXTENSIONS FOR SYSTEM RELIABILITY ESTIMATE

In principle, similar estimates are required at all other study sections of the piping system. This is manifestly infeasible, but it is possible to restrict attention to those study sections having sufficiently high design stress to govern the probability of leakage anywhere in the system. A simple empirical relationship between probability of leakage, nominal stress and wall thickness is anticipated, permitting probability values to be obtained readily for the design values of any pipe study sections.

The probability of leakage anywhere in the system, in 40 years, is found by appropriately summing over the piping system the probabilities at each of the study sections; these probabilities are given by the structure of conditional probabilities dictated by the piping probability diagram. The summation problem must recognize lack of independence among actual stress levels at successive study sections; but all variables at one study section, and the other variables at successive study sections, appear to be independent.

Extensions are also required to both initiation and growth for other mechanisms, such as stress corrosion and fatigue-assisted stress corrosion; to more extensive critical size definitions for severance, and to knowledge of conditions favoring reaching critical size for severance without prior detectable leakage.

References

1. For further details on this paper, see GEAP-10452, "Estimating Pipe Reliability by the Distribution of Time-to-Damage Method," S. A. Wilson, March 1972. For details on the pipe rupture study, see GEAP-10205, "Status of Pipe Rupture Study at General Electric Company, Part III," S. R. Vandenberg, June 1970; or "Summary, Reactor Primary Coolant Pipe Rupture Study," E. Kiss, to appear in Proceedings, ANS national topical meeting on water reactor safety, Salt Lake City, March 26-28, 1973; and quarterly progress reports in GEAP-series.
2. "The Determination of the Probability of Failure by Stress-Strength Interference Theory," R. L. Disney and J. J. Sheth, in Annals of Assurance Sciences; Proceedings of the Annual Symposium on Reliability, Boston, Mass., Jan. 16-18, 1968, pp. 417-422.
3. "Designing for Reliability Based on Probabilistic Modeling Using Remote Access Computer Systems," G. E. Ingram, C. R. Hermann, E. L. Welker, in Proceedings, Seventh Reliability and Maintainability Conference, ASME et al., San Francisco, July 14-17, 1968.
4. GEAP-5680, "Reactor Primary Coolant System Rupture Study," Quarterly Progress Report No. 13, April-June, 1968.
5. "Statistical Models in Engineering," G. J. Hahn and S. S. Shapiro, Wiley, 1967, p. 229ff.
6. "Importance Sampling in Monte Carlo Analyses," Charles E. Clark, Operations Research, September-October, 1961, p. 603ff.
7. "Monte Carlo Methods," J. M. Hammersley and D. C. Handscomb, Wiley, 1964, p. 57ff.

GROUP REPORTS

On the next-to-last day of the conference, the conferees divided themselves into small discussion groups to assess the state of the art in their particular field of expertise. The following list of "Questions for Discussion" were provided:

1. What are the research needs?
    a) Theoretical
    b) Practical
2. What are the data needs?
    a) Who should collect, disseminate, and evaluate data?
    b) How can a meaningful effort be financed?
3. Should the Universities participate by
    a) Giving special courses or degrees?
    b) Doing research? If so, what type of research.
4. Is there a commonality in reliability engineering as practiced in diverse fields?
    a) Are the definitions uniform?
    b) Are the basic equations common?
    c) Are we drifting further apart of pulling closer together?
5. Where is the field? How are reliability concepts applied?
    a) Does industry use reliability concepts and data in design? In operation?
    b) Do companies employ reliability specialists per se, or is this field considered a subset of maintenance, safety, and operations?
6. Where is the field going? How will reliability concepts be applied in the future?
    a) Are any trends in evidence? Can they be extrapolated?
7. Can equipment reliability specifications be imposed and reliability engineers licensed?

In the summary of the group discussions which follow, it is to be noted that the first five reports (Optimization and Mathematical Techniques, Nuclear Power, Electronics and Control, Mechanical and Structural Reliability, and Human Factors) are keyed to the list of questions while the other three are in narrative form.

GROUP I: OPTIMIZATION AND MATHEMATICAL TECHNIQUES -
Leader: R. E. Evans

Participants: J. Larisse, D. P. Sen Gupta, P. Seifert, G. Reina

1. Research Needs
   a) Theoretical: None are pressing, but they should not be discouraged, except for "crank-turning" exercises.
   b) Practical: Ways to handle the present, virtually intractable optimization problems which face engineers. Handling of uncertainties in the problem parameters; exploring the regions around local maxima; and limited knowledge of available hardware now cause many difficulties and expensive computations.

2. Data Needs
   There are no special needs, except those facing everyone for more data on hardware. Most present techniques make the poor but tractable assumption that reliability parameters are known exactly.

3. University Participation
   We offer no direct suggestions. Special degrees seem not to be necessary.

4. Commonality
   Mathematical and computational techniques, of course, show much commonality in virtually all fields. There is a little confusion due to terminology problems among various fields. Of course, the clear expression of assumptions and terminology is most important in any problem.

5. Present Applications
   Probably very little of the theoretical literature is directly applied to problems. Theoreticians rarely make realistic engineering assumptions. This is true regardless of the exact meaning of reliability being used (eg. safety, maintainability, forced outage).

6. Future
   There seems to be no appreciable evidence that theoreticians are changing. Tractability and esthetics seem to be more important to theoreticians than direct applicability.

Possibly more attention should be given to solutions which are in the form of simple computational algorithems rather than closed-form solutions.

NOTE: Reliability optimization is much different from performance optimization. Many process industries (chemical, electric power, telephone) have enough good data to make very efficient use of optimization techniques - usually for maximizing some measure of profit.

GROUP II: NUCLEAR POWER - Leader: G. Volta

Participants: A. Carnino (written comments), A. Colombo, G. Freneix, J. Fussell, G. Hensley

1. What are the research needs?
   The participants agreed upon three main lines of theoretical research:
   a) Implementation of the fault-tree technique in the direction of computer aided fault-tree extraction of hazard analysis. Use of new graph techniques (like GERT), extension of existing fault-tree techniques, for dealing with elements characterized by more than two parameters, time dependent repair rates, and control dependent elements.
   b) Research on models for failure and repair rates with reference mainly to mechanical components. Application of Bayesian techniques.
   c) Development of risk analysis methodology in connection with design theory.

   As practical research needs these points have been emphasized:
   a) The need of pilot or demonstration testing; prototype software; reliability assessment of unique, large equipment.
   b) Personnel training.
   c) Data collection for mechanical components.

2. What are the data needs?
   Everybody agreed on the principle that data banks should be set-up by non-profit organizations, governmental and/or international groups.

3. Should the University participate?
   Yes. But the main need is more for minor courses in the engineering curriculums than of a specific degree.

4. _Commonality_
   The language and definitions as shown during this meeting are not uniform. Some effort is needed in the future for standardization and clarification of the concepts and terminology.

   There is a general impression that we are going in the right direction.

5. _Where is the field?_
   The use of reliability techniques (redundancy, etc.) is common in the nuclear field.

   There is a lag in the use of concepts. It is a problem of education (see point 3).

   In the nuclear field reliability is a subset of "safety". This fact creates difficulties in defining the "objective function" for reliability analysis (human risk).

6. _Where is the field going?_
   Generally forward. The use of reliability techniques appears inevitable in the long term. In the nuclear field political and economic constraints hamper a regular and smooth development.

   Technical and practical difficulties also exist mainly in data collection and analysis.

7. _Specifications_
   The problem, however, is the demonstration of the reliability. Specifications can be of help in fixing goals, and in orienting designers and manufacturers.

GROUP III: ELECTRONICS AND CONTROL - Leader: A. Carnino

Participants: R. C. Bennetts, A. R. Eames (part time), G. J. Dasani, J. Humphreys, J. Lynn, W. Schneeweiss

1. _Research Needs_
   Improvement of software reliability in all areas is required. Methods for quantitatively establishing software reliability should be further investigated.

   Methods of practical validation of theoretical predictions of systems and equipments reliability and failure rates need further development.

A new idea for theoretical research (Dr. Lynn): the relationship between entropy and reliability should be investigated for large systems, if such a relationship exists.

2. Data Needs

More data is required on specific types of components (with manufacturers' names) so that the best components can be selected for sensitive applications. The tests should be organized by an independent international data bank. The collection, dissemination and evaluation of the data should be carried out by a body independent of the manufacturers and users, and who is supported by government and user subscriptions.

3. Role of the Universities

The Universities should provide a basic Masters course to encompass reliability and feasibility techniques. The Universities should provide theoretical research and industry, in conjunction with Universities, should provide the practical research for validation of theory.

4. Commonality

From our observations during the conference it appears that we are drifting apart, but we would hope that in the future we will pull together.

5. Where is the Field?
   a) In general, industry does not apply reliability techniques. They have, by trial and error, or past experience applied safety factors which have produced products which the customers have been prepared to accept.

   Only with the onset of highrisk situations have manufacturers and users been forced into using reliability concepts. In many cases, reliability concepts have been initiated as a result of accidents or financial loss.
   b) Again, only in the high risk areas do companies generally employ reliability specialists. These specialists are often considered as a separate entity from the maintenance, safety and operation areas. However, it should be stated that these disciplines ought to be closely related.

6. Where is the Field Going?

Reliability and testability will be increasingly applied in the design of future equipment, and in the improvement of equipment, and system maintainability and availability, in the framework of the overall economic aspects.

a) There are trends in evidence but it is not clear how they can be extrapolated.

7. <u>Can Specifications be Imposed?</u>
Yes.

GROUP IV: HUMAN FACTORS - Leader: R. Weaver

Participants: F. Y. Hjelte, F. P. Lees, J. Rasmussen, R. T. L. Regulinsky

1. <u>Research Data Needs</u>
Research at both theoretical and practical levels should be aimed at the taxonomy or structure of relevant data in human performance. Synergistic input from engineers, psychologists and to some extent management personnel should be recruited. Conferences of the Gordon and Engineering Foundation variety can prove to be an invaluable catalyst in this direction.

There appear to be generic components of the human factors in the spectrum of industrial and organization frameworks which are unequally developed and which should be the discipline for conceptual research in "human modeling". The engineers' role in this effort is properly that of the interpreter for the machine in the man/machine complex-rather than that of number cruncher, at least at this point in time.

The A.I.R. Data Bank effort with all its deficiences reveals both a baseline and timeliness to this attention to the structure of data to be taken (before any massive Data Bank program is proposed).

2. <u>Relevant University Activity and Professional Preparation</u>
The Universities could offer training at three levels:
a) "Awareness courses" in human factors psychology (motivation, personality traits (vigilance, communicability), intellectual traits, dexterity, physiological traits (fatigue), etc.) to be taught at the upperclassman level.
b) Generic reliability technology courses.
c) Practicum programs, probably at the master's level.
d) Post-graduate short courses.

It is felt that an engineer/psychology or business management/psychology program leads to a more employable and functional man than one in pure psychology. It is also felt that any university program should be flavored with on-site experience.

3. Interface with Other Reliability Technology

Stochastic and simulation methodology are and will remain to be a common ground between physical system reliability assessment and human factors analysis.

4. Place of the Discipline in Industry

The discipline appears ill-represented in industry between the purely clerical level effort in some safety and public relations posts on one hand, and its absorption at high responsibility levels in the myriad of talents and responsibilities expected of management. So often, human factors personnel are used at a cosmetic level. However, the situation is somewhat different in military and "critical mission" efforts. Also the trend in industry is to turn toward more sophisticated consideration of human factors - sometimes as a buffer between management responsibility and labor or public opinion. Human factors personnel it would appear should have a good background in practical psychology and serve in a staff function providing both special case expertise and tutorials armed at enhancing the general level of attention to human factors by the engineering staff at large.

5. An additional question was posed by a university faculty member, to-wit-- what can be done to enhance the reliability of medical treatment and explicit attention to it by practitioners and medical students?

Clearly some of the human factors considered earlier are germane to medical practice (the suggested "awareness course" could well be a common course in psychology for medical students). Potentially, some medical clinicians could profit from exposure to reliability technicians per se.

More importantly, progress in physiological simulation and computer-aided diagnosis, which are just beginning to reach a level of reasonable usefulness, deserve to be followed by those interested in assessing human factors.

GROUP V: MECHANICAL AND STRUCTURAL RELIABILITY -
Leader: S. A. Wilson

Participants: A. B. Bofors, G. Garriba, P. Martin, J. Munro, G. Rydman

1. Research Needs
   a) Theoretical - Studies dealing with the "Physics of Failure".
   b) Practical - General design methods which inherently improve reliability.

2. Data Collection
   The proprietary nature of equipment failure rates suggests that the technology of data collection of this type of reliability data needs further exploration. The role of trade associations, and the possibility of separate banks for materials and components should be explored.

3. University Participation
   Reliability concepts should be taught in each discipline. No BS curriculum is needed; an MS in reliability is a possibility. Continuing education and a stronger relationship between graduate schools and industry to assure that "real problems" are understood by academicians are recommended.

4. Commonality
   The word "reliability" should be reserved for the statistical concept of the probability of successful life. Other terms should be reserved for other concepts. For example, system performance often needs to be characterized by some measure other than reliability alone; especially if the system is complex.

   Reliability is only one element in systems engineering, which encompasses operations research, operational analysis, engineering economics, optimization, risk analysis, etc.

5. Status of the Reliability Field
   a) It needs continued promotion and technical public relations in the form of seminars, workshops, courses, etc.
   b) It is increasingly used to satisfy economic objectives in industry, safety objectives in nuclear and aerospace activities, performance objectives of military equipment, etc.
   c) It is of importance in warranty cost demonstration.

6. Future of the Reliability Field
   It will be of increasing importance due to the increased size of the risk in all items in 5b.

7. Reliability Specifications? Licensing?
   Reliability specifications can and are being written, but the field is not broad enough to become separately licensable.

GROUP VI: CHEMICAL ENGINEERING - Leader: G. Powers

Participants: B. Bulloch, E. Henley, G. Hensley (part time), G. Pearson

The following report expresses the views of the Chemical Engineering Group on the status of safety and reliability techniques in areas related to chemical engineering and the chemical industry.

We see several trends which make these areas of utmost importance to the chemical process industries. These trends are:

1. Single Train Plants (Economies of Scale).
2. Larger Units.
3. Extremes in Operating Conditions (Pressures, Temperatures, Hazardous Species, etc.).
4. Operating Closer to Safety Constraints (Explosion Limits, Equipment Limits).
5. Large Scale Integration (Makes the plant more difficult to analyze, ie. heat recovery ties the complete plant together).
6. Higher Maintenance Costs.
7. More Pollution Control Constraints.
8. Government Safety Regulations.
9. Union's Concern with Safety.
10. Insurance Companies Concern with Potential Process & Human Losses.

These trends when viewed against the state-of-the-act of reliability and safety analysis in the chemical process industries indicate the following needs:

Theory of Reliability & Safety Analysis

1. <u>Rare Event Prediction</u>: we need to become aware of or develop techniques for predicting rare events. Gathering data will obviously be important here.

2. <u>Sequential or Batch Operations deserve a more detailed consideration</u>. The techniques used in the detection of faults in logic circuits are most likely useful here.

3. <u>Synthesis of Fault-Trees or Reliability Diagrams for Chemical Plants</u>. This very time-consuming activity should be automated.

4. <u>Objective Function Definition</u>: It is important to define the proper objective function when discussing the worth of

reliability and safety methods. The objective function must include:
a) <u>The standard cost items</u>. raw material costs, capital costs, labor costs, utility costs, etc.
b) <u>Maintenance costs</u> as a function of the care in designing the system. A simple percentage of the installed capital cost will not reflect the worth or reliability and safety studies.
c) <u>Operating Factor</u>. the availability of the plant must be considered. The fact that overcapacity exists for a certain period in a plant's life must also be included.
d) <u>Uncertainties in plant costs</u>. The objective function should reflect the possible uncertainties that exist in the costs which are included. Techniques for handling these distributed parameters should be used.
e) <u>Safety costs</u>. The potential costs of plant hazards are extremely important, many times a profitable plant will not be built due to safety problems.

5. <u>The Gathering of Reliability Data</u> should be made a part of a more general management information system. This will reduce the direct cost of the information gathering for the reliability effort. Data should also be gathered on plants which have had extensive safety and reliability design efforts.

6. <u>Universities can supply several valuable services</u>:
a) An overview of the general techniques used in reliability and safety studies.
b) More studies on the behavior of failed systems.
c) Process design courses which directly include the safety and reliability of the plant as a design objective.
d) Development of the basic mathematical tools and techniques required in this area.
   1) Probability and Statistics.
   2) Simulation Methods (Monte Carlo, etc.).
   3) Combinations.
   4) Modeling of Non-Numerical Systems.

7. <u>Investigate the Writing of Process Reliability and Safety Specifications</u>. Very few plants currently have explicit quantitative specifications pertaining to safety and reliability. If a method for building these specifications into each design contract could be developed, a plethora of reliability and safety studies would result.

8. <u>Development of a Common Language with Other Disciplines</u>:
We as chemical engineers need to become aware of the techniques available in other fields. Logic circuit and power systems design have developed several powerful techniques for detecting and repairing faults. Studies are needed to

translate these ideas into methods useful for the chemical process industries. Hopefully a common language will result which will reflect the commonality of all of our methods. That is, fault-trees are similar to reliability diagrams; fault detection using Boolean differences of D-cubes is similar to availability studies, etc. A possible scheme for categorizing these systems would be:
a) Design Methods (Continuous/Batch Systems)
   Availability Studies
   Maintenance System Design
   Safety Analysis
   Synthesis Methods for Inherently Safe Systems
b) Operations Planning and Control
   Diagnosis of Faults (Real time)
   Operator Displays and Training
   Automatic Shutdown Procedures
   Fault Interception
   Data Gathering

Summary

In general, we see the beginning of an increased application of reliability and safety analysis and synthesis methods in the chemical process industries. We now know how to perform a standard reliability and safety analysis of a chemical plant. What is required is a program for:
a) Conveying the idea to managers of chemical companies that safety and reliability can be quantified in chemical systems.
b) Educating future generations of graduates in these techniques.
c) Developing new analysis and synthesis techniques which will reduce the time required to perform the safety analysis.
d) Gather and analyze the data required to assess plant safety and reliability.

We must remember that chemical processes tend to be robust and not everyone of them will be either hazardous or complicated enough to warrant a rigorous reliability and safety analysis. A major part of our problem is to develop our understanding of this field so that we can tell what degree of effort is justified for each given plant.

When this stage is achieved we will be much better able to discuss the true safety and reliability of chemical processing systems.

GROUP VII: POWER SYSTEMS - Leader: R. Bilington

Participants: S. L. Surana, A. Brameller, L. H. Burgesse.

There has been considerable development and application of reliability engineering concepts in this area within the last few years. This has resulted in an apparent gap between the theoretical and practical applications. Some research is needed at the implementation level and on the difficulties of communicating reliability concepts to non-specialist groups. Research is also required in the development of system reliability definitions as opposed to component definitions. One area which does not seem to have developed sufficiently is that of mechanical system reliability appraisal and further research is needed in this field. The biggest stumbling block to the general reliability evaluation is in the availability of adequate and consistant data. Any data collection system must be of a dynamic nature and should be capable of producing individual element and component data. The most desirable, though not necessarily always possible way to collect this data, would be through an independant body established for this purpose. This organization should be capable of communicating directly with other computer bank data gathering agencies in other countries. Research is also needed on data needs and the most suitable forms of data collection.

Universities should actively participate in reliability education by implementing graduate level courses on general and specialized reliability concepts. In addition, fundamental reliability engineering concepts should be introduced into the undergraduate programs in connection with existing design systems and economic analysis courses. The awareness of reliability implications should also be made available to the practising engineer through appropriate extension classes and programs. University research should be of both speculative and application orientations. Both types will add considerably to the teaching function. Contract research is desirable as it will add a much needed realistic feature to the program. The research should also study the financial implications of reliability or unreliability.

The present definitions used in the power field appear to be diverging somewhat from those originally adapted in the general reliability area. The present definitions appear to be more physically related to the system and therefore more suitable from an applications viewpoint. In general, the specialist areas are tending to drift apart as more expertise is built up in each discipline. Conferences such as this aid immeasureably in pulling together the different disciplines by increasing the overall awareness of common areas of application.

The concepts and mathematical models are considerably ahead of the practical applications. Utilities are very slow to adopt new concepts, but there has been considerable advancement over the past five years. Industry in general is on the threshold of application and analysis in the design and operation areas. In general, utilities do not employ reliability specialists, though this may come about as applications increase. Engineers within a company are often given specialized training for work within their particular area in the company. Many specialists in industrial applications will be in a consulting role, rather than as company employees.

There will be increased application in the future in the use of reliability concepts to economic appraisals of alternate facilities and expansion programs. Optimum utilization of available capital and an appreciation of the ability of reliability concepts to aid in consistent appraisal of design, operation and maintenance programs is evident in future applications. The optimum utilization of resources both physical and economic will dictate extensive utilization of basic reliability concepts in future system design. One obvious trend is in the decreased use of approximate or "rule of thumb" techniques and their replacement by consistent quantitative reliability indices. Further educational and research effort is required in this regard.

It was concluded that there is no need for the licensing of reliability engineers, and in fact it may not be desirable to attempt to produce a specialist known as a reliability engineer. It was believed that equipment reliability specifications can and should be imposed in the power field. Reliability implications and contractual obligations should be included in specifications and agreements. This may be difficult to implement, but reliability is worth paying for as is efficiency and decreased power losses, etc. In conclusion the group would like to extend their appreciation to the organizing committee for this opportunity to discuss these topics.

GROUP VIII: DATA ORGANIZATION AND TREATMENT -
Leader: M. L. Shooman

Participants: H. Asher, L. Calderola, I. Bazovsky, A. R. Eames, (Part time), T. Heimly, J. Salmons, R. A. Reid, W. Vinck

1. <u>Goals of Data Banks</u>: What do we need data banks for, and what do we hope to accomplish by consulting them?
   a) To produce data from which parameters can be deduced for use in prediction of system reliability, availa-

bility, maintainability, and other probabilistic measures. These measures are important in assessing system performability, safety, and economics.
   b) By virtue of its sample size it serves to:
      1) Detect specific modes and mechanisms of primary failure (including design and process defects) more quickly than other means.
      2) Yields smaller sample variances.
      3) Allow us to subdivide the data into environmental groups to compare environments and to isolate and study the significant environmental parameters.
      4) To detect differences among manufacturers, designs, and processes for the same or similar components.
      5) Allows us to analyze and detect early failures and wearout as well as differences in successive generations of items.

2. <u>What Input Data Do We Need To Reach These Goals for Each Failure</u>: What input data must go into the bank and what output data should be supplied?
   a) Identification of item and source.
   b) Symptoms of failure.
   c) Past history of environment and maintenance.
   d) Results of past failure analysis.
   e) Significant item lines:
      1) Repairable: past history of up and down times, time of present repair.
      2) Replaceable: time to failure, number of items on test.

3. <u>What Kind of Output Data Do We Need?</u>
   a) Raw data for a limited period into the past.
   b) Raw data segregated into like environments.
   c) Parametic models of failure rates (hazard).
   d) Nonparametic (trend) analysis of hazard.
   e) Failure rate (hazard) analyzed by mechanism of failure.
   f) Parameters for appropriate stochastic models for repairable items.

4. <u>Do the Existing Systems Satisfy These Needs</u>: from a users point of view are we satisfied with the existing data banks, what is lacking?
   a) Raw data - some banks do not keep raw data, only processed data.
   b) Raw data segmented into like environments - some do, some don't.
   c) By and large, only constant failure rates.
   d) ------------------------not done.
   e) ------------------------not done.
   f) ------------------------not done.

g) ------------------not done.
h) Accuracy of data collection (human factors).
i) Some data banks have only laboratory or only field failure data.

5. <u>What Must Be Done to Improve Things</u>: from a user's point of view what appears to be a reasonable set of approaches to improving uses.
   a) Improvement of human factors: control, education and motivation.
   b) Specialized data banks can lead to improved technical expertise, and may be more easily funded.
   c) If specialized data banks are to grow then we must be certain that provisions are made in some manner to collect data on missing components as well as data on system interconnections such as (sockets, welding, wire wrap, etc.). Also, there is the problem of a system manufacturer having to join many specialized data banks.
   d) Further study must be done of the helpfulness of the data supplied by independent laboratories, manufacturers (eg. field operations, or field inspector information).
   e) Better tie-ins between data which is available and theoretical analysis must be found.
   f) Possibility of using an interactive conversational, teletype oriented, computer program or optical card or document scanners to improve the data input phase.

ATTENDANCE

Mr. H. E. Ascher, Naval Research Laboratory, Washington D.C. U.S.A.

Dr. D. H. Allen, Dept. of Chemical Engineering, University of Nottingham, U.K.

Mr. S. A. Austin, U. S. Army Electronics Command, Fort Monmouth, New Jersey, U.S.A.

Dr. W. Bastl, Technische Universitat, Munchen, Germany.

Mr. I. Bazovsky, Jr., Sherman Oaks, California, U.S.A.

Mr. R. G. Bennetts, Dept. of Electronics, the University of Southampton, U.K.

Mr. B. C. Bullock, I.C.I. Runcorn, Cheshire, U.K.

Prof. R. Billinton, Dept. of Electrical Engineering, University of Saskatchewan, Canada.

Mr. A. J. Bourne, Systems Reliability Service, U.K.A.E.A., Culcheth, U.K.

Dr. A. Brameller, Dept. of Electrical Engineering, University of Manchester Institute of Science and Technology, U.K.

Mr. L. Calderola, Institute fur Systemtechnik und Reaktorphysik Karlsruhe, W. Germany.

Madame A. Carnino, Saclay, Gif-Sur-Yvette, France.

Dr. F. J. Charlwood, The City University, London, U.K.

Dr. C. Chopping, C.E.G.B. London, U.K.

Mr. A. Colombo, Euratom, Ispra-Varese, Italy.

Mr. A. R. Eames, Systems Reliability Service, U.K.A.E.A., Culcheth, U.K.

Dr. R. Evans, Durham, North Carolina, U.S.A.

Mr. D. N. Farnan, U. S. Army Electronics Command, Fort Monmouth, New Jersey, U.S.A.

Mr. C. D. H. Fothergill, U.K.A.E.A., Risley, Lancs., U.K.

Mr. J. Freneix, Framatome, Courbevoie, France.

Dr. J. B. Fussell, Aerojet Nuclear Co., Idaho Falls, U.S.A.

Prof. S. Garribba, Polytechnic Institute of Milano, Italy.

Mr. S. B. Gibson, I.C.I. Ltd., Runcorn, Cheshire, U.K.

Mr. A. E. Green, Systems Reliability Service, U.K.A.E.A. Culcheth, Warrington, U.K.

Mr. T. Heimly, Det norske Veritas, Oslo 6, Norway.

Prof. E. J. Henley, Cullen College of Engineering, University of Houston, Texas, U.S.A.

Mr. G. Hensley, Nuclear Installations Inspectorate, Liverpool, U.K.

Mr. J. A. Hess, U.S. Army Electronics Command, Fort Monmouth, New Jersey, U.S.A.

Mr. F. Y. Hjelte, Royal Institute of Technology, Stockholm 70, Sweden.

Mr. P. Humphreys, U.K.A.E.A. Risley, U.K.

Mr. J. Larisse, Euratom, Ispra-Varese, Italy.

Dr. F. P. Lees, Dept. of Chemical Engineering, Loughborough University, U.K.

Dr. M. A. Loughton, Dept. of Electrical Engineering, Queen Mary College, London, U.K.

Dr. J. W. Lynn, Dept. of Electrical Engineering, University of Liverpool, U.K.

Dr. G. Manzoni, ENEL-CREL, Milano, Italy.

Dr. P. Martin, Dept. of Mechanical Engineering, University of Liverpool, U.K.

Mr. J. McGowan-Docherty, U.K.A.E.A. Risley, Lancs, U.K.

Dr. J. Munro, Dept. of Civil Engineering, Imperial College of Science & Technology, London, U.K.

Dr. J. D. Murchland, London, U.K.

Prof. P. K. McPherson, The City University, London, U.K.

Mr. S. Panichelli, ENEL, Rome

Mr. G. Pearson, Dept. of Chemical Engineering, Nottingham University, U.K.

Mr. R. H. Pope, C.E.G.B., Cheltenham, U.K.

Prof. G. Powers, Dept. of Chemical Engineering, M.I.T., Cambridge, Massachusetts, U.S.A.

Mr. J. Rasmussen, Atomic Energy Commission, Research Establishment Riso, Denmark.

Prof. T. L. Regulinski, Dept. Electrical Engineering, Air Force Institute of Technology, WPAFB, Ohio, U.S.A.

Dr. R. A. Reid, Philips, Eindhoven, Holland.

Dr. G. Reina, Soc. A.R.S. Sp.A., Milano, Italy.

Mr. R. Roughley, U.K.A.E.A. Risley, Lancs., U.K.

Mr. G. Rydman, A.B. Bofors, Sweden.

Mr. J. F. Salomons, Netherlands Ministry of Defence, the Hague, Netherlands.

Mr. S. Scalcino, ENEL, Roma, Italy.

Dr. W. Schneweiss, Siemens A.G., Karlsruhe, Germany.

Dr. D. P. Sen Gupta, Dept. of Electrical Engineering, Indian Institute of Science, Bangalore, India.

Prof. M. L. Shooman, Dept. of Electrical Engineering, Polytechnic Institute of Brooklyn, New York, U.S.A.

Dr. D. J. Siddons, C. E. G.B., Banwood, Gloucester, U.K.

Mr. P. Siefert, Technische Hochschule, Darmstadt, Germany.

Mr. S. L. Surana, Dept. of Electrical Engineering, University of Manchester Institute of Science & Technology, Manchester, U.K.

Dr. A. D. Swain, Sandia Laboratories, Albuquerque, New Mexico, U.S.A.

Prof. D. R. Towill, University of Wales, Institute of Science & Technology, Cardiff, U.K.

Mr. W. Vinck, E.E.C., Brussels.

Mr. G. Volta, Euratom, ISPRA, Italy.

Prof. R. Weaver, Tulane University, New Orleans, Louisiana, U.S.A.

Mr. S. A. Wilson, Nuclear Energy Division, General Electric Co. San Jose, California, U.S.A.

Mr. E. R. Woodcock, U.K.A.E.A. Risley, Lancs., U.K.